Sustainable Land Sector Development in Northern Australia

Sustainable Land Sector Development in Northern Australia

Indigenous Rights, Aspirations, and Cultural Responsibilities

Edited by
Jeremy Russell-Smith, Glenn James, Howard Pedersen,
and Kamaljit K. Sangha

(Courtesy of Duwun Tony Lee, Birrimilanga Nadine Lee, Ian Lee, Glenn James)

CRC Press
Taylor & Francis Group
Boca Raton London New York

CRC Press is an imprint of the
Taylor & Francis Group, an **informa** business

CRC Press
Taylor & Francis Group
6000 Broken Sound Parkway NW, Suite 300
Boca Raton, FL 33487-2742

First issued in paperback 2020

ISBN 13: 978-0-367-57119-1 (pbk)
ISBN 13: 978-1-138-60020-1 (hbk)

Library of Congress Cataloging-in-Publication Data

Names: Russell-Smith, Jeremy, editor.
Title: Sustainable land sector development in Northern Australia : indigenous rights, aspirations, and cultural responsibilities / edited by Jeremy Russell-Smith, Glenn James, Howard Pedersen, and Kamaljit K. Sangha.
Description: Boca Raton, FL : Taylor & Francis, 2018. | Includes bibliographical references.
Identifiers: LCCN 2018013164 | ISBN 9781138600201 (hardback : alk. paper)
Subjects: LCSH: Aboriginal Australians--Australia--Northern Territory--Social conditions. | Aboriginal Australians--Land tenure--Australia--Northern Territory. | Sustainable development--Australia--Northern Territory. | Land tenure--Australia--Northern Territory.
Classification: LCC GN667.N95 S87 2018 | DDC 338.994--dc23
LC record available at https://lccn.loc.gov/2018013164

Visit the Taylor & Francis Web site at
http://www.taylorandfrancis.com

and the CRC Press Web site at
http://www.crcpress.com

Contents

List of Text Boxes

Preface

This book provides a synthesis of a suite of research projects addressing the broad theme, 'Building community resilience in northern Australia', focused especially on remote, predominantly Indigenous (Aboriginal) communities. This research was contracted by the Bushfire and Natural Hazards Cooperative Research Centre (BNHCRC), Melbourne, in late 2013, and undertaken under the auspices of the BNHCRC's 'northern hub' administered through Charles Darwin University (CDU) in partnership with the North Australian Indigenous Land & Sea Management Alliance Ltd (NAILSMA).

The book incorporates findings from BNHCRC component sub-projects that address a range of related topics (community resilience and governance, economic opportunities especially through payments for environmental services, appropriate bushfire and natural hazards training, savanna fire management, management of high biomass flammable grassy weeds). For reporting and wider readership purposes, we have intentionally placed these matters in a broader cultural, historical, environmental, economic, development and governance context. At an early project workshop, we asked colleagues what issues they considered necessary to illuminate the broad project theme, and this book is the result. The views expressed are those of respective contributing authors, and do not purport to represent those of the BNHCRC.

—Jeremy Russell-Smith, Howard Pedersen, Glenn James, and Kamaljit K. Sangha

Acknowledgements

The book comprises chapter and text box contributions from many people, including the inputs of independent reviewers, who have given generously of their time and expertise. We wish to acknowledge the contributions of our colleagues who have been involved with and helped guide the undertaking of the research presented here, especially (1) our project end-users – Steve Rothwell (Northern Territory Fire & Rescue Service), Mark Ashley and Collene Bremner (Northern Territory Government), Suellen Flint (Western Australia Department of Fire & Emergency Services), Naomi Stephens (New South Wales Department of Environment; (2) BNHCRC colleagues – Dr Richard Thornton, Dr Michael Rumsewicz, Nathan Maddock, Dr Desiree Beekharry; and CDU and NAILSMA administrative support – Prof Andrew Campbell, Melissa George, Prof Karen Gibb, Roanne Ramsey, Janely Seah, Tahlia Timms, Emmylou Trombley, Cameron Yates.

Finally, we would like to acknowledge the enthusiastic assistance provided by Alice Owen and her team at CRC Press in the production of this book.

Editors

Jeremy Russell-Smith is an ecologist who, since the early 1980s, has worked and published extensively on fire ecology, the carbon market, and Indigenous land management issues in northern Australia, and related issues elsewhere in the tropics. He is a professor of tropical fire ecology and management at Charles Darwin University, Darwin, where he leads the applied research program of the Darwin Centre for Bushfire Research.

Howard Pedersen has been working with Indigenous leaders in northern Australia as a policy advisor and speechwriter since the early 1980s. He has worked in government and Aboriginal organisations and his professional experience spans the breadth of the big picture policy issues which have shaped northern Australian development including land rights and native title, economic empowerment and cultural protection and renewal. In 1995 his award-winning book, *Jandamarra and the Bunuba Resistance*, written in collaboration with storyteller Banjo Wurunmurra, was released and subsequently promoted through children's books, an award-winning TV drama/documentary, and theatre productions. Howard currently works as policy advisor to the Yawuru native title holders in Broome.

Glenn James is a program and policy manager with the North Australian Indigenous Land and Sea Management Alliance Ltd in Darwin. He has lived in remote Central and North Australia for almost 30 years. Over this period his work has spanned a number of interrelated areas including Indigenous community arts, anthropology and land management enterprise; in recent years focussing on tools to identify and track the health of local Indigenous values as they transition to land management-based enterprise. Glenn's formal education in political science, philosophy and anthropology, combined with his equally formative lived experience in remote communities and many personal and professional relationships in remote Australia, informs his long-term support for the aspirations and achievements of Indigenous people.

Kamaljit K. Sangha works as an ecological economist at the Darwin Centre for Bushfire Research at Charles Darwin University. Currently she is involved in developing mechanisms for Payments for Ecosystem Services (PES), highlighting the role of natural systems for human well-being especially in Indigenous and local communities, and valuing ecosystems for their services (ES) to inform policy decision making.

Contributors

Ricky Archer
North Australian Indigenous Land & Sea Management Ltd
Darwin, Northern Territory, Australia

Otto Bulmaniya Campion
Aboriginal Research Practitioners Network
Charles Darwin University
Darwin, Northern Territory, Australia

and

Bushfire & Natural Hazards Cooperative Research Centre
Melbourne, Victoria, Australia

Danny Burton
Yalu Marnggithinyaraw Indigenous Cooperation
Galiwinku, Northern Territory, Australia

and

Bushfire & Natural Hazards Cooperative Research Centre
Melbourne, Victoria, Australia

Ian Chambers
Crawford School for Public Policy
Australian National University
Canberra, Australian Capital Territory, Australia

Peter Christophersen
Kakadu Native Plants Pty Ltd
Jabiru, Northern Territory, Australia

Peter Cooke
14 Lantana Street, Nightcliffe
Darwin, Northern Territory, Australia

Robert Costanza
Crawford School for Public Policy
Australian National University
Canberra, Australian Capital Territory, Australia

Julian Cribb
Australian National Centre for the Public Awareness of Science
Australian National University
Canberra, Australian Capital Territory, Australia

Neville Crossman
School of Biological Sciences
University of Adelaide
Adelaide, South Australia, Australia

Allan Dale
Cairns Institute
James Cook University
Cairns, Queensland, Australia

Cherry Daniels
Aboriginal Research Practitioners Network
Charles Darwin University
Darwin, Northern Territory, Australia

and

Bushfire & Natural Hazards Cooperative Research Centre
Melbourne, Victoria, Australia

Grace Daniels
Aboriginal Research Practitioners Network
Charles Darwin University
Darwin, Northern Territory, Australia

and

Bushfire & Natural Hazards Cooperative Research Centre
Melbourne, Victoria, Australia

Jenny Davis
College of Engineering, IT & Environment
Charles Darwin University
Darwin, Northern Territory, Australia

Jeremy Dore
Aboriginal Carbon Fund Ltd
Melbourne, Victoria, Australia

Andrew Edwards
Darwin Centre for Bushfire Research
Charles Darwin University
Darwin, Northern Territory, Australia

and

Bushfire & Natural Hazards Cooperative Research Centre
Melbourne, Victoria, Australia

Murray Garde
School of Culture History & Language,
College of Asia and the Pacific,
Australian National University, Canberra,
Australian Capital Territory, Australia

Mark R. Gardener
Research Institute for the Environment and Livelihoods
Charles Darwin University
Darwin, Northern Territory, Australia

Melissa George
North Australian Land & Sea Management Alliance Ltd
Darwin, Northern Territory, Australia

Rolf Gerritsen
Northern Institute
Charles Darwin University
Alice Springs, Northern Territory, Australia

Jacky Green
Waanyi Garawa Rangers
Northern Land Council
Borroloola, Northern Territory, Australia

Hmalan Hunter-Xenie
Aboriginal Research Practitioners Network
Charles Darwin University
Darwin, Northern Territory, Australia

and

Bushfire & Natural Hazards Cooperative Research Centre
Melbourne, Victoria, Australia

Bentley James
North Australian Land & Sea Management Alliance Ltd
Darwin, Northern Territory, Australia

Glenn James
North Australian Land & Sea Management Alliance Ltd
Darwin, Northern Territory, Australia

and

Bushfire & Natural Hazards Cooperative Research Centre
Melbourne, Victoria, Australia

Seán Kerins
Centre for Aboriginal Economic Policy Research
Australian National University
Canberra, Australian Capital Territory, Australia

Ida Kubiszewski
Crawford School for Public Policy
Australian National University
Canberra, Australian Capital Territory, Australia

Paul Lane
Kimberley Institute
Broome, Western Australia, Australia

Elaine Lawurrpa
Yalu Marnggithinyaraw Indigenous Cooperation
Galiwinku, Northern Territory, Australia

and

Bushfire & Natural Hazards Cooperative Research Centre
Melbourne, Victoria, Australia

Dominique Lynch
Darwin Centre for Bushfire Research
Charles Darwin University
Darwin, Northern Territory, Australia

and

Bushfire & Natural Hazards Cooperative Research Centre
Melbourne, Victoria, Australia

Alan Maratja
Yalu Marnggithinyaraw Indigenous Cooperation
Galiwinku, Northern Territory, Australia

and

Bushfire & Natural Hazards Cooperative Research Centre
Melbourne, Victoria, Australia

Erica McCreedy
North Australian Land & Sea Management Alliance Ltd
Darwin, Northern Territory, Australia

Joe Morrison
Northern Land Council
Darwin, Northern Territory, Australia

and

North Australian Land & Sea Management Alliance Ltd
Darwin, Northern Territory, Australia

Douglas Paton
School of Psychological and Clinical Sciences
Charles Darwin University
Darwin, Northern Territory, Australia

and

Bushfire & Natural Hazards Cooperative Research Centre
Melbourne, Victoria, Australia

Frank Peacocke
Herron Todd White (Northern Territory) Pty Ltd
Darwin, Northern Territory, Australia

Howard Pedersen
Nyamba Buru Yawuru Ltd
Broome, Western Australia, Australia

Stuart Phillpot
School of Business
University of New South Wales
Canberra, Australian Capital Territory, Australia

Natalie Rossiter-Rachor
Research Institute for the Environment & Livelihoods
Charles Darwin University
Darwin, Northern Territory, Australia

and

Bushfire & Natural Hazards Cooperative Research Centre
Melbourne, Victoria, Australia

Jeremy Russell-Smith
Darwin Centre for Bushfire Research
Charles Darwin University
Darwin, Northern Territory, Australia

and

Bushfire & Natural Hazards Cooperative Research Centre
Melbourne, Victoria, Australia

Kamaljit K. Sangha
Darwin Centre for Bushfire Research
Charles Darwin University
Darwin, Northern Territory, Australia

and

Bushfire & Natural Hazards Cooperative Research Centre
Melbourne, Victoria, Australia

Samantha Setterfield
School of Biological Sciences
University of Western Australia
Perth, Western Australia, Australia

and

Bushfire & Natural Hazards Cooperative Research Centre
Melbourne, Victoria, Australia

Bevyline Sithole
Aboriginal Research Practitioners Network
Charles Darwin University
Darwin, Northern Territory, Australia

and

Bushfire & Natural Hazards Cooperative Research Centre
Melbourne, Victoria, Australia

Natalie Stoeckl
Centre for Tropical Water and Aquatic Ecosystem Research
James Cook University
Townsville, Queensland, Australia

Stephen Sutton
Darwin Centre for Bushfire Research
Charles Darwin University
Darwin, Northern Territory, Australia

and

Bushfire & Natural Hazards Cooperative Research Centre
Melbourne, Victoria, Australia

Peter Whitehead
Darwin Centre for Bushfire Research
Charles Darwin University
Darwin, Northern Territory, Australia

Dean Yibarbuk
Aboriginal Research Practitioners Network
Charles Darwin University
Darwin, Northern Territory, Australia

and

Bushfire & Natural Hazards Cooperative Research Centre
Melbourne, Victoria, Australia

Peter Yu
Nyamba Buru Yawuru Ltd.
Broome, Western Australia, Australia

and

North Australian Land & Sea Management Alliance Ltd
Darwin, Northern Territory, Australia

CHAPTER 1

2-Way Country – Challenges for Inclusive, Equitable, and Prosperous Development in North Australia*

Joe Morrison, Peter Yu, and Melissa George

CONTENTS

The North Australian Indigenous Land and Sea Management Alliance (NAILSMA) is a not-for-profit company that advances the sustainable use and management of northern Australian lands and waters based on Indigenous cultural, social, and economic imperatives. NAILSMA has been operating since 2001, representing Traditional Owners and Native Title Holders throughout the North. The authors of this chapter are and have been either the chair or CEO of NAILSMA since its inception. We share a collective Indigenous view that North Australia development should be based on a decolonized land tenure system with an associated economic use that is both compatible and inclusive of Indigenous rights, knowledge and practices.

1.1 INTRODUCTION

From our perspective, northern Australia is the mega-landscape of the river systems across the top of Australia, the region covered by the North Australian Indigenous Land and Sea Management Alliance Ltd. (NAILSMA). This was the area under study by the Northern Australia Land and Water Taskforce established by Prime Minister John Howard in 2007 and continued under a modified composition by the Rudd Government. To ensure there is no confusion about what we consider the North in geographical terms, if one drew a line from Townsville to Broome, we are referring to all the land north of that line.

* This chapter is based on a keynote address by Joe Morrison, delivered at the Northern Australian Development Conference, Cairns, June 2017.

The Northern Australia Strategic Partnership comprising the Prime Minister and First Ministers of Queensland, Western Australia and the Northern Territory decided at the Council of Australian Governments (COAG) meeting in Hobart in June 2017 to establish a ministerial forum to focus on key priority matters concerning Australia's northern region. The joint media statement announcing the forum said, 'given the significant land holdings of Indigenous people in Northern Australia, the first meeting of the forum will also bring together ministers with responsibility for Indigenous affairs and senior Indigenous representatives from northern Australia'.

This initiative is strongly supported by northern Australian Indigenous leaders, although it must be said that the decision to include Indigenous interests in the national commitment to develop northern Australia has been a long time coming. 'The significant landholdings' that the joint media release refers to is not just those held by Indigenous landowners who directly own or control 40% of northern Australia – either through statutory land rights laws, native title exclusive possession, pastoral lease ownership, Aboriginal reserves or management of national parks – it also extends to coexisting native title rights to pastoral leases. This means that various Traditional Owner groups across northern Australia have rights to more than 90% of northern Australia, including its coastline.

It is not just landownership that highlights the essential importance of Indigenous inclusion; it is also demography. The Indigenous population of the North constitutes the majority of the region's permanent population. The vast majority of non-Indigenous settlers are not born or live out their lives in the North. Indigenous people in North Australia by virtue of landownership and demography must be seen as primary stakeholders and not a special interest group. Indigenous interests do not engage with governments about our region's future development as advisors or marginal participants. We engage as equity stakeholders with a commitment to inclusive and sustainable development of North Australia.

1.2 TRANSCENDING THE OLD POLITICS OF CONFLICT

Indigenous interests will not participate in a formal engagement process with governments on the basis of securing benefit for our communities from the Northern Development White Paper.[*] Our commitment is to change the existing policy framework for the North's development. The White Paper does not address, in any form, the substance of the Indigenous political position. Long-held Indigenous North Australian policy agenda positions were largely ignored or simply paid lip service to in the White Paper and the federal government's policy response. Those positions are well known because they have been the basis of contested politics in North Australia for more than six decades.

From the Yolgnu Bark petition in 1963; to the Gurindji walk off in 1966; to Aurukun and Mornington Island self-government claims in 1978; to Noonkanbah assertion of self-determination in 1980; to the Burunga Statement of 1988; Mabo in 1992; and Wik four years later, Indigenous people have consistently and courageously asserted our inherent rights. In the negotiations with governments following the Mabo and Wik High Court judgments[†] and the 2009 Northern Australia Land and Water Taskforce Report,[‡] the Indigenous political positions relating to northern Australia are embedded in universal principles:

Firstly: All Traditional Owners view the protection of their *country* and continuation of their culture and languages as paramount.

Secondly: The capacity for Indigenous people to participate in their regional and local economies, and to be independent from government welfare, is critically important to all northern Indigenous groups.

[*] Australian Government (2015). *Our North, Our Future: White Paper on Developing Northern Australia*. Commonwealth of Australia, Canberra.

[†] Recognition of inherent Indigenous rights in common law as part of Australia's legal framework.

[‡] Northern Australia Land and Water Taskforce (2009). *Sustainable Development of Northern Australia*. Department of Infrastructure, Transport, Regional Development, and Local Government, Canberra.

Thirdly: Every Indigenous group aspires to transcend the legacy of domination and exclusion from the authoritarian colonial era and live decent lives in good houses supported by services of a First World nation.

Fourthly: Indigenous people have sought recognition of their customary governance structures within the governance and land planning system that administers the North.

Many royal commissions, parliamentary enquiries, official reviews and judicial hearings have heard the evidence of countless Indigenous and other people about these matters. In a number of notable reports and determinations, bold and visionary recommendations have been put to governments so that Indigenous people's rights can be embraced, and people's economic and social potential realised. With the exception of the Woodward Royal Commission into Aboriginal Land Rights in the Northern Territory, where the recommendations were accepted by the Whitlam Government and translated into law with the unanimous support of every member of federal Parliament in 1976, state, territory and national governments have generally responded negatively to Indigenous assertions concerning the principles outlined earlier.

The White Paper on Developing Northern Australia continues government practice of ignoring the legitimate and reasonable Indigenous claims for recognition and inclusion. Indigenous leaders come to the engagement process not as naysayers and grandstanders but as representatives of people who want to make northern development work for the many who live here, not the power elites who are motivated to extract wealth from the North with little or no return for those who have and will always live here.

1.3 THE CURRENT DEVELOPMENT PARADIGM – RISK AND UNCERTAINTY

In a keynote address to the 2016 Northern Australian Development Conference in Darwin, Peter Yu described the White Paper and the federal government's response to it as flawed because it is based on the false logic of trickledown economics.* There is an inherent assumption that government support for investment in mines and large-scale irrigation works supported by transport, dam, and port infrastructure development will create opportunities for Indigenous employment and transform the North's underdeveloped status. As Peter Yu argued, this targeted support for big project development won't work because it is not accompanied by a range of structural reforms to support Indigenous economic inclusion and Traditional Owners' informed consent for development.

In a situation where Indigenous people are fundamentally excluded from the economy, our people continue to live in structural impoverishment and treated as mendicants waiting for the crumbs in the form of limited employment in specific resource-rich locations to fall from the development table; a pervasive culture of anguish and resentment festers. This is the social and economic reality of North Australia, and in this environment, proposals for development will always involve some level of community opposition on the grounds of environmental, social, and cultural impact. This situation therefore presents a huge investment risk for northern development. How many examples do government and industry need to understand this fundamental point about the risk of investing in the north? James Price Point, Coronation Hill, Jabiluka, Muckaty, Century Mine, McArthur River, Daintree, Carmichael ...† the list goes on.

Peter Yu's central point that the White Paper is defective because it fails to recognize and address this fundamental reason for northern Australia's continuing underdevelopment is supported by the broad spectrum of Indigenous leaders. Whatever political spin one wants to describe competing interests and political conflict surrounding resource development in the North, there is no hiding

* Peter Yu, Keynote Speech, Developing Northern Australia Conference, Darwin, 20 June 2016.

† Major development battles involving significant cultural and environmental issues.

from the fact that entrenched distrust on the part of Traditional Owners to development proposals by government and industry is a critical factor concerning uncertainty and risk. The national government must take care that its approach to northern development does not entrench that risk.

1.4 IDEOLOGICAL DRIVEN OPPOSITION TO INDIGENOUS RIGHTS

Since the passing of the Aboriginal Land Rights (Northern Territory) Act in 1976, the North has been engulfed in political conflict about the rights and capacity of Indigenous people to own, manage and live on their traditional *country*. It has been a grim history of opposition to Indigenous rights by the mining and pastoral industries and southern Australian political power committed to continuing the structures of northern Australian settler colonialism. It has been an ideological war waged against Indigenous society to maintain the power structures of colonial subjugation so that wealth can be extracted from the North in a manner blindly entrenched since the nineteenth century. The simple truth is 'you reap what you sow'.

Whilst Indigenous people have consistently won court battles concerning the recognition of our common law rights and determinations through statutory land claim processes, those victories have not translated into public policy and institutional reform required to decolonize northern Australia and to ensure that development is inclusive of the majority permanent population. For the sake of sustainable development and reliable prosperity for those who live in the North, there comes a time when this ideological war must end. We hope that time is now.

Since the release of the Finkel Report on Australian energy policy in early June 2017,* Prime Minister Turnbull has talked about ending the ten-year partisan political standoff over climate change policy. He has asked for a bipartisan approach without ideology and with economic imperatives to be at the heart of future energy policy development – although it would appear that his biggest political hurdle may be his own party room.

Indigenous leaders are asking that the same principles apply to North Australia's future development. Indigenous leaders come to the formal engagement process about northern development without partisan politics, without ideology. Our interest is to make northern development work for the benefit of the people who 'live' in the North and for the nation as a whole. This is no longer an argument about Indigenous rights. That has well and truly been won and established in Australian law.

The Native Title Act and the Aboriginal Land Rights (Northern Territory) Act have enabled the legal determination of traditional ownership and connection to lands throughout the North. This provides a fundamental basis for investment certainty in the North. But what is missing is a policy framework of economic and social inclusion that can unlock the North's development potential. Politicians, industry leaders and economists often refer to the underlying fundamentals when describing the health of an economy. Indigenous leaders say that the underlying economic fundamentals concerning northern Australia's development must be addressed or this region will remain in structural underdevelopment. The Indigenous argument is that the White Paper did not deal with underlying fundamentals and because of this its development objectives are unrealistic. In truth they are fantasy.

1.5 NORTH AUSTRALIAN INDIGENOUS POLICY AGENDA

North Australia's Indigenous leaders have identified key underlying economic fundamentals requiring policy development and reform as a necessary pre-condition for northern development. In a sense they could be described as development pillars for a new northern Australian development

* Finkel A, Moses K, Munro C, Effeney T, O'Kane M (2017). *Independent Review into the Future Security of the National Electricity Market: Blueprint for the Future*. Commonwealth of Australia.

paradigm – an interwoven framework for sustainable development comprising land reform, access to finance, fiscal policy incentives for social development and Indigenous governance capacity building.

Firstly, the land title system in North Australia needs to be comprehensively overhauled. Across the north's three jurisdictions, pastoral leases dominate the region's land tenure. It is a fragmented mess of nineteenth century titles that needs urgent modernisation.

The High Court Wik decision in 1996 revealed the truth of the pastoral lease in Australian law. It is not a lease in the English common law but simply a license to graze livestock that can coexist with Traditional Owners' rights to use the land.

The Wik decision also revealed a political truth. The loudest and most virulent cries of opposition to Wik and demands for the Commonwealth to wind back Native Title holder's rights to negotiate came from the mining industry. Northern Australia's pastoral rangelands have been viewed by the mining industry as their domain to explore and mine resources on lands covered by pastoral leases. Attempts to restrict the mining industry's access have been fiercely resisted by the miners and their political supporters. Because the mining industry and pastoralists, who for the most part are absentee landlords, enjoy their privileged control over this huge northern Australian landscape at the exclusion or marginalisation of Traditional Owners, reform of this land tenure system has hardly been on the government's policy agenda.

Yet if Australia is to seriously consider the potential of transformational development of North Australia and the potential for generating other uses of the pastoral rangelands such as horticulture, carbon farming and sequestration, conservation and tourism, reform of the pastoral land tenure system has to be addressed. If the nation is serious about reform and inclusion of Traditional Owners and Native Title Holders, creative and innovative thinking should be encouraged.

One significant idea is for governments to transfer the ownership of pastoral leases from the Crown to Native Title Holders who can then lease them with clear conditions for their use. This would be a major transformational change, which could guarantee income for many Traditional Owner groups as well as structurally embed Aboriginal people into the economic and political institutional fabric of the North.

The statutory land title system recognising Aboriginal ownership is also fragmented and inconsistent across the North. The Northern Territory is the high-water mark for statutory land rights in Australia. Queensland Aboriginal land rights law, passed in 1990, is a far more limited form of Indigenous land tenure, whilst Western Australia has no statutory land title recognising traditional ownership whatsoever. Aboriginal communities in Western Australia live on Aboriginal reserves owned by the state, which leases reserve land to Aboriginal corporations. In Western Australia and Queensland, native title determinations often cover existing Indigenous landholding arrangements that governments entered into in during the pre-Mabo *terra nullius* era.

Governments preside over confusion and uncertainty regarding Indigenous landholding status in many instances, particularly in Western Australia and Queensland, which contributes hugely to the North's continuing underdevelopment. A critical issue for North Australia's future development is that much of the Indigenous land estate is not fungible and therefore is dead capital in a classic economic sense. Indigenous owners can't go to the banks to get loan finance to develop their lands like other property owners. This impacts negatively not just on indigenous landowners but the whole region.

The White Paper highlighted the importance of Indigenous land title reform, but little has come of it. There is a critical need to create a fungible land title which does not extinguish Indigenous rights such as what the Aboriginal Land Rights (Northern Territory) Act provides.* This would have an impact on the capacity of Indigenous landowners to generate income and importantly embrace Indigenous people as key equity partners in the North's economic future.

* S 19 and S19(a) of the Aboriginal Land Rights Act, 1976.

Secondly, Indigenous agencies have got to be in the planning wheelhouse: They've got to be able to plot the course of their own development, rather than having to negotiate with external research organisations who are assisting third parties to access and develop Indigenous land. Governments must acknowledge and respect the knowledge and capacity of Indigenous bodies to lead the research agenda concerning Indigenous lands. There is considerable capacity and experience in Indigenous organisations such as NAILSMA, land councils and native title corporations which have well established formal relationships with universities and research bodies. That governments continue to ignore this reality and direct public investment for research and development concerning Indigenous lands to external bodies such as the Commonwealth Scientific and Industrial Research Organisation (CSIRO); this is highly inefficient and not conducive for producing agreed developmental approaches.

It must be recognized that Indigenous people have been colonised and fundamentally excluded from the regional economy that has dispossessed them and usurped their traditional authority meaning that we do not operate from a level playing field. Governments should invest in a comprehensive Indigenous-led research and development strategy concerning ideas for development on Indigenous lands. Associated with this research strategy there should be serious consideration for concessional finance to be made available to Traditional Owners and Native Title Holders under a loan facility possibly modelled on the Northern Australian Infrastructure Facility.*

Thirdly, the Commonwealth should be looking at exploring a fiscal policy framework to encourage development investment on Indigenous properties in the North involving rebates, subsidies, and tax incentives. Indigenous leaders appreciate the constitutional limitations regarding a special economic zone for northern Australia, but there should be serious collaborative investigation between Indigenous bodies and governments about the potential for fiscal incentives for development on Indigenous land in a similar fashion to the way government have supported Australian industries in the past such as car manufacturing or the wine industry.

Fourthly, governments – Commonwealth and State and the Territory, in partnership with Indigenous people, local shires, and industry – should develop a comprehensive social advancement strategy based on a place-based development model that incorporates housing and social services. The current fragmented model of social and community investment across the three northern jurisdictions is a disaster and needs to be totally revamped.

Fifthly, there needs to be a comprehensive investment in Indigenous governance development in terms of decision making concerning Indigenous land management and investment and community social development. The White Paper also highlighted this as an important issue for northern development, but little action has followed. Traditional Owner management of significant land estates in the North means that Indigenous governance capacity must be prioritised as a critical dimension of North Australia's future development. But we must look not just at improving the capacity of Indigenous corporate governance but at the development of North Australia's governance systems. In this regard, the relationship between Indigenous landowning entities and local government is critical to future local and regional governance including harmony with existing Indigenous structures.

1.6 THE EMERGING ECONOMY OF NORTH AUSTRALIA

There is an emerging northern economy based on traditional owners' connection with and knowledge of their *country*. The management and conservation of these lands are of strategic importance to this nation. Conservation is becoming a critical part of the northern economy in terms of employment, tourism, and income generation from carbon farming and sequestration.

* The Northern Australia Infrastructure Facility (NAIF) was established in 2016 as an initiative of the Northern Development White Paper. The NAIF offers up to $5 billion over five years concessional finance to encourage and complement private sector investment in infrastructure that benefits Northern Australia.

About 25% of the North is already recognised as formal conservation reserves, of which half – 154,000 km^2 – are Indigenous Protected Areas (IPAs). In the last two decades there has been significant growth in the northern conservation economy. There are more than 650 Indigenous rangers employed, some of whom work on 33 IPAs across the region. There are also many other Indigenous conservation workers employed in national parks or other aspects of land and sea management. A large proportion of the North is already under formally registered savanna burning projects, reflecting a huge development since 2012.

At the time of writing there are 79 savanna burning projects in the 600 mm rainfall zone and above earning millions of dollars per year. In the Northern Territory alone there are 13 Traditional Owner fire projects with a total value of $40M. There are 32 Indigenous carbon projects delivering 1.9 million carbon credits from savanna burning alone. The burgeoning savanna burning industry shows that promotion of an ecosystem services-based economy will have big benefits for remote indigenous communities across the North. For example, savanna burning projects in Arnhem Land last year earned approximately $10M. Tourism, the native fruits industry and sustainable horticulture, are growing economic uses of northern Australia's lands, a potential that unfortunately is currently retarded. These are not mega-economies but nonetheless are important where orthodox economies don't exist.

1.7 CONCLUSION

A national commitment for the inclusive and sustainable development of North Australia must consider ideas about how best to implement the structural reform that is required. We are talking about transformational change which will need serious engagement and negotiation and the informed consent of Traditional Owners across North Australia.

The key to reform and its implementation is through regional and local agreements that have cultural, geographic and societal integrity and validity. There should be consideration to establishing a comprehensive settlements process to negotiate local and regional agreements so that matters such as land tenure, economic investment strategies, and social development, governance and compensation matters can be negotiated and formalized within a relevant cultural framework. The Yawuru and Mirriwung Gajerrong Indigenous Land Use Agreements in the Kimberley region, and the Noongar Agreement in the southwest of Western Australia, give some indication of what is possible in terms of regional agreement making.

It could be argued that these Indigenous Land Use Agreements within the Native Title Act are akin to a 'domestic treaty', which are well known in North America. It has often been said that the Noongar Indigenous Land Use Agreement is a form of, and the closest example we have, of the type of settlement found in North America. These ideas have been advocated by some groups in their submission to the Senate concerning the Native Title Act Amendment (Indigenous Land Use Agreements) Bill 2017, which was passed in the national Parliament in June 2017. Given the strong commitment of Indigenous people for agreement making expressed in the 2017 Uluru Statement concerning Indigenous recognition in Australia's Constitution,* the potential for engaging Indigenous people about a comprehensive settlement process appears a very sensible and logical approach for transformational development of North Australia.

* See *Uluru Statement from the Heart*; https://www.referendumcouncil.org.au/sites/default/files/2017-05/Uluru_Statement_From_The_Heart_0.PDF.

CHAPTER 2

Change and Continuity

The North Australia Cultural Landscape

Ricky Archer, Jeremy Russell-Smith, Seán Kerins, Robert Costanza, Andrew Edwards, and Kamaljit K. Sangha

CONTENTS

Our story is in the land ...
It is written in those sacred places.
My children will look after those places,
That's the law.
When that law started?
I don't know how many thousands of years.
Europeans say 40,000 years,
But I reckon myself probably was more because ...
It is sacred.*

2.1 THEMES ADDRESSED BY THIS BOOK

This book sets out a vision for developing North Australia that differs substantially from a business-as-usual preoccupation with large infrastructure development, 'food bowls', and unfettered exploitation of mineral and pastoral resources. Despite an overlay of just 150 years of European settler colonialism and sporadic attempts at European-style economic development, North Australia supports resilient Indigenous cultures and knowledge systems spanning tens of thousands of years. Outside small numbers of urban centres scattered across the sparsely settled North, the population is substantially Indigenous. Forms of legal title over much of the landscape and seascape belatedly has been, or is in the process of being, implemented. Despite many Indigenous groups owning vast areas of land they have been excluded from policy development processes that shape land and resource use, resulting in Indigenous peoples experiencing significant social and economic disadvantage.

* Bill Neidjie, Stephen Davis, Allan Fox (1985). *Kakadu Man ... Bill Neidjie*. Mybrood, NSW.

Indigenous land and sea management, or what Indigenous Australians term Caring for *Country*, is an Indigenous development strategy. As such, it is a life project, seen elsewhere in similar settler colonial contexts like Canada, the United States, and Aotearoa/New Zealand, which is 'embedded in local histories and encompassing visions of the world and the future that are distinct from those embodied by projects promoted by the state and markets'.[1] The dominant ideas and practices about Indigenous 'development' promulgated by the settler states are built solely around problems and deficits, and are rejected by Indigenous peoples engaged in caring for *country*. Instead, what we are seeing across North Australia, as is occurring in other places around the world,[2] are distinct Indigenous forms of land tenure built around Indigenous peoples' common property, along with the cultural and social capital of Indigenous peoples being used as the building-blocks for a new development pathway. Together these can bring both social benefits to remote Indigenous communities and environmental benefits to wider Australia, through the protection of natural systems and the ecosystem services they provide.

Some Indigenous peoples frame this as a struggle for environmental justice which is part of a global Indigenous movement to resist 'companies, governments, policies, and other forces that threaten to fragment, displace, assimilate, or drive them toward cultural disintegration'.[3]

The United Nations Declaration on the Rights of Indigenous Peoples supports Indigenous life projects. Article 23 of the Declaration determines that 'Indigenous peoples have the right to determine and develop priorities and strategies for exercising their right to development'. Article 32 states that 'Indigenous peoples have the right to determine and develop priorities and strategies for the development or use of their lands or territories and other resources'.[4]

Vital to these new development pathways is Indigenous ecological knowledge that plays an essential role in identifying and solving environmental problems.[5] Its use in problem solving (along with scientific knowledge) is crucially important in recalibrating the power relationship between Indigenous peoples and government agencies – past models that placed Indigenous ecological knowledge and scientific knowledge as 'opposing extremes of a scale of accuracy and therefore, authority'[6] need to be rejected. The use of both Indigenous ecological knowledge and science is important because it facilitates ongoing relationships between Indigenous peoples, scientists and government officials working together to solve environmental problems; often resulting in more effective and efficient natural resource management outcomes.[7]

The industrialized world is finally rediscovering the interdependent relationship between humans and the rest of nature that Indigenous cultures have always known. The industrial revolution and some religious traditions have emphasized the distinctions between humans and 'nature' – that humans are somehow above, apart from, or fundamentally different from the rest of nature. In fact, the more we learn about the way the world and its complex interconnected systems function, the more we recognize that *Homo sapiens* is, and has always been, an integral component of the ecosystems it is embedded within. Humans are not apart from nature but are a part of the natural world, and their health and well-being cannot be understood or managed separate from that complex and evolving context. This is very consistent with Indigenous Australians' concept of *country*.

The concept of *ecosystem services* makes this interdependence with the rest of nature more apparent and quantitative. It does this by analysing, modelling, quantifying, and valuing the degree to which humans are connected with and benefit from the ecosystems that enclose them. Ecosystems provide a range of services that are of fundamental importance to human well-being, health, livelihoods and survival.[8–11]

The idea that preserving the environment *as an asset*, rather than an impediment to economic and social development, is both very old and very new. For most of human history, at least until the start of the Industrial Revolution, the benefits humans derived from the rest of nature were well recognized and embedded in various cultural rules and norms. Parts of forests, lakes, wetlands or mountains were often deemed sacred and off limits. But it is no coincidence that these sacred natural assets also supplied essential life-support services for the communities involved. This is in stark contrast to the post-industrial view in much of the Western world that nature is merely a pretty picture – nice to enjoy if you can afford it but not essential to the more important business of 'growing the economy'. Too often, when the issue of conservation of the environment has entered public or political discussions, it has been purported to come at a cost, and the discussion has been framed as 'the environment versus the economy'.

Probably the most important contribution of the widespread recognition of 'ecosystem services' is that it reframes the relationship between humans and the rest of nature to be more consistent with what we know, and what Indigenous cultures have always known. A better understanding of the role of ecosystem services emphasizes our natural assets as critical ingredients to inclusive wealth, well-being, and sustainability. Sustaining and enhancing human well-being requires a balance of all of our assets – individual people, society, the built economy and ecosystems. This reframing of the way we look at 'nature' is essential to solving the problem of how to build a sustainable and desirable future for humanity and for northern Australia – a goal that we all share.

The ecosystem services concept makes it abundantly clear that the choice of 'environment versus economy' is a false choice. If the environment contributes significantly to human well-being, then it is a major contributor to the *real* economy, and the choice becomes how to manage *all* our assets, including natural and human-made capital, more effectively and sustainably.[12]

Drawing on case studies from across the North we focus particularly on developing a culturally appropriate, ecologically sustainable, and economically diversified North Australia land sector economy that supports both Indigenous people's responsibilities to *country* and Indigenous aspirations, as well as enhancing development opportunities for society as a whole. It is no secret that a plethora of top-down policy agendas have and continue to fail Indigenous peoples living in remote regions. Our purpose is to help breach that gap – not by attempting to revisit and revitalise tried and failed approaches but by acknowledging and harnessing Indigenous cultural strengths and knowledge systems for looking after *country* and its people as part of an inclusive, responsive, innovative, diversified ecosystem services economy. This purpose is not about advocating for mutual assimilation. Rather, it is about recognising that there are now two major coexistent cultural traditions which, from time to time and place to place, may intersect constructively to provide mutual benefit.

Building on contemporary challenges identified in the opening chapter, here we set out to establish an introductory framework for addressing key themes and chapters which follow. We focus first on giving voice to ongoing connections to, and the fundamental importance of, maintaining law, culture, and *country*, and the aspirations of Indigenous people across the North. Such connections and aspirations can be seen to differ essentially from those of non-Indigenous residents – for a start, one's *country* is not real estate which can be simply sold or traded. The remainder of the chapter characterizes the North Australia focal region addressed by this book – its defining geographic features, the long history of human occupancy in the context of significant landscape change, and contemporary patterns of settlement, land use and socio-political organization.

Succeeding chapters describe: Chapter 3 – the grim, recent history of Indigenous dispossession and, in response, the ongoing assertion of Indigenous rights; Chapter 4 – a critique of recurring historical and contemporary short-sighted regional socio-economic misadventures; Chapter 5 – the case for developing a diversified land and sea sector economy that recognises the generally poor economic and environmental performance of the predominant land use sector, the pastoral industry, and which takes advantage of novel, emerging ecosystem services (e.g. carbon farming) markets and Indigenous land management enterprise development opportunities.

The final three chapters address associated governance challenges to sustainable development: Chapter 6 – bottom-up cultural requisites for empowering community-based resilience for developing robust and prosperous enterprises; Chapter 7 – a regional case study addressing the magnitude of the challenges involved and potential solution pathways; Chapter 8 – priorities for governance reform, from effective grass-roots processes to genuine cross-jurisdictional cooperation on policy and investment priorities.

CONTEMPORARY INDIGENOUS VOICES

My name is Ricky Archer. I am a Djungan man; that is where I start. My *country* is in the Western Tablelands region of North Queensland and encompasses land and waters north to the Mitchell River, east to Hahn Tableland, south along the Walsh River, and west to Elizabeth Creek. At the heart of Djungan *country* lies Nguddabulgan (Mount Mulligan: Figure 2.1), and our home is Wungu, a small block of land on the outskirts of Dimbulah, on the Walsh River. *Country*, culture, and kinship are values that drive me (Figure 2.2). *Country* is home, an un-tradeable commodity, a birth right and a deep pool of resource. Culture gives me identity, my connectivity to place and systems of knowledge. Kinship covers family, continuity and succession, and is the platform of my being. Before I get to stories across the North, I will begin with mine. *Country* has different meanings and provides *country* men* and women with different opportunities dependant on available resources from *country*.

I value richness of traditional knowledge, biodiversity and healthy *country* over mineral resources, production value of grasses, or commercial value of water allocations. At the intersection of tradition and commerce lies potential opportunities in culture-based economies, emerging markets encompassing provision of ecosystem services, and tried and trusted caring-for-*country* initiatives such as the establishment of Indigenous Rangers by Traditional Owner groups for managing cultural and natural capital. *Country* is an enabler to success; people are sometimes inhibitors.

To gain an understanding of on-*country* perspectives and contemporary Indigenous voices, I spoke to countrymen across North Australia and asked what *country* means to them. I wanted to understand how traditional and or contemporary values could be used as enablers rather than inhibitors in the Northern Development context. Tales and topics from

Figure 2.1 Nguddabulgan (Mount Mulligan), Queensland.

* Generally used as a gender-neutral term of respect used by various Indigenous peoples across the north to signify cultural connections with their traditional estates (*country*).

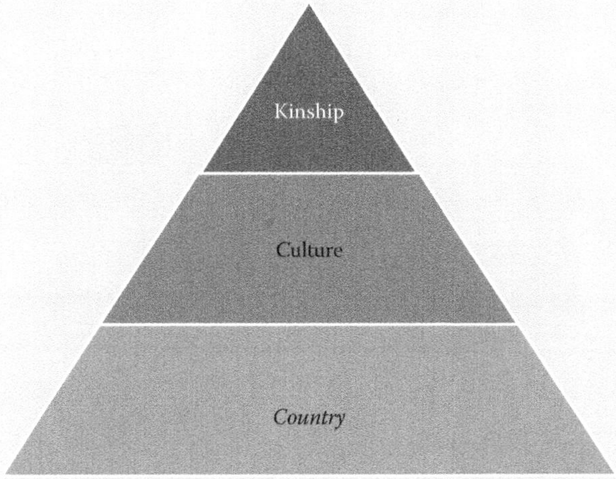

Figure 2.2 *Country*, culture and kinship relationship.

countrymen resonated across the North. Shared visions and aspirations were unveiled through people yarning about what their *country* does for them and what they do for their *country*.

RIGHT PEOPLE NEED TO SPEAK FOR *COUNTRY*

We need strong people, and we need strong *country*. People need *country*, and *country* needs people. Leaders and *country* are alike: they are cyclical, where some cycles are better than others. The functions of a strong leader generally remain the same; what changes are the evolving social and cultural issues to which leaders must adapt and respond. The functions of *country* generally remain the same; what can be argued is how environmental factors such as an altering climate affect its functions.

In a time where these two core themes, people and *country*, are continuously cycling there is an apparent need for good leadership to influence outcomes in the Northern Development context. We need to recognise and respect cultural protocols which outline terms of leadership. Leadership is also dependent on place, on *country*. The leadership style for Bininj in West Arnhem land differs to the leadership style of mob in the Gulf of Carpentaria or countrymen in the Kimberley.

A common reoccurring question asks what makes a good leader. The question we should be asking is how we can support our leaders. We need leadership to influence succession and establish pathways for practitioners of the future. We must build and strengthen structures to ensure future generations have opportunities to develop as leaders. Leaders come and go, but their legacy and passion are ingrained in the people they have influenced. Everyone has their own interpretation of good and bad leadership; let's learn from the past, nurture the present, and hope for the future. At the least we can follow this general rule – right people need to speak for *country*.

GOVERNANCE

Phil Rist, Nywaigi man, from North Queensland Wet Tropics/Cardwell, stresses the importance of good governance. Rist is the executive officer of Girringun Aboriginal Corporation and is advocating approaches where business-on-*country* is driven by cultural authority (Figure 2.3). Girringun's governance framework supports numerous Traditional Owner groups to engage in economic development. For this to work, Rist states Girringun's constituents need to have, 'trust and faith in Girringun', to act in the best interest of countrymen.

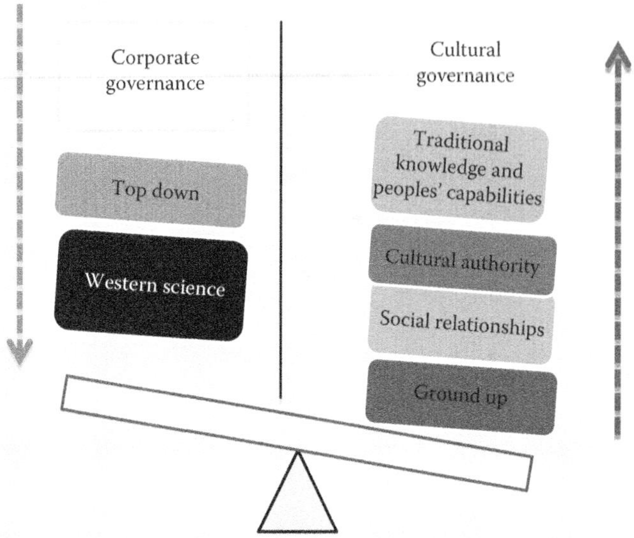

Figure 2.3 Cultural and corporate governance systems.

This is a continual work in progress and has taken many years to develop. The current economic and political appetite for northern development does not pay consideration to the value and importance of such well-established, 'trust based' relationships.

Sights fixed 700 kilometres (km) towards the north in Cape York is Chuulangun, Kaanju *country*. David Claudie is a senior Kaanju man with the cultural obligation of looking after his *country*. Claudie echoes the need for good governance. He has established strategic partnerships with research and academic institutes and is a published author and also spins a good yarn in the plight to share his successes and challenges. Kaanju operate with the intricacies of cultural/traditional indifferences which somewhat promote the continuation of customary systems in place on their *country*. 'Right people must speak for *country*', Claudie amplifies this statement in a way that suggests there is no alternative.

Follow a dusty stock route 500 km south-west and you end up in cattle *country* in the Gulf of Carpentaria. Joseph Rainbow is a Senior Kurtijar man, based in Normanton. Rainbow sits on the Morr Morr Pastoral Company Board which oversees the management of one of the most successful Indigenous owned and managed pastoral properties in the *country*, Delta Downs Station. This mob has developed a three-tiered governance framework using a 'grassroots' approach to ensure the requirements of business, culture, and *country* are met. This framework supports a successful beef enterprise, manages legislative requirements, and achieves cultural and environmental aspirations of Kurtijar people.

VALUES

The tangible value of *country* is great; the intangible value of *country* is greater. Measuring the assets of one's *country* can be, rightly so, overshadowed through the indescribable feelings *country* gives you. Spiritual connection to *country* is difficult to explain; Big Bill Neidjie,* Bunitj Elder (now deceased) takes a crack at it.

* Bill Neidjie (1989). *Story about Feeling*. Magalaba Books, Broome.

You listen to my story and you will feel im
Because spirit e'll be with you.
You cannot see but e'll be with you e'll be with me
This story, just listen careful.

Opportunities for economic development exist where economic scales can find balance between the two value sets (Figure 2.3). There is a perceived risk that inappropriate development can result in some losses to *country*. Once *country* loses its ability to heal its people, it is no good. *Country* can become high in commercial value but low in spiritual value. This trade-off must be carefully thought to minimise the negative impacts, by applying balanced and managed approaches before the next steps are taken.

Beware the third-party syndrome – it starts with an individual, the individual identifies an opportunity; enter a third party. Business-on-*country* is a sector that almost always requires third party investment or support. The 'third-party syndrome' changes the dynamics of an opportunity through devaluation of the individual. Further explained, a traditional owner can identify an opportunity on *country* which incorporates their profound knowledge of *country* and culture. To pursue this opportunity the traditional owner will most likely engage with a third party for initial investment or support. This step immediately places the third party in the 'driver's seat' and elevates the importance of a third party rather than highlighting the value of the Traditional Owner knowledge base which created the opportunity. The treatment of the third-party syndrome is a lengthy process, and the symptoms of a cure are when you can see knowledge of *country* and knowledge of culture are valued higher than financial investment.

POTENTIAL OF *COUNTRY*

With attention given to leadership, governance, and values, the northern landscape has somewhat untapped potential. What is now made clear are tapped resources through mining which are unsustainable and non-renewable. However, *country* offers many sustainable resources and services, and related economic options (Figure 2.4). Traditional Owners could harness this potential and be in the driver's seat of northern development. We just need the keys.

Leila Nambidja is a Gorrgoni senior Traditional Owner of Nangak. Her outstation overlooks the Tomkinson River Floodplain, south of Maningrida, West Arnhem Land. Her passion for *country* is entangled between knowledge of *country*, knowledge of culture, and aspirations for the future. Leila has identified potential and initiated opportunities through an established platform of in-depth traditional ecological knowledge. Culture and conservation-based tourism opportunities have been initiated at various stages; native plant nurseries are progressing; engagement with the education and academic sector is ongoing. However, something is always missing. There are numerous opportunities for potential on Leila's *country*, which all come with limitations.

The opportunities on *country* for Leila come and go and are ever present. What always remains and is growing every day is her knowledge of *country* and knowledge of culture. The economic values of an industry or sector fluctuate with simple market trends such as supply and demand. In her culture and her society, the knowledge Leila possesses is an invaluable commodity where the stocks are continually rising.

Helen Williams is a Wurnal woman from Saltwater *country* of Maningrida, West Arnhem Land. Helen advocates for biodiversity and environmental conservation with a strong focus on cultural governance and ensuring *country*, language, and culture are maintained for the benefit of future generations. Helen is the senior landowner for Entrance Island and Second Island, which are important nesting sites for migratory birds and endangered sea turtles. There are also numerous sacred sites that have been documented in attempts to improve management

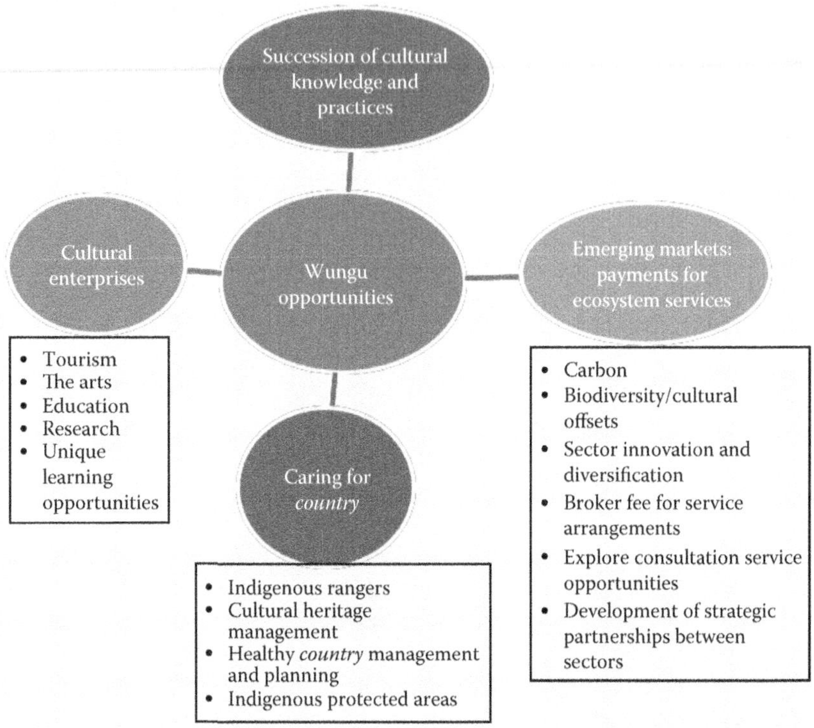

Figure 2.4 Potential of Wungu, Djungan *country*.

initiatives and increase awareness. Helen is a staunch opponent of economic influences outside of her core ideologies. She proudly waves 'Frack Off' banners and promotes anti-coal steam gas initiatives and urges her community to prioritise land and sea health over economic promises. Helen is working closely with Djelk Rangers to develop collaborative opportunities that deliver cultural, environmental, and economic opportunities.

Bart Pigram is a Yawuru man from Broome, with ties to Wadjarri *country* encompassing the Murchison River, south in the Pilbara region. Pigrim owns and operates a small-scale culture-based tourism operation on Yawuru *country*. He stresses the importance of economic diversification to enable opportunities for other Yawuru people. Pigrim highlights the notion, in reference to the pastoral sector, that 'it's a one-eyed industry' and calls for collaboration and support to grow potential opportunities which promote the region's rich cultural and environmental assets. Yawuru themselves manage a successful pastoral operation, Roebuck Plains, and have also engaged with the broader pastoral industry to negotiate access to Yawuru *country*. Pigrim claims there are opposing views about *country* between the sectors: 'the pastoral industry limits access for our people … gates are locked'. To find the balance between business and culture, Yawuru have to 'keep a minimal footprint on our own *country*, which sounds wrong to me'. Pigrim advocates to maintain cultural connection and strengthen cultural identity through better access to *country* but states, 'Diversification of leases needs to happen', to achieve this.

There are two components Pigrim highlights as priorities: being on *country* with Elders and maintaining the health of *country*. In the economic development context, these pillars should be supported with good leadership and frameworks which enable social return to Yawuru people. 'There needs to be cultural shift in leadership', Pigrim reiterates, so as to ensure the negative traits of leaders do not filter down to younger generations. Pigrim advises fellow youth to 'remain rooted in what you're doing, and maintain your vision', while keeping close counsel with Elders. Pigrim's latter priority stresses the importance of healthy *country*, and he delivers a one-liner that perfectly encapsulates the collective views of countrymen across the north, 'It's in our DNA to be looking after *country*'.

The Yiriman Project was established in 2000 as an on-*country* program developed by Nyikina, Mangala, Karajarri, and Walmajarri Elders of the Kimberley region. The project initiated out of a need for Elders to support and improve the social and cultural health of younger generations. Annie Miljin and Martina Watson are both Nyikina women working in the Yiriman project.

The old lady,* Annie Miljin, proclaims, 'We never had no funding to start Yiriman'. The project has grown since then and has gained widespread support and recognition for their achievements and successes. Martina Watson, Annie's daughter, explains: 'Every plant, animal, they got a story. They connected to dreaming, they connected to land, they connected to *country*'.

The cultural context of business on *country* is one small aspect of what Yiriman does. The program integrates Caring-for-*Country* initiatives based on strong traditional ecological knowledge, with social outcomes for Indigenous youths. Yiriman efforts also enable the importance of Elders and their knowledge to shine. The program encourages participants to engage in cultural economies whilst delivering social outcomes for the improvement of future generations.

The Yiriman women call for more secure funding to deliver longer-term outcomes. 'We'd like to see this more for our young people', Martina explains. She continues to urge future generations to 'realise the importance of having the knowledge of culture, law, and *Country*'.

Through listening to their stories and successes, these women exemplify the Yiriman logo written on their shirts: 'Building stories in our young people'. The future of Indigenous peoples in the North is in the hands of present and future generations; if future generations mirror the sentiments, values, and aspirations of young leaders such as Yawuru's Pigrim or Nyikina's Watson, the North will succeed.

The context of northern development needs to be valued case by case. The perspectives of countrymen and women will differ regionally, and I think we need to make priorities to fully inform countrymen, so they can make decisions on their own and for their own *country*. Regardless of opportunities, I will lean towards whatever will sustain my *country*; whatever will support my *country* to be there for my children and their children. I welcome the potential economic opportunities but acknowledge that what *country* gives us we should repay tenfold. After all, *Country*, culture, and kinship is what drive me. I am a Djungan man, and that is where I finish.

Ricky Archer

* Old lady and old man are terms of respect for seniors.

2.2 A LONG HISTORY OF OCCUPATION

Well-dated archaeological sites at least 40,000 to 65,000 years old are known from the Kimberley,[13,14] the Top End,[15–17] and Cape York.[18] At the time of this early migration, Australia was joined to New Guinea by an extensive exposed continental shelf, and sea levels would have been around 60–70 meters (m) lower than they are today[19,20] (Map 2.1). Once these early immigrants had made the relatively short sea crossing from Timor, or through the Moluccas to present-day New Guinea, genetic evidence suggests that people spread rapidly around the Australian coastline and then into the interior.[2]

after 6k YBP	Sea-level stabilised at ~present level,[21] with increasing rainfall variability associated with ENSO after 4k YBP.[22–24] Population increased exponentially from ~2k YBP,[25] associated especially with expansion of freshwater floodplain resources.[26] This whole period marked by very substantial technological, artistic, cultural developments.[26–30]
6–16k YBP	Rapid sea-level rise from 18k YBP flooding continental shelf; New Guinea finally cut-off from Australia 9k YBP; stabilising at present level ~6 kYBP.[21] Regional climate wetter especially from 9–6k YBP.[22–23] Regional population beginning to increase after 10k YBP.
22–18k YBP	Last glacial maximum (LGM): sea-level 120 m below present,[31,21] climate cold and arid.[32] Regional population low and scattered.[25]
40–30k YBP	Widespread evidence of habitation across the north,[33] likely sparsely distributed and population in low thousands.[25] Sea-level ~70–80 m below present.[31] Climate getting cool and arid across the north[32,34]. Peak megafaunal extinction around this time.[35]
55–60k YBP	People entered Australia from the north, rapidly spreading around the coast.[33,36–38] Small founding population.[25,37] Sea-level ~60 m below present.[31] Climate cooler but likely much wetter than present across the north.

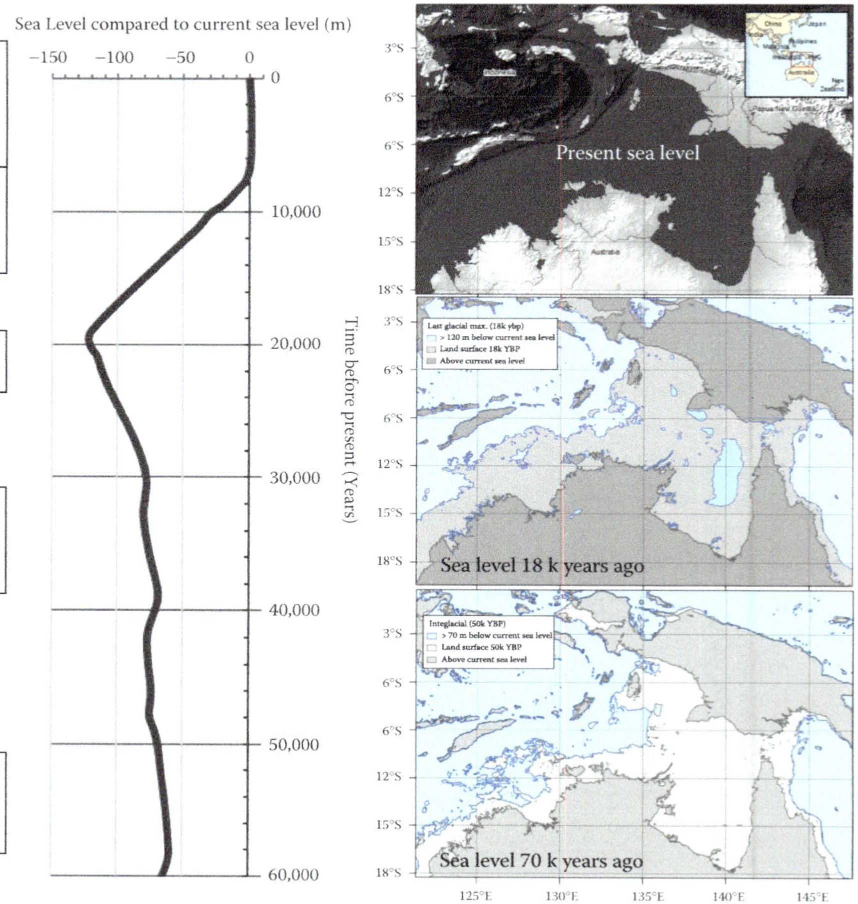

Map 2.1 The long prehistory of Indigenous occupation as currently understood from palaeorecords. For numerical citations referred to in text boxes, refer to the references section for Chapter 2.

Current archaeological and demographic models suggest that relatively low population numbers (measured in tens of thousands of people) fluctuated throughout the early period of occupancy, were especially low in the arid, colder period known as the Last Glacial Maximum (or Ice Age) around 20,000 years ago when sea levels had fallen to 120 m below current levels (Map 2.1), and remained low until the start of the modern epoch, known as the Holocene, about 10,000 years ago.[25] By the time of European colonization in the late eighteenth century, it is estimated that rapid expansion of the Aboriginal population in the late Holocene had resulted in a continental population of the order of one million people, spread across all habitat types including Australia's dominant semi-arid and arid core.[25,39,40]

After the very rapid rise in sea-level following the Last Glacial Maximum and the associated submergence of the northern continental shelf, New Guinea was finally cut-off from the Australian mainland around 9,000 years ago, and sea-level stabilized at around present levels some 6,000 years ago[21] (Map 2.1). A rich extant Indigenous oral tradition from around Australia describes this 'great inundation' (Box 2.1).

BOX 2.1 ORAL TRADITION OF THE GREAT INUNDATION

BENTLEY JAMES

NORTH AUSTRALIAN LAND & SEA MANAGEMENT ALLIANCE LTD DARWIN

Aboriginal oral traditions describing sea-level rise and coastal flooding show an enormous depth of cultural continuity that corroborates empirical evidence of postglacial sea-level rise.[1–3] Dating Aboriginal stories from 21 locations has been calculated against the sea-level envelope for Australia.[4] These traditions appear to have endured over a period spanning 13,000 years BP.[3]

The extraordinary longevity of trans-generational oral traditions is continued in current Indigenous bilingual education curricula in the Northern Territory. Yirrkala School deputy principal, M. Ganambarr-Stubbs, in two-way Yolngu curriculum teaching of oral traditions, describes an inundated 'island called Gunyipinya: That water came pouring, like a tidal wave, from the sea, broke open the land'.[5] This is one of many such examples of oral traditions about coastal inundation among Yolngu people from north-east Arnhem Land.

Neville White[6–8] details genetic evidence for a theory of westerly migration from the east into north-east Arnhem Land by patrilineally organised Pama Nyungan (PN) language speakers following rising sea levels during the early Holocene. The migration of PN speakers, with distinctive patrilineal attachment to land has implications for transmission of genes through kinship networks along economic, ritual, and connubial lines.[7–10] White[7] observes this westerly displacement supported in Yolngu religious beliefs: 'The Djangkawu and his sisters, principal ancestral beings of the Dhuwa moiety … are believed to have come from the east in a bark canoe from the island Baralku far out to sea'. Anthropologists Berndt and Berndt reported that these principal ancestral beings 'caused floods which formed the sea between Elcho Island and the mainland'.[7]

The following Yan-nhaŋu story from Elcho Island recounts the travels of the three named *waŋarr* (ancestral/creator-beings) that paddle a sacred canoe to the west singing the songs of the sacred sites in the sea, as they lay down the language and the law. As the Yan-nhaŋu waŋarr paddle the sacred canoe, they give names to the tides, winds, birds, fish, and other land and sea phenomena.

(*Continued*)

BOX 2.1 (Continued) ORAL TRADITION OF THE GREAT INUNDATION

Wayanha balay ŋurrunnaŋgalbu bnaŋ'ku ŋurrunaŋgalbu nyeninha balay manana waŋanha 'galiyana ŋalili gurrku baŋ'ku Bral-bralŋa ŋali gurrku wanha galiyana. Yan-nhaŋu wäŋana wanha dhanthanaba balay wänbalili. Nyeninha balay mananha ŋali gurrku manaba galiyana ŋunha dhakalŋuru wäŋana balay dhakal waŋgala balay waŋgala yakarra yirrpana balay wäŋana' Dunhuba ŋali gurrku dhabakthun yo wanha galiyanaba garanha galiyana garanha ŋurrunaŋgalbu ga ŋurrunaŋgalbu galiyana balay garanha yirrpana baŋ'ku dhakal waŋgala ŋali galiyana baŋ'ku garanha dhakal waŋgalaŋa wäŋana go nhaŋu waŋgala ŋumunyu baŋ'ku mananha galiyana yakarraway lipalipa ŋurrunaŋgalbu ga ŋurrunaŋgalbu galiyananha baŋ'ku garanha yo galiyanha ba garanha ŋumungarra dhula madayinŋagu wäŋanalaba nhaŋuba riya.

The two named sisters paddle across the sea. They spoke in the language of Yan-nhaŋu as they travelled towards the sunset. As they paddled they created the island and they spoke out the names of places. The two sisters and their brother paddle to the west from the island in the sacred canoe; they sing and paddle the sacred canoe, and they sing and name the sacred places in the sea. They paddle on and put more sacred named places in the sea. (translated by Baymarrwaŋa)

The following illustration depicts a few of the numerous named Dhuwa and Yirritja Yan-nhaŋu ancestors and their travels in the Crocodile Islands called dreamings. These journeys are linked by songs to named sites created by the ancestors. These ancestors, sites, and songs share ancestral essences with the people who inherit their names, ceremonies, and responsibility for their care and intergenerational reproduction (Figure B2.1.1).

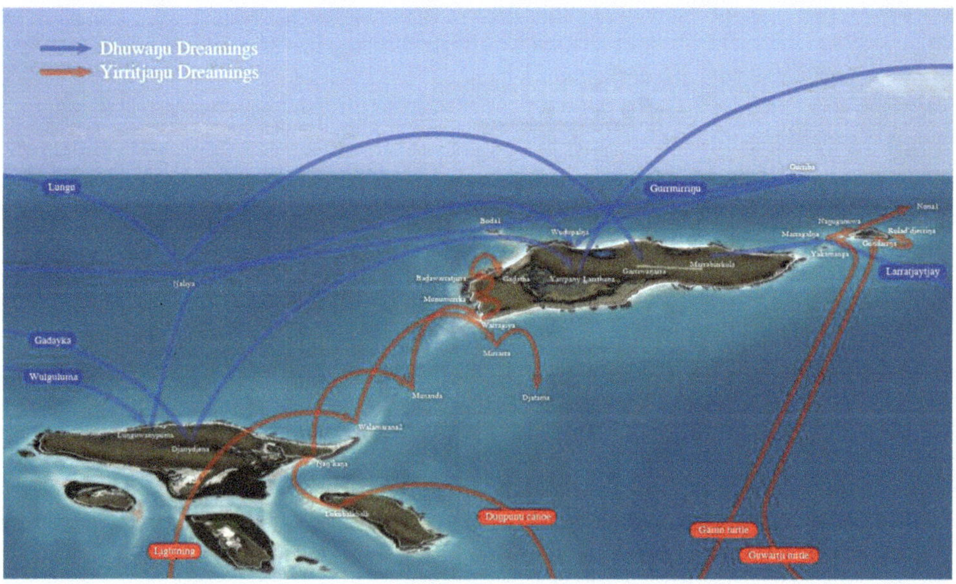

Figure B2.1.1 Dhuwa and Yirritja Yan-nhaŋu dreamings. (From Baymarrwaŋa, L., Senior Australian of the Year 2012. https://www.australianoftheyear.org.au/honour-roll/?view=fullView&recipientID=780.)

(Continued)

BOX 2.1 (Continued) ORAL TRADITION OF THE GREAT INUNDATION

Indigenous oral traditions of extraordinary longevity reflect local adaptation to continent-wide catastrophic environmental changes with distinctive local continuities in associated linguistic, genetic, economic, and cultural patterns.[7,10–13]

REFERENCES

1. Blake BJ (1991) Woiwurrung, the Melbourne language. In *Handbook of Australian languages* (Eds Dixon RMW, Blake BJ), pp. 30–122. Australian National University Press. Canberra, Canada.
2. Flood JM (1995) *Archaeology of the Dreamtime: The Story of Prehistoric Australia and Its People (revised edition).* Angus and Robertson, Sydney, Australia.
3. Nunn DN, Reid NJ (2016) Aboriginal memories of inundation of the Australian coast dating from more than 7000 years ago. *Australian Geographer* **47**, 11–47.
4. Lewis SE, Sloss CR, Murray-Wallace CV, Woodroffe CD, Smithers SG (2013) Post-glacial sea-level changes around the Australian margin: A review. *Quaternary Science Reviews* **74**, 115–138.
5. Ganambarr-Stubbs M (2017) http://mobile.abc.net.au/news/2017-05-25/indigenous-education-in-a-modern-world/8555368.
6. White NG (1976) A preliminary account of the correspondence among genetic, linguistic, social and topographic divisions in Arnhem Land. *Mankind* **10**, 240–247.
7. White NG (1979) The use of digital dermatoglyphics in assessing population relationships in Aboriginal Australia. *Birth Defects: Article Series* **15**, 437–454.
8. White NG (1997) Genes, languages and landscapes in Australia. In *Archaeology and Linguistics: Aboriginal Australian in Global Perspective* (Eds McConvell P, Evans N), pp. 207–236. Oxford University Press, Melbourne, Australia.
9. Allen H (1997) The distribution of large blades: Evidence for recent changes in Aboriginal ceremonial exchange networks. In *Archaeology and Linguistics: Aboriginal Australian in Global Perspective* (Eds McConvell P, Evans N), pp. 357–377. Oxford University Press, Melbourne, Australia.
10. McConvell P, Evans N (1997) Clues to Australia's human past: Pulling together the strands. In *Archaeology and Linguistics: Aboriginal Australian in Global Perspective* (Eds McConvell P, Evans N), pp. 1–16. Oxford University Press, Melbourne, Australia.
11. Keen ID (1997) A continent of foragers: Aboriginal Australia as a 'regional system'. In *Archaeology and Linguistics: Aboriginal Australian in Global Perspective* (Eds McConvell P, Evans N), pp. 357–377. Oxford University Press, Melbourne, Australia.
12. James B (2009) *Time and Tide in the Crocodile Islands: Change and Continuity in Yan-nhaŋu Marine Identity.* Ph.D. Dissertation. Australian National University Canberra.
13. Baymarrwaŋa L, James B (2014) *Yan-nhaŋu Atlas and Illustrated Dictionary of the Crocodile Islands.* Tien Wah Press, Singapore.

The resultant drowning of coastal river valleys was followed by the 'big swamp' phase, characterized by initial development of extensive mangrove systems and then, over the past few thousand years, ongoing replacement by resource-rich freshwater wetland systems.[26,41] In the early Holocene rainfall conditions across northern Australia were likely more humid than today, but increased rainfall variability and substantially more arid conditions have subsequently characterized much of the Holocene.[22,27,42] In line with these changing environmental conditions, the Holocene has also

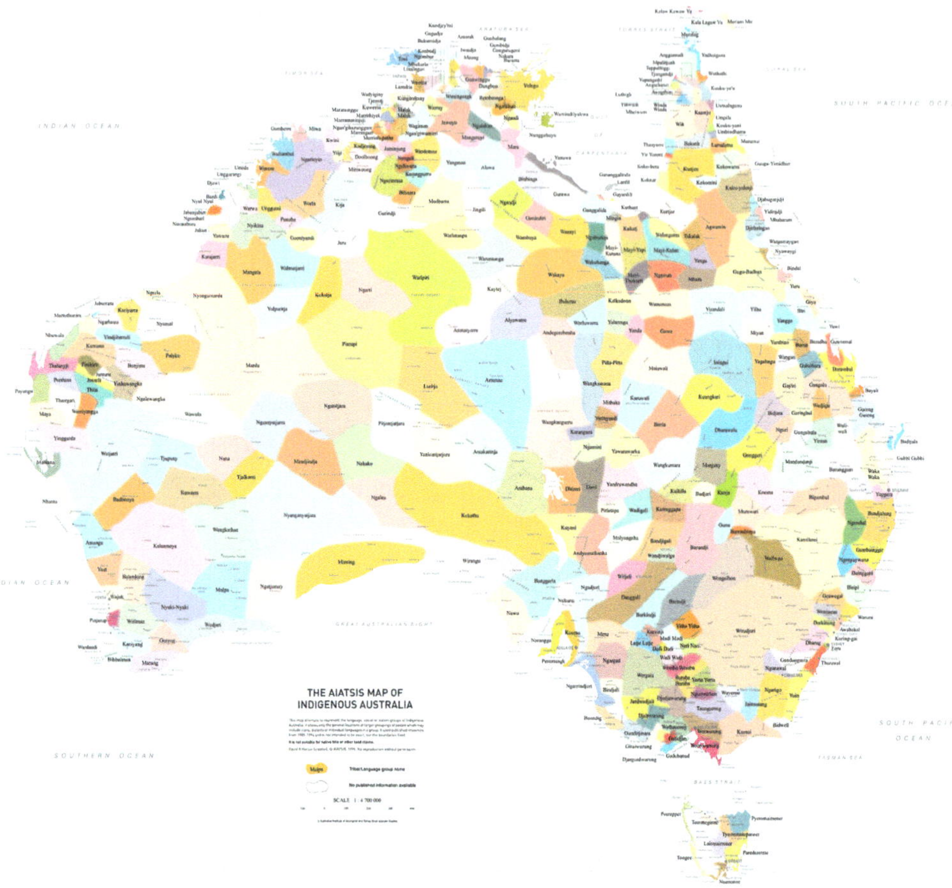

Map 2.2 Indigenous language map compiled by David Horton. (From Australian Institute of Aboriginal and Torres Strait Island Studies [AIATSIS], reproduced with permission.)

witnessed significant cultural developments, including the spread of languages (Map 2.2[30]), artistic styles[27,28] (Figure 2.5), material technologies,[26,29] and ongoing changing occupation patterns.[39,40]

Given this long and varied period of occupancy, there has been considerable speculation about the impacts of early Indigenous people's burning and hunting activities. With respect to burning, the admittedly sparse and geographically biased charcoal record 'provides little support for large-scale impacts of human activity on Australasian fire regimes'.[43] Rather, lack of correlation between archaeological and charcoal records points to a much stronger association with external climatic forcing.[44] However, interpreting landscape-scale changes in the charcoal record (including from respective soot, microcharcoal and macrocharcoal fractions) remains highly challenging.[44–46]

More contentious has been the implication that people's burning and associated hunting activities resulted in the demise of much of Australia's unique megafauna (reptiles, birds, and marsupials over forty kilograms).[35,47–51] Key assumptions advanced by proponents of this 'ecosystem collapse' hypothesis are that megafauna extinctions occurred mostly in close association with human arrival (with a peak extinction probability at approximately 41,000 BP[21]), with 'little evidence for exceptional climate change around [that] time'.[48] However, conflicting evidence indicates that (a) substantial

(a)

(b)

Figure 2.5 Arnhem Land prehistory rock-art record illustrated by (a) a now extinct thylacine from the 'pre-estuarine' period, before sea-level attained present levels 6,000 years BP (Courtesy of Peter Cook.) and (b) a modern barramundi fish (foreground), in the contemporary 'X-ray' style which became fully developed over the last few thousand years after sea-level stabilized (Courtesy of David Hancock.). (From Chaloupka, G. *Journey in Time: The World's Longest Continuing Art Tradition*. Reed Books, Chatswood, Australia, 1993.)

megafaunal (and smaller faunal) extinction had occurred prior to the arrival of people, associated with longer-term processes of aridification and habitat change[52,53]; and (b) very extensive landscape drying occurred across the Australian interior from about 50,000 years ago.[34] Nevertheless, against a background of precarious climate variability it is plausible that hunting and burning activities undertaken by small, localized human populations could have incurred landscape-scale impacts on surviving megafauna with characteristically low reproductive rates[54] in remnant habitat refugia.

An important corollary of the earlier discussion is that Aboriginal burning practices, and the scales over which these were applied, doubtless have changed markedly through time. Palaeorecords attest that pre-European patterns of occupation, technological development, population size, and mobility have changed markedly even within the late Holocene.[26–29,39,40] It is therefore likely that intensive Aboriginal patch-mosaic burning practices evident at the time of European settlement developed only over the past few, especially the last 1,500 years.[25,40,55] By the time European colonisers invaded, however, it is evident that systematic landscape-scale fire management was being undertaken in the western deserts,[56] and substantial productive areas of south-east and south-west Australia[57] as part of intensive agricultural practice.[58] Fire-stick farming is still practised in various locations across the north, especially Arnhem Land.[59–61]

2.3 NORTH AUSTRALIA

For the purposes of this book, our geographic focus is concerned especially with the vast (~1.2 M km^2), very sparsely populated savanna landscapes extending across northern Australia, from around Broome on the west coast, to the Burdekin region south of Townsville on the east coast but excluding north Queensland's relatively well-watered Wet Tropics. North Australia as defined here is broadly comparable with the 'Northern Australia' region (1.2 M km^2) defined by water catchments,[62] and the 1.9 M km^2 Australian 'tropical savannas' based on a biogeographic regionalization (Map 2.3).

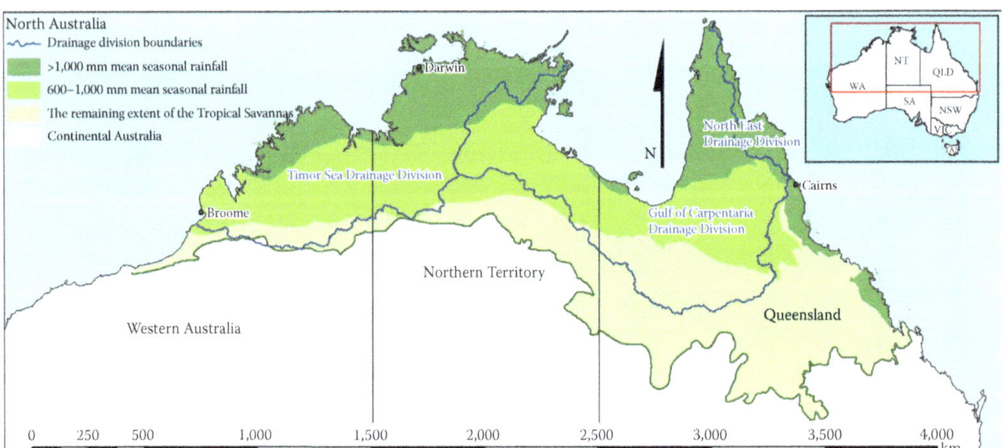

Map 2.3 'Northern Australia' region as defined by the Northern Australia Land and Water Taskforce (2009), and the 'Tropical Savanna' region shown in different shades of green as defined by the Cooperative Research Centre for Tropical Savannas. (From http://www.savanna.org.au/savanna_web/index.html.)

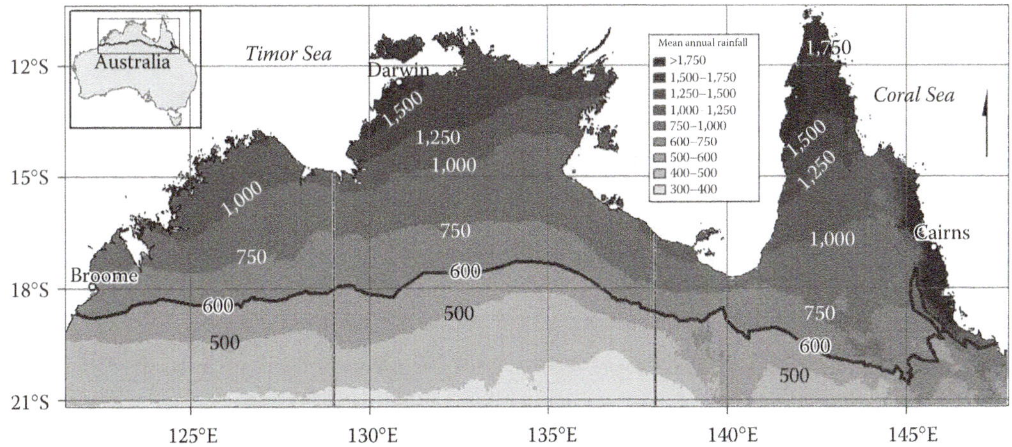

Map 2.4 Mean annual rainfall, based on rainfall data from 1970 to 2012. (From Jones, R. Hunters in the Australian coastal savanna. In *Ecology in Savanna Environments* (Ed Harris D) pp. 107–146. Academic Press, London, UK, 1980.)

North Australia thus defined encompasses an apparently little modified region receiving an average of at least 600 mm annual rainfall predominantly during the wet season months November–March (Map 2.4). The 600 mm rainfall isohyet is significant here in that it distinguishes between generally more settled, productive and viable cattle *country* to the south (especially in the eastern sector), and, to the north, rapidly developing novel ecosystem services opportunities on internationally significant biodiverse, but generally less pastorally productive, lands (Chapter 5).

Most of the region is of low elevation and topographically subdued (Map 2.5) but note the 'ridge' of relatively high *country* extending from the eastern Kimberley to Cape York. This higher country comprises mostly rugged terrain derived from low fertility sedimentary and metamorphosed rocks,

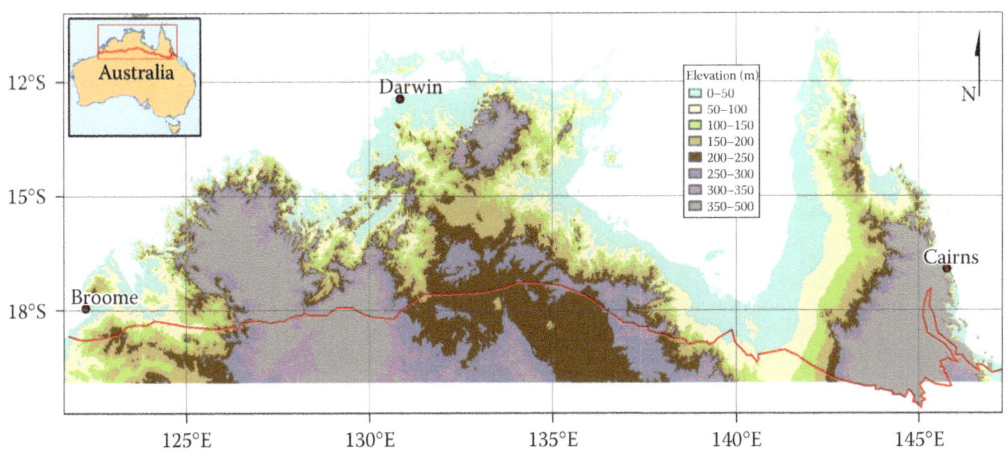

Map 2.5 Elevation. (From Geoscience Australia: SRTM-derived 3 Second Digital Elevation Models Version 1.0RTM DEM; http://www.ga.gov.au/metadata-gateway/metadata/record/gcat_72760.)

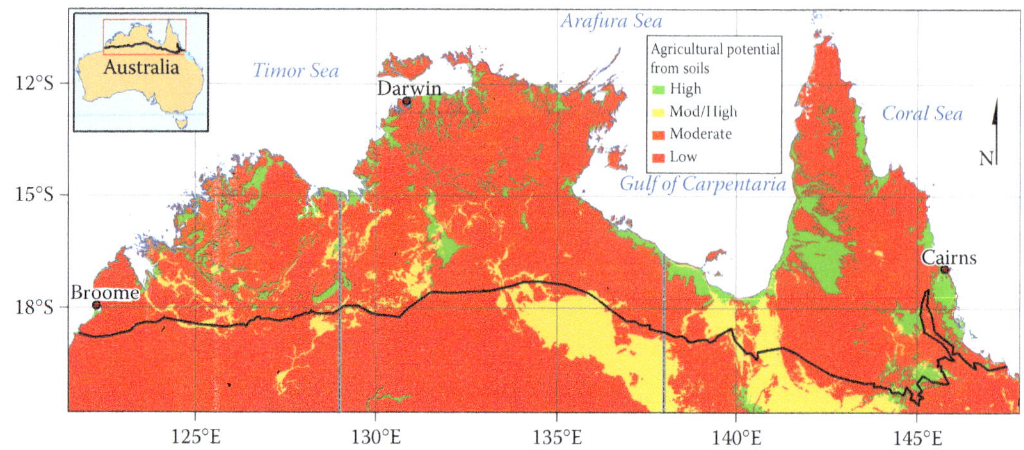

Map 2.6 Soils and derived agricultural potential. (From Australian Soil Resource Information System; http://www.asris.csiro.au/themes/Atlas.html.)

supporting high biodiversity values. Soils are typically of low agricultural potential (Map 2.6). Related pastoral potential and stock carrying capacity maps are presented as Figure 5.4a, b, in Chapter 5. A simplified vegetation map, based on the 1:2,000,000 vegetation map for the tropical savannas region,[63] is presented as Figure 5.8d in Chapter 5.

Today, the major economic activity at landscape scales over the majority of our focal region is extensive (as opposed to intensive, grain-fed) beef cattle pastoralism (Map 2.7), typically occurring on very large property sizes as reflected in the contemporary cadaster (Map 2.8). A highly significant proportion of the region (28%) comprises legally recognised Indigenous interests in land (Map 2.9). The formal conservation estate (e.g. National Parks, Indigenous Protected Areas), including

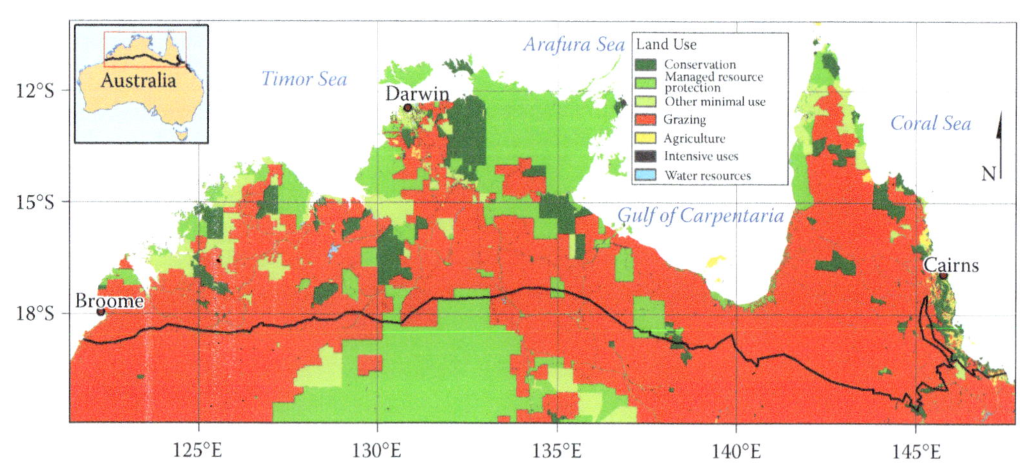

Map 2.7 Land use categories following land tenure. (From http://nationalmap.gov.au.)

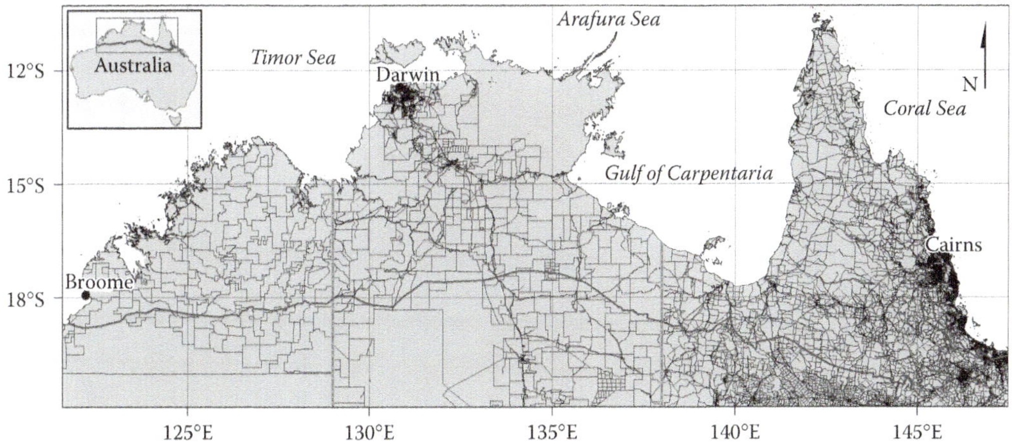

Map 2.8 Cadastre using land tenure data from each of the jurisdictions in Western Australia, Queensland and the Northern Territory. (From State and Territory Spatial Data Portals: WA – https://data.wa.gov.au/; Qld – http://qldspatial.information.qld.gov.au/catalogue/custom/search.page?q=cadastre; and the NT – https://nt.gov.au/property/land/get-land-information-online.) Note the relatively small number, but typically large property sizes across the north.

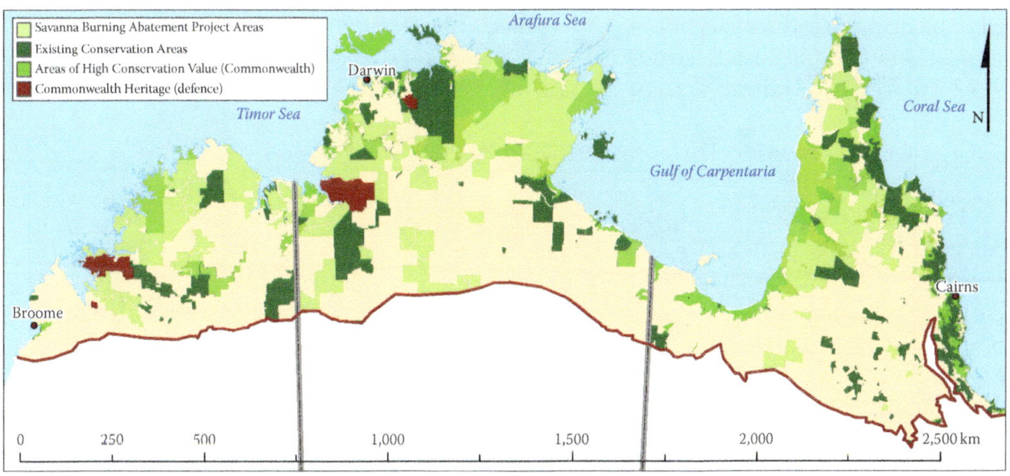

Map 2.9 Conservation estate. (From CAPAD (Collaborative Australian Protected Area Database) 2014 dataset – http://www.environment.gov.au/fed/catalog/search/resource/details.page?uuid=%7B4448CACD-9DA8-43D1-A48F-48149FD5FCFD%7D; World Heritage sites – http://www.environment.gov.au/metadataexplorer/full_metadata.jsp?docId=%7B6C54FE6C-2773-47C6-8CBC-4722F29081EF%7D&loggedIn=false; Arnhem Land Aboriginal Land Trust – https://www.legislation.gov.au/Details/F2015L00344); Riparian areas using Topographic data from Geoscience Australia; Wetlands – http://www.environment.gov.au/metadataexplorer/full_metadata.jsp?docId=%7B3F179472-DE1F-4C6C-B7CA-535DF2896656%7D&loggedIn=false; and land tenure information from each state on Defence land.)

Indigenous-owned lands, also comprises 28% of the region (Map 2.10). Although confined to relatively few and small sites, mining (Map 2.11) is the major regional land sector industry, generating at least $6 billion annually[64]; however, multipliers (measures of the extent to which money circulates and stimulates additional activity in the economy) from mining are low, because local communities are often poorly placed to provide the goods or services sought by these industries, including labour.[65]

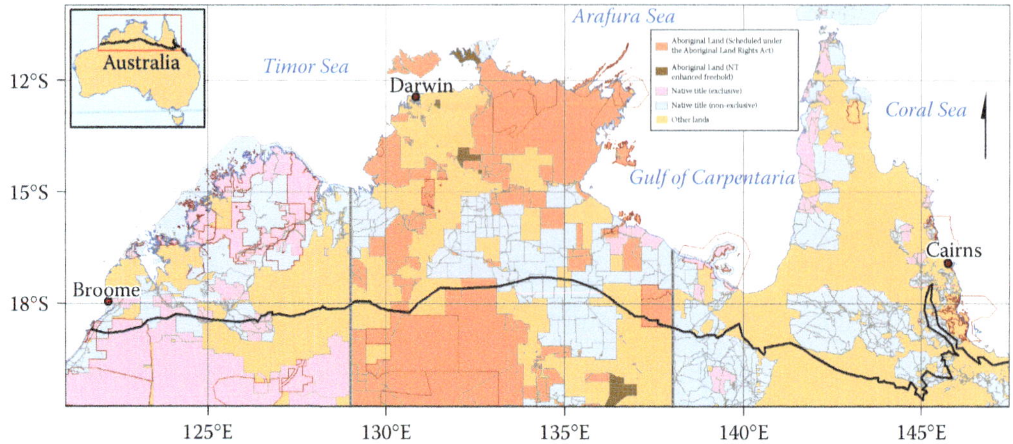

Map 2.10 Indigenous interests in land. (From http://www.nntt.gov.au/assistance/Geospatial/Pages/Spatial-aata.aspx, as at January, 2017.)

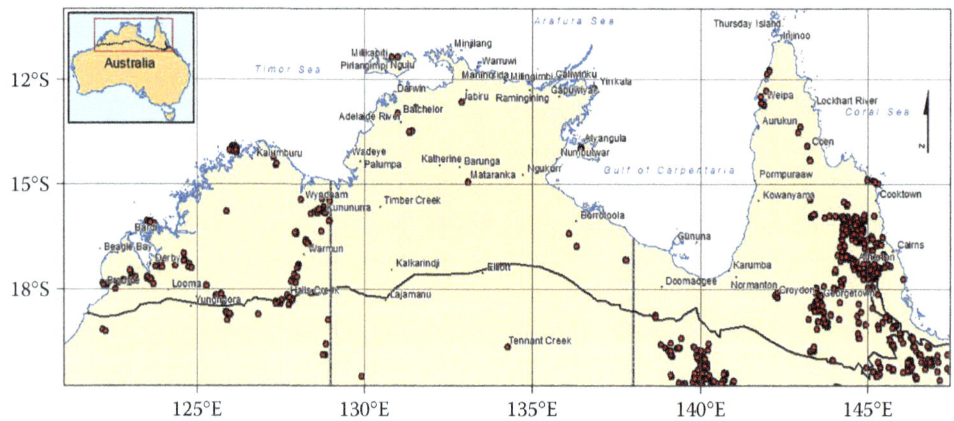

Map 2.11 Current and historic mining sites in Western Australia, Queensland and the Northern Territory. (From WA – http://geodownloads.dmp.wa.gov.au/downloads/dasc/Static/Documents/DMP_DataAndSoftwareCentre_DownloadURLs.pdf; Qld – https://data.qld.gov.au/dataset?q=mines; NT – http://nrmaps.nt.gov.au/;jsessionid=1mlhbmfurfnxj1l82g2ulrr6oq.)

Although regional savannas are little modified by international standards, losses of ecological function, chiefly evidenced in biodiversity declines, are increasingly observed at a number of levels and across large areas.[66,67] More reliably productive environments (like wetlands and riparian habitats) are valued highly by many different interests or sectors. Rivers are mostly unregulated. Most rivers carrying high wet season flows cease flowing during parts of the annual dry season so that many water-dependent ecosystems are maintained by springs or other near-surface groundwater. Primary industries (agriculture and mining) rely heavily on groundwater extraction even in the high rainfall areas, and so can readily impact these systems.

North Australia is characterized also by frequent fires (Map 2.12), especially extensive wildfires in the late dry season (Map 2.13). In some areas, but especially in regions receiving >1000 mm

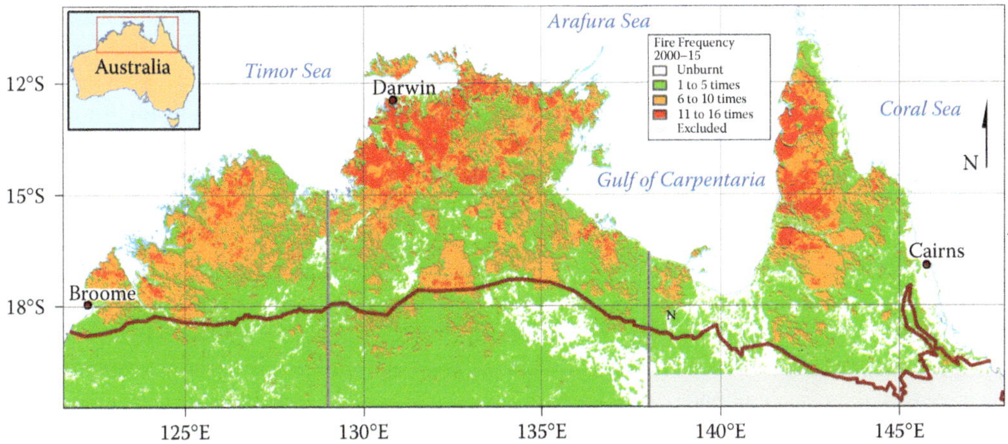

Map 2.12 Fire frequency, 2000–15. (From North Australia Fire Information website (www.firenorth.org.au) (Fire frequency classes: unburnt, burnt 1 to 5 times, 6 to 10 times and 11 to 16 times, represent 10%, 46%, 35%, and 9% of total high and low rainfall regions, respectively. In the high rainfall region the proportions are 9%, 30%, 41%, and 20%, respectively; whilst in the lower rainfall region the proportions are 11%, 56%, 31%, and 2% respectively.)

annual rainfall, fires currently recur at frequencies greater than once every two years.[68,69] In absence of marked grazing pressure and other grassy fuel-limiting conditions (e.g. closed canopy vegetation, rock platforms), fires have been observed to recur annually in areas receiving at least 600 mm rainfall.[70] This recurrent fire management problem has resulted in novel economic opportunities becoming available in recent years through the development of market-based, landscape-scale, 'savanna

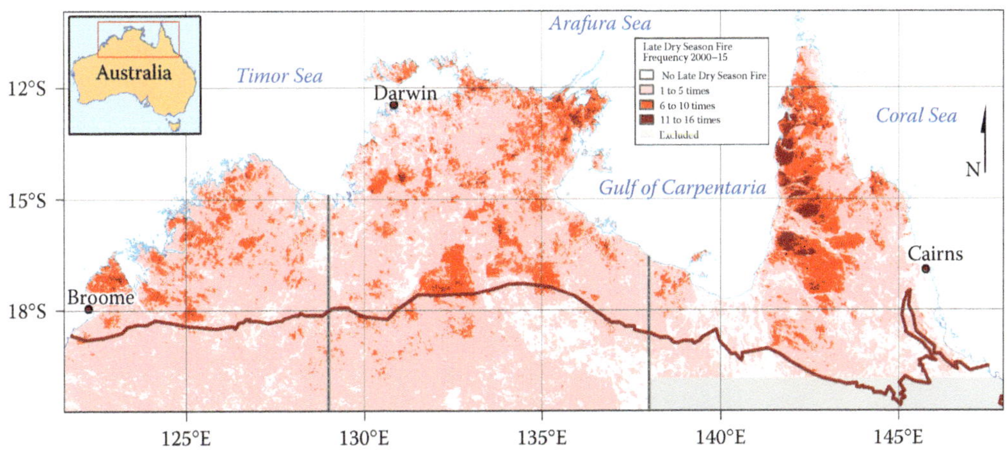

Map 2.13 Late dry season fire frequency, 2000–15 Late dry season fires are those occurring after July. (From North Australia Fire Information website www.firenorth.org.au.) (Late dry season fire frequency classes: unburnt, burnt 1 to 5 times, 6 to 10 times, and 11 to 16 times, represent 14%, 66%, 19%, and 2% of the total high and low rainfall regions, respectively. In the high rainfall region the proportions are 13%, 57%, 26%, and 4%, respectively; whilst in the low rainfall region proportions are 14%, 71%, 14%, and 0%, respectively.)

burning', carbon farming projects. Importantly, savanna burning projects serve as an illustration of the regional potential of the ecosystem services industry sector generally, and especially for remote Indigenous land owners and managers.[71]

The North Australia population (including major towns such as Cairns, Ingham, Townsville, Mt Isa, Ayr and Home Hill [Queensland], Darwin, Palmerston, Katherine [Northern Territory], and Broome [Western Australia]), comprised 901,000 people in the 2016 census, of whom 14% were Indigenous. Outside of towns, the North Australia savanna population was around 170,000, with Indigenous people comprising a much greater proportion of the population. In the Kimberley and Top End savannas, about half of the population is Indigenous, and in very remote regions generally, more than 90 per cent. Nationally, 45 per cent of the population living in areas classified by the Australian Bureau of Statistics as very remote is Indigenous.[72] Average population density of about 0.75 persons/km^2 is low by Australian (3.0 persons/km^2) and global standards. Excluding major town centres, there is about one person for each 700 ha (0.14 persons/km^2). By any standard, North Australia is sparsely populated (Map 2.14). Given its land area, regional populations are also poorly politically represented in state/territory and national legislatures (Maps 2.15 and 2.16).

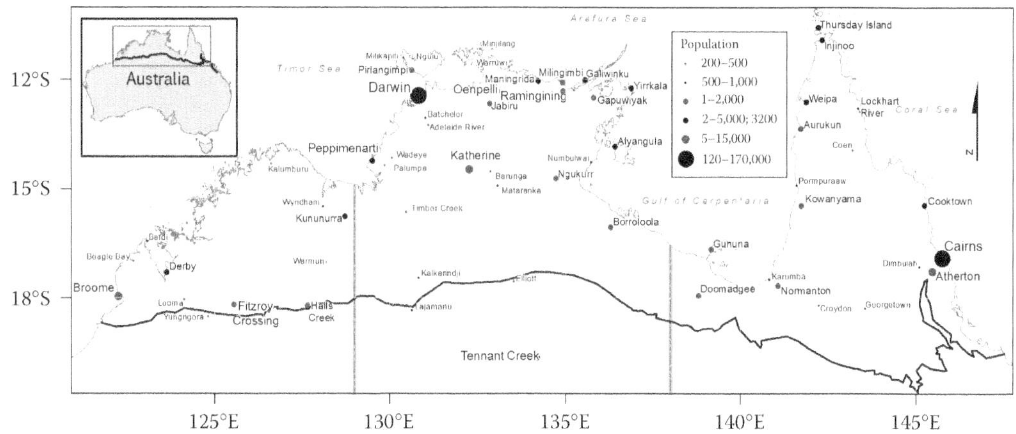

Map 2.14 Population distribution. (From Australian Bureau of Statistics, census data 2011.)

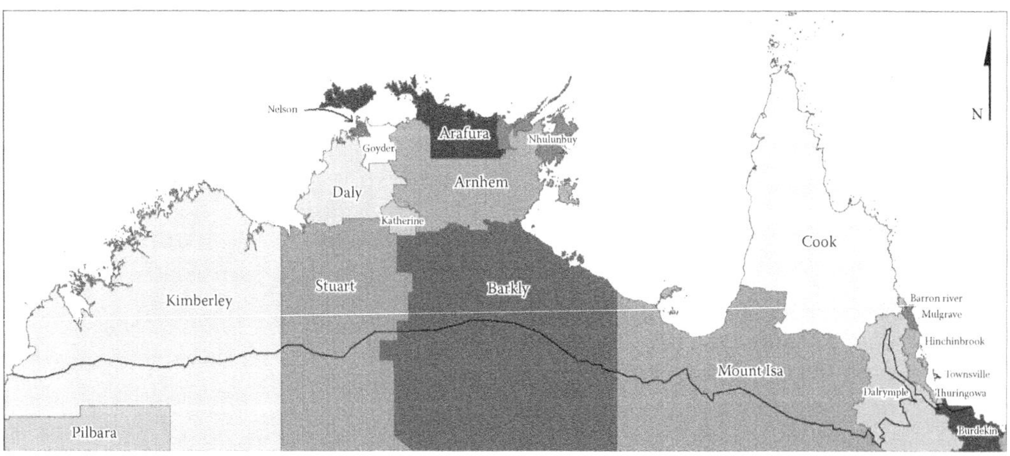

Map 2.15 State/territory electorates covering North Australia. (From http://nationalmap.gov.au.) Noting that political representation at the State/Territory is particularly low in Western Australia (with the Kimberley region included as part of one mega-electorate) and Queensland (with just three savanna region electorates).

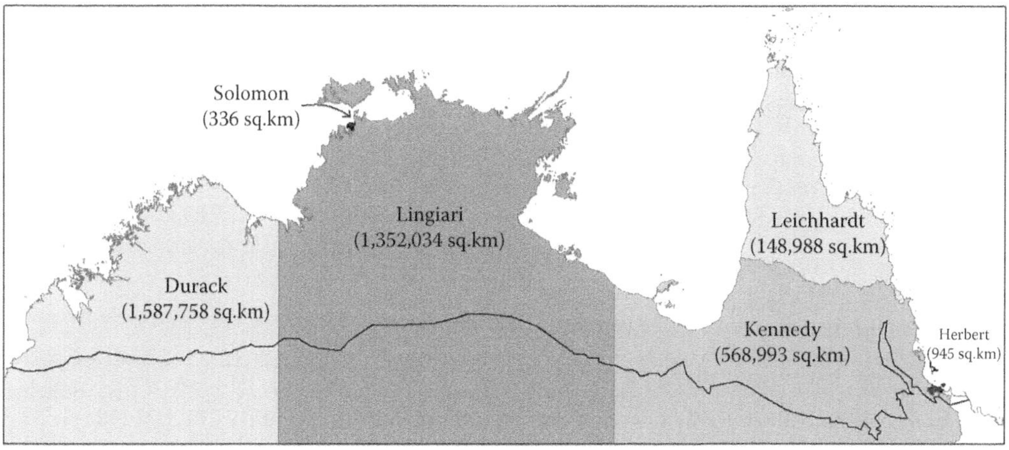

Map 2.16 National electorates covering North Australia. (From http://nationalmap.gov.au.) Noting that political representation of North Australia savanna interests at the national scale is restricted to just four electorates.

REFERENCES

1. Blaser M (2004) Life projects: Indigenous peoples' agency and development. In *In the Way of Development, Indigenous Peoples, Life Projects and Globalization* (Eds Blaser M, Feit HA, McRae G). Zed Books, London, UK.
2. Escobar A (2011) *Encountering Development: The Making and Unmaking of the Third World.* Princeton University Press, Princeton, NJ.
3. Schlosberg D, Carruthers D (2010) Indigenous struggles, environmental justice, and community capabilities. *Global Environmental Politics* **10**(4), 12–35.
4. United Nations (2007) *United Nations Declaration on the Rights of Indigenous Peoples.* United Nations, New York. Retrieved from http://www.un.org/esa/socdev/unpfii/documents/DRIPS_en.pdf.
5. Whyte KP (2013) On the role of traditional ecological knowledge as a collaborative concept: A philosophical study. *Ecological Processes* **2**. doi:10.1186/2192-1709-2-7.
6. Blaser M, Feit HA, McRae G (2004) Indigenous people and development processes: New terrains of struggle. In *In the Way of Development, Indigenous Peoples, Life Projects and Globalization* (Eds Blaser M, Feit HA, McRae G). Zed Books, London, UK.
7. Kerins S (2012) Caring for country to working on country. In *People on Country Vital Landscapes Indigenous Futures* (Eds Altman J, Kerins S). Federation Press, Sydney, Australia.
8. Costanza R, d'Arge R, de Groot R, Farber S, Grasso M, Hannon B, Limburg K et al. (1997) The value of the world's ecosystem services and natural capital. *Nature* **387**, 253–260.
9. Daily GC (1997) *Nature's Services: Societal Dependence on Natural Ecosystems.* Island Press, Washington, DC.
10. de Groot R, Brander L, van der Ploeg S, Costanza R, Bernard F, Braat L, Christie M et al. (2012) Global estimates of the value of ecosystems and their services in monetary units. *Ecosystem Services* **1**, 50–61.
11. Millennium Ecosystem Assessment (MEA) (2005) *Ecosystems and Human Wellbeing: Synthesis.* Island Press, Washington, DC.
12. Costanza R, Daly M, Folke C, Hawken P, Holling CS, McMichael AJ, Pimentel D, Rapport D (2000) Managing our environmental portfolio. *Bioscience* **50**(2), 149–155.
13. O'Connor S (1995) Carpenter's Gap Rockshelterl: 40,000 years of Aboriginal occupancy in the Napier Ranges, Kimberley, WA. *Australian Archaeology* **40**, 58–59.
14. Balme J (2000) Excavations revealing 40,000 years of occupation at Mimbi Caves, south central Kimberley, Western Australia. *Australian Archaeology* **51**, 2–5.

15. Roberts RG, Jones R, Smith MA (1990) Thermoluminescence dating of a 50,000-year-old human occupation site in northern Australia. *Nature* **345**, 153–156.
16. Roberts RG, Jones R, Spooner MA, Head MJ, Murray AS, Smith MA (1994) The human colonisation of Australia: Optical dates of 53,000 and 60,000 years bracket human arrival at Deaf Adder Gorge, Northern Territory. *Quaternary Science Reviews* **13**, 575–583.
17. Clarkson C, Jacobs Z, Marwick B, Fullager R, Wallis L, Smith M, Roberts RG et al. (2017) Human occupation of northern Australia by 65,000 years ago. *Nature* **547**, 306–310.
18. David B, Roberts R, Tuniz C, Jones R, Head J (1997) New optical and radiocarbon dates from Ngarrabullgan Cave, a Pleistocene archaeological site in Australia: Implications for comparability of time clocks and for human colonization of Australia. *Antiquity* **71**, 183–188.
19. Lambeck K, Chappell J (2001) Sea level change through the last glacial cycle. *Science* **292**, 679–685.
20. Siddall M, Chappell J, Potter E-K (2007) Eustatic sea level during past Interglacials. *Developments in Quaternary Sciences* **7**, 75–92.
21. Lewis SE, Sloss CR, Murray-Wallace CV, Woodroffe CD, Smithers SG (2013) Post-glacial sea-level changes around the Australian margin: A review. *Quaternary Science Reviews* **74**, 115–138.
22. Wyroll K, Miller GH (2001) Initiation of the Australian monsoon 14,000 years ago. *Quaternary International* **83–85**, 119–128.
23. Griffiths ML, Drysdale RN, Gagan MK, Zhao JX, Ayliffe LK, Hellstrom JC, Hantoro WS et al. (2009) Increasing Australian-Indonesian monsoon rainfall linked to early Holocene sea-level rise. *Nature Geoscience* **2**, 636–639.
24. Shulmeister J, Lees BG (1995) Pollen evidence from tropical Australia for the onset of an ENSO-dominated climate at c. 4000 BP. *The Holocene* **5**, 10–18.
25. Williams AN (2013) A new population curve for prehistoric Australia. *Proceedings of the Royal Society Series B* **280**, 20130486.
26. Brockwell S et al. (2011) Holocene settlement of the northern coastal plains, Northern territory, Australia. *The Beagle* **27**, 1–22.
27. McGowan H, Marx S, Moss P, Hammond A (2012) Evidence of ENSO mega-drought triggered collapse of prehistory Aboriginal society in northwest Australia. *Geophysical Research Letters* **39**. doi:10.1029/2012GL053916.
28. Chaloupka G (1993) *Journey in Time: The World's Longest Continuing Art Tradition*. Reed Books, Chatswood, Australia.
29. Mulvaney J, Kamminga J (1999) *Prehistory of Australia*. Allen & Unwin, St. Leonards, Sydney, Australia.
30. McConvell P, Bowern C (2011) The prehistory and internal relationships of Australian languages. *Language Linguistics Compass* **5**, 19–32.
31. Lambeck K, Chappell J (2001) Sea level change through the last glacial cycle. *Science* **292**, 679–685.
32. Dodson JR (1989) Late Pleistocene vegetation and environmental shifts in Australia and their bearing on faunal extinctions. *Journal of Archaeological Science* **16**, 207–217.
33. O'Connell JF, Allen J (2015) The process, biotic impact, and global implications of the human colonization of Sahul about 47,000 years ago. *Journal of Archaeological Science* **56**, 73–84.
34. Cohen TJ, Jansen JD, Gliganic LA, Larsen JR, Nanson GC, May J-H, Jones BG, Price DM (2015) Hydrological transformation coincided with megafaunal extinction in central Australia. *Geology* **43**, 195–198.
35. Saltré F, Rodríguez-Rey M, Brook BW, Johnson CN, Turney CSM, Alroy J, Cooper A et al. (2016) Climate change not to blame for late Quaternary megafauna extinctions in Australia. *Nature Communications* **7**, 10511.
36. Roberts RG, Jones R, Smith MA (1990) Thermoluminescence dating of a 50,000-year-old human occupation site in northern Australia. *Nature* **345**, 153–156.
37. Tobler R, Rohrlach A, Soubrier J, Bover P, Llamas B, Tuke J, Bean N et al. (2017) Aboriginal mitogenomes reveal 50,000 years of regionalism in Australia. *Nature* **544**, 180–184.
38. Clarkson C, Jacobs Z, Marwick B, Fullagar R, Wallis L, Smith M, Roberts RG et al. (2017). Human occupation of northern Australia by 65,000 years ago. *Nature* **547**, 306–310.
39. Smith MA (2013) *The Archaeology of Australia's Deserts*. Cambridge University Press, Cambridge, UK.
40. Williams AN, Mooney SD, Sisson SA, Marlon J (2015a) Exploring the relationship between Aboriginal population indices and fire in Australia over the last 20,000 years. *Palaeogeography, Paleaoclimatology, Palaeoecology* **432**, 49–57.

41. Woodroffe CD, Thom BG, Chappell JMA (1985) Development of widespread mangrove swamps in mid-Holocene times in northern Australia. *Nature* **317**, 711–713.
42. Donders TH, Haberle SG, Hope GS, Wagner F, Visscher H (2007) Pollen evidence for the transition of the eastern climate system from the post-glacial to the present-day ENSO mode. *Quaternary Science Reviews* **26**, 1621–1637.
43. Mooney SD, Harrison SP, Bartlein PJ, Stevenson J (2012) The prehistory of fire in Australia. In *Flammable Australia* (Eds Bradstock RA, Williams RJ, Gill AM) pp. 3–25. Cambridge University Press, Cambridge UK.
44. Williams AN, Ulm S, Turney CSM, Rohde D, White G (2015b) Holocene demographic changes and the emergence of complex societies in prehistoric Australia. *PLoS One* **10**, e0128661. doi:10.1371/journal.pone.0128661.
45. Lynch AH, Beringer J, Kershaw AP, Marshall A, Mooney S, Tapper N, Turney C, van der Kaars S (2007) Using the palaeorecord to evaluate climate and fire interactions in Australia. *Annual Review of Earth and Planetary Sciences* **35**, 215–239.
46. Singh G, Kershaw AP, Clark R (1981) Quaternary vegetation and fire history in Australia. In *Fire and the Australian Biota* (Eds Gill AM, Groves RH, Noble IR) pp. 23–54. Australian Academy of Science, Canberra, Australia.
47. Flannery TM (1994) *The Future Eaters*. Reed Books, Sydney, Australia.
48. Johnson CN, Alroy J, Beeton NJ, Bird MI, Brook BW, Cooper A, Gillespie R et al. (2016) What caused extinction of the Pleistocene megafauna of Sahul. *Proceedings of the Royal Society Series B* **283**, 20152399.
49. Miller GH, Fogel ML, Magee JW, Gagan MK, Clarke JS, Johnson BJ (2005a) Ecosystem collapse in Pleistocene Australia and a human role in megafaunal extinction. *Science* **309**, 287–290.
50. Miller GH, Mangan J, Pollard D, Thompson S, Felzer B, Magee J (2005b) Sensitivity of the Australian monsoon to insolation and vegetation: Implications for human impact on continental moisture balance. *Geology* **33**, 65–68.
51. Rule S, Brook BW, Haberle SG, Turney CSM, Kershaw AP, Johnson CN (2012) The aftermath of megafaunal extinction: Ecosystem transformation in Pleistocene Australia. *Science* **335**, 1483–1486.
52. Price GJ, Webb GE, Zhao J, Feng Y, Murray AS, Cooke BN, Hocknull SA, Sobbe IH (2011) Dating megafaunal extinction of the Pleistocene Darling Downs, eastern Australia: The promise and pitfalls of dating as a test of extinction hypotheses. *Quaternary Science Reviews* **30**, 899–914.
53. Wroe S, Field JH, Archer M, Grayson DK, Price GJ, Louys J, Tyler Faith J, Webb GE, Davidson I, Mooney SD (2013) Climate change frames debate over the extinction of meagafauna in Sahul (Pleistocene Australia-New Guinea). *Proceedings of the National Academy of Sciences USA* **110**, 8777–8781.
54. Johnson C (2006) *Australia's Mammal Extinctions: A 50,000 Year History*. Cambridge University Press, New York.
55. Bliege Bird R, Bird DW, Codding BF, Parker CH, Jones RH (2008) The 'fire stick' farming hypothesis: Australian Aboriginal foraging strategies, biodiversity and anthropogenic fire mosaics. *Proceedings of the National Academy of Sciences USA* **105**, 14796–14801.
56. Burrows ND, Burbidge AA, Fuller PJ, Behn G (2006) Evidence of altered fire regimes in the Western Desert region of Western Australia. *Conservation Science Western Australia* **5**, 272–284.
57. Gammage W (2012) *The Biggest Estate on Earth: How Aborigines made Australia*. Allen & Unwin, Sydney, Australia.
58. Pascoe B (2014) *Dark Emu, Black Seeds: Agriculture or Accident?* Magabala Books, Broome, Australia.
59. Jones R (1980) Hunters in the Australian coastal savanna. In *Ecology in Savanna Environments* (Ed Harris D) pp. 107–146. Academic Press, London, UK.
60. Garde M, Nadjamerrek LB, Kolkiwarra M, Kalarriya J, Djandjomerr J, Birriyabirriya B, Bilindja R, Kubarkku M, Biless P (2009) The language of fire: Seasonality, resources and landscape burning on the Arnhem Land Plateau. In *Culture, Ecology and Economy of Savanna Fire Management in Northern Australia: Rekindling the Wurrk Tradition* (Eds Russell-Smith J, Whitehead PJ, Cooke PM) pp. 85–164. CSIRO Publications, Melbourne, Australia.
61. Yibarbuk D, Whitehead PJ, Russell-Smith J, Jackson D, Godjuwa C, Fisher A, Cooke P, Choquenot D, Bowman DMJS (2001) Fire ecology and aboriginal land management in central Arnhem Land, northern Australia: A tradition of ecosystem management. *Journal of Biogeography* **28**, 325–344.

62. NLaWT (Northern Australia Land and Water Taskforce) (2009) *Sustainable Development of Northern Australia*. Department of Infrastructure, Transport, Regional Development and Local Government, Canberra, Australia.
63. Fox ID, Neldner VJ, Wilson GW, Bannink PJ (2001) *The Vegetation of the Tropical Australian Savannas*. Environment Protection Agency, Brisbane, Australia.
64. Brereton D, Klimenko V, Cote C, Evans R (2009) The minerals industry and land and water development in northern Australia. In *Northern Australia Land and Water Science Review: Full Report*. Chapter 8: 8.1–8.50. CSIRO Sustainable Agriculture Flagship, and Northern Australia Land and Water Taskforce, Canberra, Australia.
65. Welters R (2010) Atypical labour markets require atypical policy solutions. In *North Australian Political Economy: Issues and Agendas* (Ed Gerritsen R) pp. 55–69. Charles Darwin University Press, Darwin, NT.
66. Franklin DC, Whitehead PJ, Pardon G, Matthews J, McMahon P, McIntyre D (2005) Geographic patterns and correlates of the decline of granivorous birds in northern Australia. *Wildlife Research* **32**, 399–408.
67. Woinarski JCZ, Legge S, Fitzsimons JA, Traill BJ, Burbidge AA, Fisher A, Firth RSC et al. (2011) The disappearing mammal fauna of Australia: Context, cause, and response. *Conservation Letters* **4**, 192–201.
68. Felderhof L, Gillieson D (2006) Comparison of fire patterns and fire frequency in two tropical savanna biomes. *Austral Ecology* **31**, 736–746.
69. Edwards AC, Russell-Smith J, Meyer CP (2015) Contemporary fire regime risks to key ecological assets and processes in north Australian savannas. *International Journal of Wildland Fire* **24**, 857–870.
70. Russell-Smith J, Yates CP, Brock C, Westcott V (2010) Fire regimes and interval-sensitive vegetation in semi-arid Gregory National Park, Northern Territory. *Australian Journal of Botany* **58**, 300–317.
71. Whitehead PJ, Russell-Smith J, Yates C (2014) Fire patterns in north Australian savannas: Extending the reach of incentives for savanna fire emissions abatement. *The Rangeland Journal* **36**, 371–388.
72. Taylor J (2006) *Population and Diversity: Policy Implications of Emerging Indigenous Demographic Trends*. CAEPR Discussion Paper No. 283/2006. Centre for Aboriginal Economic Policy Research, Australian National University, Canberra, Australia.

CHAPTER 3

North Australian History – Dispossession, Colonisation, and the Assertion of Indigenous Rights

Howard Pedersen and Stuart Phillpot

CONTENTS

3.1 INTRODUCTION

This chapter describes the colonisation of North Australia from 1860 to the present day. Historical context is important to support any political and economic analysis of a society or region. An understanding of North Australia's history is particularly relevant, not simply because of repetitive failures of major projects and policies that define Western development of the North since the nineteenth century, but more importantly the region's contemporary underdevelopment, impoverishment, and dependence on southern Australia is historically embedded. North Australia's past is not a foreign *country*. Contemporary northern Australia continues to live in its history.

The chapter is written from a colonised perspective, which is not to say that it is written with an Aboriginal voice. Rather, the chapter's underpinning analysis is influenced by dependency theory concerning underdevelopment and an understanding of coloniality of power; a Eurocentric system of knowledge, in which race is seen as 'naturalisation of colonial relations between Europeans and non-Europeans'.[1]

In the 1970s, dependency theory was widely taught and read at universities as a way to understand the disparity of wealth and power between the industrialised, developed global north and the impoverished, underdeveloped global south. The central contention of dependency theory is that poor states are impoverished, and rich ones are enriched by the way poor states are integrated into the global economic system. Leftist scholars and thinkers such as Andre Gunder Frank and Water Rodney developed the theory in response to the widely accepted modernisation theory of the Washington Consensus which dominated Western global policy-making in the Cold War era.

To many non-Indigenous people in the 1970s who were trying to make sense of post-colonial Australia in the period where Indigenous people had gained citizenship rights and were asserting their inherent cultural and traditional land ownership rights under the political banner of self-determination, dependency theorists and decolonisation writers such as Franz Fannon and Albert Memmi were guiding literature. From the 1980s and into this century, dependency and decolonisation theory lost favour in the face of neo-liberal economic and political thinking, which came to dominate domestic and international policy setting in the era of globalisation. It was not just the ascendant post-Cold War triumph of neo-liberalism that marginalised dependency and decolonisation theory, it was also the obvious failures of these theories to explain economic globalisation's apparent successes, particularly in Asia.

Consequently, modernisation theory infused with neo-liberal economic approaches has provided a guiding philosophical and policy setting backdrop to rampant economic globalisation over the past three decades. That dominant global ideology concerning economic development has significantly influenced Australian Indigenous public policy with concerted application to North Australia. Closing the Gap initiatives, welfare reform, the Northern Territory Emergency Response (Intervention), and the North Australian White Paper – hallmarks of Australian bipartisan Indigenous public policy over the past decade or so – have largely been shaped by modernisation theory.

The failure of these approaches to change the underdeveloped status of colonised people has led to a resurgence of Indigenous-led critique of modernisation theory which embraces an understanding of continuing colonialism, increasingly referred as 'settler colonialism'.[2] Whilst the critique of modernisation theory by Australian Indigenous leaders and scholars draws from a tradition of Indigenous opposition to colonisation, new thinking, within this approach, advanced by Aboriginal land councils, native title holding entities, and the North Australian Indigenous Land and Sea Management Alliance (NAILSMA), argue that structural issues within the Indigenous and Australian Nation State relationship must be addressed to enable the emergence of inclusive and sustainable development. This political objective has a vision of decolonised economic and political relationships as opposed to a development paradigm where Indigenous people's standard of living is assumed to improve through their acceptance and inclusion in a western society based on capitalist modes of production and associated social values. Proposals for investment in new economic endeavours and more effective land and natural resource use based on reformed land titles, which recognises the traditional ownership and Indigenous culture, are fundamental to this vision.[3]

A key tenet of dependency theory is its attention to the role of history and, in particular, the role of colonialism in constructing the positions of different countries within the global economy. This chapter directs that attention to North Australia while understanding that northern Australian colonisation should be seen in the historical context of empire and the contemporary context of global commodity trade and investment. Critically important to the Indigenous critique of modernization theory is that it is non-historical and that its policy prescriptions target the victims of colonisation and ignores entrenched structural relationships that are embedded in colonialism.

North Australia's nineteenth century colonial character remains fundamentally intact. This assertion is highlighted by key institutional, economic, and political dimensions of contemporary North Australia; the pastoral land tenure regime; marginalised but numerically dominant Indigenous populations living outside a handful of concentrated European settlements; wealth extracted from the north without an equitable reinvestment; and high rates of Indigenous imprisonment. Importantly, the real decision-making power over North Australia's administration is located in southern Australia. The chapter outlines the historical setting for the broad political narrative of this book: inclusive and sustainable North Australian development must go hand-in-hand with changing the laws, institutions, and policies that have been inherited from colonial history.

The chapter is not a social history but rather a broad political and economic account of North Australia's colonial and more recent development. It overviews the frontier period of conflict, the takeover of Indigenous people's lands, and the subsequent shaping of a colonial society which enabled European occupiers to subjugate and control the colonised majority well into the second half of the twentieth century. It describes the collapse of that colonial society in the late 1960s and its political consequences. The impacts of the breakdown of formal colonialism – end of discriminatory laws and practices, equal wages, closing of missions, extension of social welfare, and end of alcohol restrictions – engulfed the whole North. In the wake of pastoral evictions and relocation pressure following the 1966 pastoral industry labour award decision and the closure of several missions in the latter part of the 1960s and into the 1970s, North Australia became immersed in a massive population displacement crisis.[4]

That crisis coincided with the election of the Whitlam Labor Government in 1972 and subsequent Commonwealth leadership in Indigenous policy at a national level. Whilst the 1967 constitutional amendment of the Race Power provided the Commonwealth with the power to make laws concerning Indigenous people, Australia's Constitution constrained the Whitlam Government's national reach particularly concerning land titles. So it focused on its own jurisdiction, the Northern Territory, which Whitlam saw as a beacon for a bold and visionary Indigenous self-determination agenda that would eventually embrace the whole of Indigenous Australia.[5]

The Woodward Royal Commission ushered in the Northern Territory Aboriginal Land Rights Act in 1976, which first gained wide popular approval and bipartisan political support at the national level, but the potential for national consensus quickly evaporated as the mining industry supported by conservative political forces in Western Australian, Queensland, and Northern Territory opposed Indigenous land rights in a sustained clash over land ownership and use that fundamentally persists today. Much of the impetus for self-determination that has influenced the wider Australian body politic has emanated from northern Australia. And it has been in the North where much of the conflict between Western development interests and Indigenous cultural imperatives – whether it be direct political action or litigation – has taken place.

Conflicts between Indigenous groups, developers, and governments and between governments were not conducive to an agreed national effort for northern Australian development during a period of extraordinary unsettling change. Thousands of people who moved to small communities, larger settlements, and towns or took over former church missions were supported by Commonwealth-funded services and infrastructure. Commonwealth and state tensions over funding responsibility and state opposition to Commonwealth interventions in their land management regimes impeded social and economic development and transitional change.

By the beginning of the twenty-first century, conservative policy prescriptions for dealing with the social fracturing in Indigenous communities and their broader social environments had come to influence government policy making based on a widely held view that Commonwealth policies emanating from the Whitlam period had created endemic welfare dependency. Here was modernization theory at work using facile and distorted historical analysis to develop social engineering policy interventions to change behaviours and lifestyle values. The vision to expand the so-called 'real economy' in the North, where Aboriginal people would participate and adopt the 'social norms' of Western society, has driven a national bipartisan policy approach relating to Indigenous Australia northern development for more than a decade. The Commonwealth White Paper on Northern Australian Development reflects this approach.

The chapter reflects broadly on four key epochs in the history of dispossession, colonization and assertion of Indigenous rights (Table 3.1). Major legislative developments affecting governance and tenure arrangements over the same period are set out in more detail in Appendix 3.1.

Table 3.1 Governance and Land Tenure Structures 1860–1999

Period	Time Period	Governance	Land, Tenure, Rights and Reform
Establishing the Northern Colonial Frontier	1861–1901	1861–1900 Australian Colonial Governments of Queensland, South Australia and Western Australia	Colonial land tenure legislation that included grazing licences, pastoral leases and freehold title supporting closer settlement (See Appendix 3.1 for details)
Northern Australian Colonial Society	1902–1939	Queensland, South Australia and Western Australian State, Territory and the Australian Commonwealth Government after 1911 Aboriginal Protections Acts (See Appendix 3.1 for details)	As above with the addition of legislation covering the declaration of reserves, the establishment of missions and government settlements as institutions for the protection of Aboriginal people and the expansion of Land tenure acts to include soldier settlement (See Appendix 3.1 for details)
Fracturing of the Colonial System	1940–1972	State and Commonwealth Aboriginal Affairs Acts and the Northern Territory Welfare Ordinances	
Indigenous Rights and Conflict over opposing visions	1972–1998	Commonwealth and State Departments of Aboriginal Affairs/followed by Aboriginal Development Commission Act 1980 and then the Aboriginal and Torres Strait Island Commission (ATSIC) in 1990	As above with the addition of the Northern Territory Land Rights Act 1976. In the 1980s various State legislation recognised some Aboriginal Land entitlements; these applied until the recognition of Native Title in 1992 and the Commonwealth Government Native Title Act in 1994 followed by the Native Title Amendment Act in 1998 (See Appendix 3.1 for details)

3.2 ESTABLISHING THE NORTHERN COLONIAL FRONTIER 1860–1901

It should be acknowledged that European colonisers were not the first foreigners to have contact with Indigenous people in North Australia. Indonesian people have been visiting Australia's north for centuries. They came to fish mainly for *trepang* (sea cucumbers or sea slugs), which was a highly prized delicacy in much of South East Asia and China. These trepang fisher people came mainly from the trading centre of Macassar, in the southwest of Celebes, now Sulawesi. Remains of Macassarn campsites have been found on Australia's northern coast and date back at least 800 years. The Macassarns sailed to Australia in boats called *praus* and fished for trepang in dugout canoes across the northern coast from the Kimberley to Mornington Island in the east of the Gulf of Carpentaria. The Macassarn engagement with the North had a profound effect on Indigenous cultures. The visits are still remembered through oral history, songs, and dances and paintings on rock and bark. Trade dropped off towards the end of the nineteenth century because of duties imposed by Australian Governments. The last prau left Arnhem Land in 1906.[6]

Establishing settler colonial society in North Australia was a protracted violent process. In contrast to southern Australia, relatively small numbers of settlers resorted to dispossessing and controlling Indigenous populations who vastly outnumbered them. Immense areas of North Australia – north Kimberley, Arnhem Land, and Cape York – could not be conquered and were designated Aboriginal reserves under mostly mission management. The basis of long-term settler occupancy depended on pastoralism and exploited Aboriginal labour. To support northern colonial settlements, governments resourced their authority with large police numbers, gaols and a judicial system to manage labour exploitation, segregation and dissent. Even though isolated mass murders of Aboriginal people continued well into the twentieth century, colonialism's geographical footprint was firmly established in North Australia by Federation.

North Australia was colonised not by Britain but by southern Australia's settler society. By the early 1860s the governmental colonial architecture with the capacity to rule over the Australian continent and Tasmania had been fundamentally developed only 70 odd years, after the British government established its penal settlement at Port Jackson in 1788. Separate colonies had been established for New South Wales, Tasmania, South Australia, Victoria, and Queensland, each with metropolitan centres of power. Only Western Australia continued to be directly ruled from London until it joined the other colonies as a self-governing political entity in 1890.

The 1850s gold rush transformed settler Australia. In Victoria, the settler population grew from 76,000 in 1850 to 560,000 in 1860. The demand for land to accommodate the surplus immigrant population became intense, resulting in land title allocation and management dominating colonial legislatures and colonial executive decision making.[7] Following the economic boom in southern Australia from 1847 to 1860, predominantly based on gold and which saw a tenfold increase in Australia's settler population, the eyes of colonial expansionists turned north. A flurry of northern expeditions, which are etched into popular Australian history telling, roamed the northern river systems between 1860 and the late 1870s reporting to their sponsors and southern settlers about luxurious domains for pasturing livestock and growing crops. In the tradition of Anglophone settler colonialism, particularly its relentless expansion in North America, it was envisaged that the fertile well-watered lands of North Australia would support large-scale European occupation.

Early European settlement attempts were spectacular failures. Port Essington, established as a trading port by the British on the Coburg Peninsular in the 1830s, was finally abandoned in 1849. Then, in the 1860s, high-profile northern Australian colonial ventures, including Camden Harbour and Roebuck Bay in the Kimberley, Escape Cliffs in the Northern Territory, and Somerset at the tip of Cape York, all collapsed within a couple of years of their establishment.

By the 1870s European settlement had taken a firm foothold in the North. A unique form of land tenure in the British land title system – the pastoral lease – had already been created and was beginning to be used in NSW, South Australia, and Western Australia in the 1850s, to legally validate the practice of settler's 'squatting' on land with their sheep and cattle. The pastoral lease enabled pastoralists to take possession of Aboriginal land under Australian colonial law as settlement rapidly spread northward. In a period spanning 25 years, several hundred pastoral leases were issued by land title administrations in Brisbane, Adelaide, and Perth that covering more than two and half million square kilometres over a vast landscape of the northern Australian continent between the Pacific and Indian oceans. Under the legal fiction of *terra nullius*, it was an extraordinary feat of dispossession of people belonging to approximately 100 language groups by only a few hundred leaseholders. As would be determined by the Australian High Court in the 1996 Wik decision, the pastoral runs weren't leases in the English Common law sense of embodying the right of exclusive possession. Legally, they were simply licenses to graze livestock.

The first northern Australian pastoralists were initially seen as temporary frontier pioneers; the advance guard who would give way to multitudes of farmers who would come to occupy smaller Torrens Title allotments and grow food. But that migration never happened, and the tenuous epoch of pastoralist frontiersmen became fossilized as the long-term backbone of northern Australian colonial society.

Once northern land title regimes were developed, colonial governments invested in ports, roads, telegraph communications and administrative resources such as judiciary, police, and prisons to support colonial settlements. Key colonial administrative centres were established in the Northern Territory and north Queensland in the later 1860s and 1870s, whilst settlements in the Kimberley emerged a decade later reflecting Western Australia's lack of governmental capacity compared to South Australia and Queensland which were eager to promote northern development following exploration surveys. Strategically located settlements where government authority was located was fundamental to the occupation of huge expanses of pastoral rangelands throughout the North. Within fifteen years the strategic centres of Darwin (1859), Townsville (1866), Roebourne (1868), Katherine (1872), Cairns (1876),

Derby (1883), Broome (1883) and Wyndham (1885) were formally gazetted. The northern pastoral industry, based mainly on cattle and to a much lesser extent on sheep grazing, was the tenuous basis for northern colonial occupation but European settlement was significantly enhanced by localised and at times lucrative mining and pearling industries. Pearling promoted settlements at Thursday Island (1877) and Broome (1883). Whereas, short-term gold rushes established the settlements of Pine Creek (1871), Charters Towers (1872), Cooktown (1874) and Halls Creek (1886) and supported numerous other short-term settlements. In other Anglophone colonial settler experiences in southern Australia, North America, and New Zealand, short-term booms played a critical role in building a dominant settler society which overwhelmed numerically Indigenous peoples. Not so in North Australia, where short-term mining booms devastated local Aboriginal communities but did not generally leave large permanent European settlements in their wake.[8]

While northern colonial occupation and consequently the denial of Aboriginal people's authority over their traditional lands occurred incrementally over many years, the instrument for legal dispossession occurred with the stroke of an official's pen in government land title offices in Brisbane, Adelaide, and Perth. In these centres of colonial power, pastoral leases were allocated to settlers and lines drawn on maps covering huge swathes of territory. Not featured on the scantily detailed nineteenth century colonial maps was any mention that this was Aboriginal owned land that was benignly allocated to settlers for their economic use. We know now from the 1996 High Court decision in the Wik case that these parcels of land weren't leases that gave exclusive possession to the leaseholder under English common law. The High Court ruled that pastoral leases were a unique form of land tenure developed for the circumstances of nineteenth-century Australian settler frontier expansion to give governments the power to manage those lands. Until then these lands had known no other possessor than their Aboriginal inheritors but now were systematically incorporated under the political authority of Britain's Empire, whose footprint on the ground were the hooves of cattle and settler horses.[9]

The term pastoral lease was a self-serving misnomer. They were simply licenses to graze stock with a number of conditions, including the rights of the Traditional Owners to live, hunt, and continue to practice their culture. Drawing from extensive historical archives, the High Court in Wik also concluded that British and Australian colonial governments viewed pastoral leases as a temporary and transitory form of land tenure which was intended to be upgraded to a more secure and exclusive form of title once European settlement intensified and the rangelands became diversified for agricultural use. But in the transition period, the British and colonial governments intended that Aboriginal rights to their traditional *country* inside pastoral boundaries would be recognised.

The Queensland Government led the way in this transition passing a series of land acts between the 1860s and 1880s, favouring agriculture over pastoralism and reducing the land rights of Aboriginal people. Yet more than a century and half after the first pastoral leases were issued, the dominant land tenure in North Australia remains the pastoral lease with freehold or exclusive possession leasehold titles outside urban settlements constituting a negligible proportion of the region's land titles.

The so-called Aboriginal reservations in colonial land laws and inserted in the pastoral lease conditions were well known to governments and pastoral leaseholders. The system fundamentally commenced in 1851 when South Australia's governor Young insisted that all pastoral leases should include reservations in favour of Aboriginal people, allowing them access to pastoral lands. This reflected nineteenth century British colonial policy and came to apply to all Australian jurisdictions that administered pastoral leasehold rangelands. Pastoralists often complained about the 'reservation', with allegations that it was a major source of stock killing, but for the most part the reservation was simply ignored as leaseholders viewed their titles literally and asserted exclusive possession. In the frontier period of protracted violence and contested occupation pastoralists, for the most part, viewed Aboriginal people on their leases as either independent, and therefore enemies, or servitude labour.

The law of coexistence was not the reality on the Australian northern frontier. Rather, it was the Doctrine of Discovery and the legal myth of *terra nullius* that determined the violent dispossession of Traditional Owners from their land. More than a century after the pastoral frontier pushed into northern Australia, High Court judge Brennan wrote in the 1992 Mabo judgement: 'Aborigines were dispossessed of their land parcel by parcel. Their dispossession underwrote the development of the nation'.[10] Within three decades from granting the first northern settlers titles to their cattle runs in the Northern Territory and far north Queensland, a pastoral belt of several hundred pastoral leases connected Cape York to the Kimberley, the economic basis of northern Australian settlement. In an extraordinary feat of colonial occupation, the northern pastoral estate was created through the mass issuing of licenses to a few hundred settlers to graze stock over a land mass far greater than contemporary Europe.

Pastoral enterprise failure was common in the early frontier period with Aboriginal resistance, climate, distance from markets, and price fluctuations taking a heavy toll. Technological advances in the latter part of the nineteenth century such as wind-driven water pumps, barb wire, and steam-driven drilling rigs and improvements in transport allowed the pastoral industry to gain a foothold and to become relatively stable over time. These innovations not only made it easier to occupy *country* with herds of cattle and to a lesser extent sheep, but it also meant that the settlers were less reliant on Aboriginal labour.

Consistent with the historical experience elsewhere in Australia, there was no attempt by settlers to negotiate with the Traditional Owners over the use of their lands. However, what was different about northern Australia was the sustained Aboriginal struggle against European incursion. Throughout the North as in other parts of Australia, Aboriginal resistance to the encroachment on their *country* was pervasive although it varied in intensity and endurance. Terrain, settler numbers, police quasi-military support, and transport and supply capabilities were critical factors which determined both the capacity for Aboriginal people to resist and for the setters to conquer. In several areas, in conflict which could be described as asymmetrical warfare, Aboriginal resistance to European incursion persisted over several decades. In the tropical rainforests and mountains of Cape York, several hundred Aboriginal people and settlers died on bloody battlefields in a sustained conflict lasting more than twenty years.[11] The Kalkatunga (Kalkadoon) in the Mt Isa region fought one of the most celebrated resistances to European occupation anywhere in Australia.[12] In the Arnhem Land and Port Keats areas of the Northern Territory, Aboriginal resistance was so strong that it slowed or in some cases prevented European occupation. In the central Kimberley region in the 1890s, frontier conflict took the form of European warfare involving guns and organised battles.[13]

Indigenous raids on stock and random attacks on settlers became pretexts for murderous campaigns against men, women, and children by settlers and police across northern Australia. Many of these massacres have been recorded in surviving official and unofficial documents and many more figure in countless Aboriginal oral accounts. Along with advances in industrial technology, developments in European weaponry were critical to successful pastoral occupation and the subjugation of Aboriginal people to settler economic interests. These weaponry developments included cartridge breach-loading repeating rifles such as the Snider–Enfield (1866), the Martini Henry (1871), and the Winchester (1873). These guns, in the hands of people from nations experienced in organised warfare against societies with no military tradition, resulted in the mass slaughter of Aboriginal people throughout North Australia. The Queensland Native Police Force was responsible for widespread indiscriminate slaughter. Hundreds, perhaps thousands of survivors over huge distances were forcibly transported and confined in large group settlements and missions such as Palm Island. Similar police systems of violence in support of settlers were also employed in the Kimberley and the Northern Territory, albeit with less obvious murderous intent.

The last known massacres of Aboriginal men, women, and children at the hands of settlers and their police supporters were at Forrest River in the East Kimberley in 1926 and Coniston in the

Northern Territory in 1928. In five decades of violent frontier conflict the Aboriginal body count will never be known, but documentary evidence combined with recorded Aboriginal oral accounts suggests that the figure is in the thousands.[11]

By the end of the 1890s in the settled pastoral regions of North Australia, many Aboriginal people were displaced, homeless, and suffering debilitating illnesses, their lifestyle and economy destroyed. When Australia federated as a nation in 1901 the colonial architecture in terms of governance administration, land tenure, and the police and justice system was firmly in place to rule North Australia.

Although colonial authority showed ruthless determination to crush dissent with the expressed purpose of bringing Aboriginal people under permanent subjugation, there are several instances of heroic Aboriginal opposition where European occupation was effectively deterred. In much of northern Australia large Aboriginal reserves were created because Aboriginal resistance was too strong. Most of these areas are still exclusive Aboriginal domains.

3.3 NORTHERN AUSTRALIAN COLONIAL SOCIETY

Soon after federation it became clear that the vision for large-scale northern Australian European settlement would not be fulfilled. Pastoralists dealt with inconsistent climate and costly distance from markets. They survived and profited because the subjugated Aboriginal populations residing within the pastoral domains provided unpaid work in a feudal relationship that would last another three to four generations. The exploitative labour system was the basis of European northern Australian settlement, supported by a colonial government apparatus of police, courts, prisons, and native welfare administration.

In the early part of the twentieth century, the cost of northern Australian colonisation was far greater than the economic wealth extracted from the region.[14] But northern Australia's strategic importance became more pronounced following Australia's federation thereby entrenching southern Australia's subsidisation of the North as an embedded feature of Australian nation building.[15] The North has played an interesting and compelling role in Australia's history of racial discrimination with respect to the denial and abuse of Aboriginal cultural and human rights and the nation's relationship with Asia. The race power in Australia's Constitution (S51, Ss26) gave the Commonwealth a head of power to make laws for any race. As expected by the Constitutional Founders, the Commonwealth used that power to enact the national parliament's first piece of legislation: the Immigration Restriction Act (1901), which provided the legal framework for what is commonly referred to as the White Australia Policy.[16] Settler Australia's concerns about non-European people, particularly Asians, coming to live in Australia is a well-documented social and political theme in Australia's historical character. Vestiges of that national character have continued to play out in recent years through the politics of 'boat people' asylum seekers. Subsidising North Australian colonial society to support the security of Australia's northern border has been a bipartisan national political imperative since federation.

The other aspect of the constitutional race power was that Indigenous people were expressly quarantined from Commonwealth power until the constitutional amendment in 1967. The historic reasoning for Indigenous exclusion from Commonwealth constitutional responsibility was concern by some states such as Western Australia about Commonwealth interference in labour relations particularly in the North, and the pervasive acceptance of policies and practices that encouraged the decline and eventual disappearance of Indigenous people in Australia, an outcome considered best managed by the states.

In the first two decades of the twentieth century, all Australian jurisdictions passed laws giving government officials inordinate powers over the lives of Aboriginal people. Queensland at first led the way through 1897's The Aboriginal Protection and Restriction of the Sale of

Opium Act, which allowed the 'chief protector' to remove local Aboriginal people onto and between reserves and hold children in dormitories. In 1905, the Aborigines Act in Western Australia was heavily influenced by the Queensland laws and became the national high-water mark in government management of Aboriginal people, highlighted by its focus on segregating Aboriginal people from settler society. Under the WA legislation, Aboriginal people were effectively confined to reserves and banned from towns. Reserves were administered by government agencies or missionaries and every aspect of Aboriginal life was controlled, including the right to marry, guardianship of children, the right to work outside reserves, and management of personal property. The legislation was subsequently imitated by South Australia and the Northern Territory. In 1911, the Commonwealth assumed responsibility for the administration of the Northern Territory from South Australia and in that year passed the Northern Territory Aboriginals Ordinance, which made the chief protector the legal guardian of every Aboriginal and mixed heritage child under 18 years of age. The ordinance empowered the chief protector to force any Aboriginal person onto a mission or settlement and to seize any Aboriginal child from their families and place them in institutions.[17]

In the first half of the twentieth century, racial biological theory came to strongly influence Australian Indigenous public policy. The Aboriginal reserves system was pivotal to government policies of segregation and control. There were three pillars to the Aboriginal reserve system. Firstly, reserves were established on the edge of towns, which allowed government authorities to manage Aboriginal populations within colonial urban centres. The second pillar comprised reserves where large concentrated Aboriginal settlements, often under church mission management, were established for the dispossessed or as institutions for dissidents and mixed heritage children forcibly removed from their families. The three jurisdictions – Queensland, Western Australia, and South Australia/Commonwealth – used their Aboriginal protection and segregation laws to create Aboriginal reserves throughout the North. Practically every northern settler town had a reserve on its outskirts where Aboriginal people could be controlled and segregated.

Thirdly, large reserves were established in regions where Aboriginal resistance to settlement was strong and on lands that governments deemed were not worth fighting over. On these reserves legal restrictions prevented settlers, particularly miners, from entering. 'Soothe the dying pillow' was the political catch-cry that would underwrite the disappearance of Aboriginal people in very remote regions – out of sight and out of mind from settler Australia. The eugenicist and white supremacist leaders of Australian Indigenous policy, A.O. Neville, the long-standing commissioner of Native Welfare in WA, and Dr Cecil Cook, chief protector of Aborigines in the Northern Territory, insisted on strong laws to ensure there was no contact between traditional Aboriginal people living on reserves and marauding miners. Interracial sexual relationships, they concluded, would expand Aboriginal populations rather than reduce them. The legal protection they demanded gave government authorised officials effective veto over mining activity on Aboriginal reserves in Western Australia and the Northern Territory.[18]

The unintended historic consequence of this power, fuelled by eugenic ideology, was its influence on Northern Territory land rights legislation and the national toxic political debate over land rights a half a century later. In his 1974 Royal Commission report on Aboriginal land rights in the Northern Territory, Justice Woodward recommended that Aboriginal reserves in the Northern Territory be transferred to Traditional Owners along with the corresponding power over mining access. The recommendation was incorporated into the Northern Territory Aboriginal Land Rights Act of 1976, and subsequently became the pivotal tension point in the national political combat over the recognition of Aboriginal land rights which continues today.

Concurrent with the development of the government Aboriginal reserve system was the growing expansion of church missions across North Australia. Church missions played a critical role in the colonisation of the North. Numerous Christian denominations established mission settlements throughout the North and were supported by governments in that the missions were almost all located on Aboriginal reserve lands. The mission settlements offered Indigenous people a protective sanctuary from frontier violence but were also often places of authoritarian repression which attempted to expunge traditional culture and replace them with Christian practices and beliefs. However, in some notable instances, church missions were sympathetic to traditional values and encouraged an enduring coexistence of Indigenous culture and Christianity.

3.4 FRACTURING OF THE COLONIAL SYSTEM

The Second World War had a decisive impact on North Australia. It has been estimated that up to 4,000 people in North Australia served in the Australian military and/or special Aboriginal or guerrilla units. In addition, a further 8,000 Aboriginal people also supported the war effort through direct employment at military bases; under mission defence contracts; or were mobilized by the Commonwealth Manpower directorate for defence related projects. There were also numerous people who were employed as guides, trackers and maritime pilots. Some Aboriginal soldiers and auxiliary personnel saw service in Timor and Papua New Guinea, and many were critical to the success of Australian Imperial Forces units posted to defend North Australia. In addition, some 34 Aboriginal communities hosted operational or emergency airstrips and 20 radar stations.[19] Wartime military experience gave many Indigenous Australians pride and confidence in demanding their rights.

After the war, the northern labour relations system began to fracture. In 1946, several hundred Aboriginal people from pastoral stations throughout the Pilbara walked off in a concerted political action vowing never to return. In 1955, the large Aboriginal government settlement of Moola Bulla was sold to a pastoralist, and more than 200 people were forced into destitution in nearby Halls Creek and Fitzroy Crossing. In 1966, the Gurindji walked off Wave Hill in the most celebrated Indigenous political action against feudal colonialism, which sparked a national political campaign for Aboriginal land rights.

By the second half of the 1960s, global decolonisation began to have a major impact on northern Australia. The Commonwealth Conciliation and Arbitration Commission in 1966 ruled that Aboriginal stock workers should be paid the same as their non-Aboriginal counterparts. Over the next six years the exploitative pastoral industry labour relations system disintegrated. During this period, the other key plank of northern Australian colonialism – institutional segregation – also crumbled. Racial discriminatory laws and administrative practices dissolved as Australia moved to comply with United Nations treaties concerning human rights and international labour standards and voted overwhelmingly to change the nation's Constitution giving the Commonwealth the power to make laws for Indigenous people. Institutional relationships between Church and State concerning the management of missions fractured causing a number of missions to close. Mainstream schools and hospitals became accessible to all. Social security payments became commonly available, and universal voting franchise changed the North's political dynamics.

The rigid colonial system based on a feudal labour pastoral industry regime, authoritarian mission residential life, institutional segregation, race-based alcohol restrictions, and a range of citizenship restrictions collapsed rapidly and without a policy framework and plan of action by governments to manage the change. The social repercussions were traumatic. Thousands of Aboriginal people across northern Australia were uprooted from their homes on pastoral properties and missions. Town reserves became appallingly overcrowded, and hastily erected camp sites of calico and corrugated iron became common place throughout the North. Freely available alcohol and universal social security fermented a national crisis of unprecedented social turmoil. In this environment the Whitlam Government was elected with a commitment to introduce land rights and self-determination as a basis to decolonise Indigenous Australia and particularly focused on the North.

3.5 INDIGENOUS RIGHTS AND CONFLICT OVER OPPOSING VISIONS

Australia's modern Indigenous political rights movement has largely been driven from North Australia. Celebrated instances of Indigenous assertions of their rights have dotted the northern landscape. From Wave Hill and the Gove Peninsular in the top end, to Uluru in the centre, to Mornington Island, Aurukun and Murray islands in Queensland and the Torres Strait, and to Noonkanbah and Gordon Downs in the Kimberley, Indigenous people asserted their rights to be recognised as Traditional Owners of their *country* that could not be ignored by Settler Australia's body politic.

Indigenous rights soon found its way into Australian nation-building as national Labor and Conservative governments introduced Land Rights legislation in the Northern Territory and cultural heritage protection laws. The election of the Whitlam Government in December 1972 was a transformational signpost in the history of the Australian state's relationship with Indigenous peoples. In addition to Whitlam's land reform agenda inherent in the Woodward Royal Commission, his government's implementation of its self-determination policy was based upon a number of strategic elements. His government established a federal Department of Aboriginal Affairs and negotiated separate agreements with each State to ensure a national approach to implement Aboriginal policy within the federal system. To enable Aboriginal communities to implement their development agendas and manage grant funding, the government encouraged the development of Aboriginal community corporations through the Aboriginal Councils and Association Act (ACAA) which was finally passed in 1976. To support Aboriginal aspirations to acquire land, the Whitlam Government established the Aboriginal Land Fund Commission which was responsible for purchasing a number of pastoral leases for Aboriginal benefit across North Australia. The Fraser Government continued the progressive policies of the Whitlam Government concerning Indigenous development during its seven and half years in office, although it often ran into loggerheads with Western Australia and Queensland over Indigenous cultural protection, land rights, and community governance.

The Whitlam Government's development agenda was particularly relevant to North Australia. In terms of local governance, his government intended to de-institutionalise government settlements and missions by enabling Aboriginal communities to incorporate under the Councils Section of the ACAA, and to also foster the development of Aboriginal civil and commercial society under the association part of that legislation. The Indigenous local governance aim of the Commonwealth Government's agenda was firstly frustrated by the Queensland Government and then by nearly all the states and the Northern Territory. The opposition to the local government (Council) provisions of the ACAA arose in 1978, during the period of the Fraser Government, when the Queensland Department of Aboriginal and Islander Advancement attempted to assume control of what were then Aurukun and Mornington Island missions. The local community objected, and supported by the Uniting Church, attempted to incorporate under the Councils provisions of ACAA. After a highly public stand-off the Queensland Government agreed to a limited form of local government for Queensland Aboriginal Communities in the form of the Queensland (Aboriginal Land) Act 1978 which provided a Grant of Deed in Trust. This form of tenure became widely used by communities on Aboriginal reserve land in Queensland. The Commonwealth has never again attempted to use the wider local governance aspect of the ACAA which meant that the developmental agenda for Indigenous self-governance was seriously restricted.[20]

The Commonwealth Government's land reform agenda for Indigenous people has been continually frustrated by the inherent tensions of Australia's federation. Woodward presented two reports on his Northern Territory Land Right Royal Commission. The first report recommended that Aboriginal Reserve land should be recognized as Aboriginal land and transferred to Aboriginal ownership. In the second report, he recommended that Aboriginal Traditional Owners should have the right to obtain royalties and that statutory Aboriginal Land Councils should be established to make land claims, negotiate agreements with the informed consent of Traditional Owners, and administer mining royalty distribution. Importantly, the Royal Commission recommended that the power of veto over mining access to Aboriginal land, enshrined in the protectionist legislation emanating from the period of eugenic policy influence, should be transferred to Aboriginal land owners.

The Whitlam Government accepted Woodward's recommendations and introduced the Northern Territory Land Rights Bill in 1975, which was enacted the following year by the Fraser Government demonstrating the bipartisan political support at the national level for Aboriginal land rights. However, the potential for national consensus quickly evaporated as the mining industry, supported by the pastoral industry and conservative political interests in Western Australian, Queensland and Northern Territory, opposed Indigenous land rights in a sustained clash over land ownership and use.

The Northern Territory *Country* Liberal Party, which held office from the time the Territory became a self-governing jurisdiction in 1977–2001, consistently opposed land claims under the Northern Territory Land Rights Act which led to protracted litigation, political contest, and tension and community ill will. The Queensland Government has also had a fractious history concerning the recognition of Aboriginal land ownership. In 1982, the Queensland Government passed the Lands Act (Aboriginal and Islander) Amendment Act, which provided the granting of land in trust to Aboriginal people, and this was further amended to enable the resumption of land granted in trust in 1984. In 1991, the Queensland Aboriginal Land Act was passed, which recognizes Aboriginal traditional ownership but is generally seen as limited compared to the rights accorded Traditional Owners under the Northern Territory Land Rights Act.

However, it is Western Australia which has most disturbed the potential to reach a national consensus on Aboriginal land recognition. Opposition to Aboriginal land rights was highly visible in Western Australia in the Noonkanbah dispute in 1980. The dispute centred on the opposition by the Noonkanbah community to an American mining exploration company's plans to drill for oil, which was strongly supported by the Western Australian Government led by Sir Charles Court. In the high-profile conflict the Western Australian Opposition Labor Party supported the Aboriginal community and, in the process, amended its policy platform committing a future State Labor Government to implement Aboriginal land rights based on the Northern Territory model.

When the Burke Labor Government came to power in Western Australia in 1983, it established an inquiry into Aboriginal land rights prompting a massive opposition campaign led by the mining industry which felt threatened by the potential for Aboriginal landowners impeding their access to land in the resource-rich state. The mining industry's campaign was also aimed at the Federal Hawke Labor Government, which was committed to enacting a national Aboriginal land rights legislative regime. Just before the Western Australian State election in 1986, both the Burke State Government and the Hawke Federal Government abandoned their commitments to legislating for Aboriginal land rights.[21] No Western Australian major political party since has proposed Aboriginal land rights as a political commitment. Western Australia remains the only Australian jurisdiction that does not have a statute recognising Aboriginal land ownership.

National political conflict over Indigenous rights erupted again in the 1990s in response to the High Court's Mabo and Wik decisions on the recognition of native title rights in Australian common law. Following the Mabo judgement in June 1992, there was hesitancy and confusion by Indigenous interests, governments, and industry about the most effective response because of the legal uncertainty about the extent of the ruling. The High Court ruled that Indigenous inherent rights to their traditional *country* had survived colonisation except where governments had extinguished those rights by granting land titles which were inconsistent with Aboriginal use.

In the months that followed the Mabo judgement, some Indigenous leaders argued that native title meant far more than land property titles and included the full spectrum of cultural, social, economic, and political rights and thus required a comprehensive negotiated regional agreement approach – in effect, a new relationship between the Australian nation state and Indigenous colonised peoples. Sections of Australia's Indigenous leadership were influenced by the Canadian Government's Land

Claims Policy approach, which was put in place following the Canadian Supreme Court's 1973 Calder Case, which had informed the Mabo Judgement. However, before these arguments could be mounted, the Australian polity became engulfed in frenzied debate about economic certainty and the validation of mining titles. The mining industry demanded the validation of thousands of mining titles issued after the Federal Racial Discrimination Act came into force in 1975, which had the effect of legally protecting native title rights. This became the central issue in a national debate where the mining industry and state governments, such as Western Australia, demanded that native title be extinguished in Australia for the sake of economic certainty.

The national response to Mabo was the Native Title Act (NTA) passed by the federal parliament in 1993. It was a compromise law which protected native title from extinguishment by State governments and secured all mining titles. In legislating mass extinguishment of Indigenous rights, parliament suspended the Racial Discrimination Act and in return, as a special measure in the legislation, native title holders were accorded a limited right to negotiate with respect to any future grant of title which impacted on Indigenous legal rights. Despite the NTA providing security for mining and establishing a formal statutory process for Indigenous people to claim native title, opposition by industry and resource rich states was intense. Western Australia passed its own law which extinguished native title throughout the state only weeks before the federal Native Title Act came into force. Eighteen months later the High Court struck down the Western Australian law in a unanimous decision on the grounds that it was unconstitutional and breached the federal Racial Discrimination Act.

It was clear from the Mabo judgement that the granting of freehold or leasehold titles had extinguished Indigenous common law rights. But what was not clear at the time was whether native title rights could coexist on other land titles such as pastoral leases. The general legal and political view arising from the Mabo judgement was that pastoral leases had extinguished native title despite the Aboriginal use reservation within statutory land laws that regulated pastoral rangelands. If the High Court had in fact ruled that this was the case at common law, the impact of native title rights would have been restricted to lands already reserved for Indigenous use or unallocated government land, which only applied to Western Australian desert regions. When the High Court ruled in Wik three years after the passing of the Native Title Act that native title was able to coexist on a pastoral lease, Australia was once again engulfed in a toxic political debate of Indigenous land rights. The Howard Government introduced major amendments to the NTA, which had the effect of winding back Indigenous people's rights.

During this period of recognition of Indigenous communal-held rights by the Australian courts, the political debate turned away from the potential to reform the colonial land tenure system and include Indigenous people in a new and inclusive development paradigm. The political reaction to Mabo and Wik by the mining industry and resource-rich states revealed the commitment of those interests to defend the status quo concerning mineral wealth extraction from the North. The Wik judgement was seen to threaten the mining industry far more than it posed a threat to the pastoral industry. In Western Australia more than two-thirds of mining leases are located on pastoral leases.[22]

3.6 CONCLUSION

Using their 1967 acquired constitutional powers, a succession of national governments responded to Indigenous demands for their rights to be recognised by funding land buybacks, homeland communities, language and cultural initiatives, Indigenous medical services, and education and employment initiatives. In a period of a little over two decades, from the Gurindji walk off in the mid-1960s to the time of the Mabo decision, North Australia was socially and politically transformed.

A thousand discrete indigenous communities had been created in what has been a demographic revolution, and land councils and community representative organisations have substantially changed political power in the North.

The political backlash against the assertion of Indigenous rights occurred at a time when modernisation theory and neo-liberal economic thinking were resurgent. And in this context the northern Australian Indigenous leadership found it difficult to build its decolonising political agenda. Supported by mainstream media and influential conservative-orientated think tanks, modernisation theory's philosophical underpinnings – individual and family responsibility, private property, home ownership, generational wealth transfer, and participation in the western industrial economy – came to dominate Australian Indigenous public policy. The policy approach's highpoint was the Howard Government's Northern Territory Emergency Response ('Intervention') in 2007. The clear failure of this approach to improve the living standards of Indigenous people, and the growing critique of trickledown economics that is embedded in the White Paper and the modernisation framework, has rekindled the decolonising agenda of northern Indigenous politics.

This renewed agenda is fuelled by developments initiated by Indigenous communities, not by governments. Forty years of Aboriginal Land Rights in the Northern Territory and more than twenty years of native title assertion throughout the North is beginning to reshape the North's economy and political power. A range of Indigenous community economic activity incorporating land and sea management, horticultural development, ecological research, tourism, community infrastructure development, and artistic endeavour is fundamentally based on traditional rights and native title. That this potential continues to be retarded through the structural issues of land tenure and governance that remains colonial in character, and maintains the North in economic and social underdevelopment, is a significant challenge of public policy and the effectiveness of Australia's federal governmental system.

APPENDIX 3.1

Policy	Time	Australian Government: Aboriginal Affairs & Related Acts	Australian Government: Land Tenure Acts	Northern Territory: Aboriginal Affairs & Related Acts	Northern Territory: Land Tenure Acts	Queensland: Aboriginal Affairs & Related Acts	Queensland: Land Tenure Acts	Western Australia: Aboriginal Affairs & Related Acts	Western Australia: Land Tenure Acts
Relevant Australian, Colonial and State Governance and Land Tenure Legislation									
Establishing the Northern Colonial Frontier	1861–1901	Not Constitutionally applicable until after Federation in 1901 and the referendum in 1967	Not Constitutionally applicable until after Federation in 1901 and the referendum in 1967	South Australian Colonisation Act (UK) 1839 SA Royal Commission of 1895 into the Northern Territory recommends changes to land tenure to encourage permanent settlement	Wastelands Protection (SA) ACT 1842 Pastoral Lands Act 1893 Closer Land Settlement Act (SA) 1897 Northern Territory Crown Lands Act (SA) 1890, Northern Territory Lands Act (SA) 1899	Aboriginal Protection and Restrictions of the sale of opium Act 1897	Land Act (1864) Land Act (1864) Land Act (Qld) 1868 enables leases to be divided for closer settlement Crown Lands Act 1884 Crown Land Act 1897 establishes a land court	Aboriginal Protection Act (WA) 1886 enables Aboriginal reserves to be declared	Colonial Land Regulations for crown land including leases and licences 1832–1890
Relevant Australian, Colonial and State Legislation									
Northern Australian Colonial Society	1902–1939	Following the transfer of the Northern Territory to the Australian Commonwealth Government in 1911, the commonwealth became responsible for Aboriginal Affairs in the Northern Territory		Aboriginal Ordinance (NT) 1911–1918 enables land to be reserved for Aboriginal people.	Land Acquisition Act and associated Land Act 1906 makes all land in the Australian Territories leasehold	Aboriginal Protection and Restrictions of the sale of opium Act 1897 as amended in 1901, 1927, 1928 and 1934 Aboriginals Preservation and Protection Act 1939 replaced the 1897 Act	The Special Agricultural Selections Acts 1901 to 1904 (closer settlement acts) Land Act Amendment Act 1909 1922 Closer settlement for returned soldiers expands in North Queensland	1904 Royal Commission on into the administration of Aborigines commences Aborigines Act 1905 Native Administration Act 1905–1936 1911 Aborigines Act Amendment Act significantly extended the Protector's guardianship power 1936 Sections 8 and 12 of the new Native Administration Act the Chief Protector's guardianship powers were increased	Land Act (WA) 1889 Land Act (WA) 1890–1909 Agricultural Land Purchase Act (WA) 1896–1909 Crown Land (WA) Act 1909

(Continued)

APPENDIX 3.1 (*Continued*)

		Australian Government		Northern Territory		Queensland		Western Australia	
Policy	**Time**	**Aboriginal Affairs & Related Acts**	**Land Tenure Acts**	**Aboriginal Affairs & Related Acts**	**Land Tenure Acts**	**Aboriginal Affairs & Related Acts**	**Land Tenure Acts**	**Aboriginal Affairs & Related Acts**	**Land Tenure Acts**
Fracturing of the Colonial System	1940–1972	1940-Defence Force recruitment regulations revised overall 4,000 Aboriginal people served in the military 8,000 are estimated to have been employed directly by military bases; under mission military contracts; or were mobilised by the Commonwealth Manpower directorate in Qld and WA 1962 Commonwealth Electoral Act is amended giving franchise to all Aboriginal people in Western Australia, Queensland and the Northern Territory	Not Applicable	Northern Territory (Administration) Act 1947, establishes a Legislative Council Welfare ordinance (NT) 1953 Social Welfare Ordinance (NT) 1964	NT Crown Land Ordinance passed 1953 NT Housing Loans 1953 passed	Aborigines' and Torres Strait Islanders' Affairs Act, (QLD) 1965, gives the Director of Aboriginal Affairs considerable power over 'assisted Aborigines' Aborigines Act 1971 and the Torres Strait Islanders Act 1971. Both of these laws abolished the Director's power to remove children	Qld Land Act 1962 amalgamates a range of nineteenth century acts	Native Welfare Act 1963	As above

(*Continued*)

APPENDIX 3.1 (*Continued*)

		Australian Government		Northern Territory		Queensland		Western Australia	
Policy	**Time**	**Aboriginal Affairs & Related Acts**	**Land Tenure Acts**	**Aboriginal Affairs & Related Acts**	**Land Tenure Acts**	**Aboriginal Affairs & Related Acts**	**Land Tenure Acts**	**Aboriginal Affairs & Related Acts**	**Land Tenure Acts**
Indigenous Rights and Conflict over Opposing Visions	1972–1999	Aboriginal and Torres Strait Islanders (Queensland Discriminatory Laws) ACT 1975 (Cth) Racial Discrimination Act 1975 (Cth) Aboriginal Land Fund Commission Act 1975 (Cth) Aboriginal Development Commission Act 1980 (Cth) Aboriginal and Torres Strait Islanders heritage Protection Act 1984 (Cth) Aboriginal and Torres Strait Islanders Commission Act 1989 (Cth) Council of Reconciliation Act 1991 (Cth)	Aboriginal Land Rights Act (1976 (Cth) Aboriginal and Torres Strait Islanders Heritage Protection Act 1984 (Cth) Native Title Act 1993 (Cth) Native Title Amendment Act 1998 (Cth) Native Title Amendment Act 1998 (Cth) Native Title (Prescribed Body Corporate Regulations) Act 1999 (Cth)	Local Government Act (NT) 1978 Community Welfare Act (NT) 1983 NT Aboriginal Sacred Sites (NT) Act 1989 Local Government Act (NT)	Aboriginal Land Act (NT) 1978 Peninsula land and sanctuary Act (NT) 1981 Nimililuk (Katherine Gorge)national Park (NT) Act 1989 Pastoral Land Act (NT) 1992 Crown Lands Act 1992 (NT) Macarthur River Project Agreement Ratification (NT) Act 1993 Validation of the Native Title Act (NT) 1994 Land Administration Act 1997(NT)	Aborigines Act and other Acts Amendment Act 1975 (Qld) Community Services (Aboriginals (QLD) Act 1984 Community Services (Torres Strait Islanders (QLD) Act 1984	Cultural Resources (Qld & Qld Estate) Act 1987 Aboriginal Land Act 1991(QLD) Torres Strait Island land Act (QLD) 1991 Native Title (QLD) Act 1993 Land Act 1994 (QLD)	Aboriginal Affairs Planning Authority (SA) Act 1972 Aboriginal Communities (WA) Act 1972 Aboriginal heritage (WA) Act 1972	Conservation and land management Act 1984 (WA) Land (Validation and Native Titles (Traditional usage) Act (WA) 1995

REFERENCES

1. Quijano A (2000) Coloniality of power, Eurocentrism, and Latin America. *Nepantia: Views from the South* **1**, 533–580.
   Quijano writes (p. 535): 'Europe's hegemony over the new model of global power concentrated all forms of the control of subjectivity, culture, and especially knowledge and the production of knowledge under its hegemony'. This resulted in a simultaneous denial of knowledge production to the conquered peoples and repression of traditional modes of knowledge production, on the basis of the superiority/inferiority relationship enforced by the hierarchical structure.
2. Young Scholars Initiative (2013) *Dialogues on Development*. Volume 1: Dependency. Institute for New Economic Thinking. Retrieved from www.ineteconomics.org/education/young-scholars-initiative.
   Settler colonialism is a distinct type of colonialism that functions through the replacement of indigenous populations with an invasive settler society that, over time, develops a distinctive identity and sovereignty. Settler colonial states include Canada, the United States, Australia, and South Africa, and settler colonial theory has been important to understanding conflicts in places like Israel, Kenya, and Argentina, and in tracing the colonial legacies of empires that engaged in the widespread foundation of settlement colonies.
3. See for example:
   Peter Yu (2016) Keynote address, *Northern Australian Development Conference*, Darwin.
   Morrison J (2013) Indigenous participation in the developing north: A national emergency. Retrieved from www.theconversation.com.
   Morrison J (2017) Keynote address, *Northern Australian Development Conference*, Cairns.
   Oscar J (2017) Mabo address, *Native Title Conference*, Townsville, Australia. Australian Institute of Aboriginal and Torres Strait Islander Studies, Canberra, Australia.
   Uluru Statement from the Heart (2017) www.referendumcouncil.org.au.
4. Thalia A (2006) *Equal pay but no jobs*. Retrieved from http://onlineopinion.com.au/view.asp?article=4839.
5. Whitlam G (1985) *The Whitlam Government 1972–1975*. Viking Press, Ringwood, VA.
6. Berndt RM, Berndt CH (1964) *The World of the First Australians*. Ure Smith, Sydney, Australia.
7. Kiddle M (1961) *Men of Yesterday: A Social History of the Western District of Victoria, 1834–1890*. Melbourne University Press, Melbourne, Australia.
8. Belich J (2009) *Replenishing the Earth. The Settler Revolution and the Rise of the Anglo-World, 1783–1939*. Oxford University Press, Oxford UK.
9. High Court of Australia (1996) *The Wik Peoples v State of Queensland & Ors; The Thayorre People v State of Queensland & Ors*, 23 December 1996, (1996) 187 CLR 1
10. High Court of Australia (1992) *Mabo and Others v Queensland (No. 2)*, 3 June 1992, 175 CLR 1, (1992) HCA 23.
11. Reynolds H (2006) *The Other Side of the Frontier: Aboriginal Resistance to the European Invasion of Australia*. University of NSW Press, Sydney, Australia.
12. Drake J (2012) *The Wild West in Australia and America*. Boomerang Press, Australia. Retrieved from www.boomerangpress.com.
13. Pedersen H, Woorunmurra B (1995) *Jandamarra and the Bunuba Resistance*. Magabala Books, Broome, Australia.
14. Blainey G (1980) *A Land Half Won*. McMillan Publishers, London, UK.
15. McGregor R (2016) *Environment, Race and Nationhood in Australia: Revisiting the Empty North*. Palgrave McMillan, London, UK.
16. Attwood B (2003) *Rights for Aborigines*. Allan & Unwin, Melbourne, Australia.
17. Haebich A (2000) *Broken Circles*. Fremantle Press, Perth, Australia.
18. Manne R (1996) *In Denial: The Stolen Generations and the Right*. Self-published. https://www.quarterlyessay.com.au/essay/2001/04/in-denial.
19. Phillpot S, unpublished research.
20. Hiatt LR (1986) *Aboriginal Political Life. Wentworth Lecture 1984*. Australian Institute of Aboriginal and Torres Strait Islander Studies, Canberra, Australia.
21. Beresford Q (2006) *Rob Riley: An Aboriginal Leader's Quest for Justice*. Aboriginal Studies Press, Australian Institute of Aboriginal and Torres Strait Islander Studies, Canberra, Australia.
22. Russell P (2005) *Recognizing Aboriginal Title: The Mabo Case and Indigenous Resistance to English-Settler Colonialism*. University of Toronto Press, Toronto, Canada.

CHAPTER 4

Economic Development Across the North

Historical and Current Context of Possible Alternatives

Rolf Gerritsen, Peter Whitehead, and Natalie Stoeckl

CONTENTS

4.1 INTRODUCTION

North Australia inspires many competing narratives about its future: historically as an ineluctable destroyer of grand visions; more recently as a globally significant 'food bowl' irrigated by limitless water and as a rich quarry and gas producer delivering great benefits despite regulatory challenges. It is also the site of extraordinarily valuable natural and cultural assets that warrant special protection, and, outside a few affluent urban enclaves, it is a place of enduring human poverty, poor health, and social dysfunction. We will consider the implications of the current development model, as expressed in the federal government's White Paper and its policy actions. We will then link this to the conflicts between this model and the sustainable realities of northern Australia. Alternative approaches to the economic development of northern Australia are proposed, mostly to do with the improved social distribution of economic development and the better use of its resources.

We identify the policy changes needed to mitigate recurring failures and propose an approach to northern development that respects the landscape and the people who continue to live in it. Intrinsic to this is to encourage equitable participation of Aboriginal people in that development. We argue that the dominant influence on government policy of the more optimistic proponents of northern development have been particularly damaging to Aboriginal society. And current policy prescriptions appear likely to continue to compromise Aboriginal roles and participation in northern economic and social development, despite official commitment to 'closing the gap'. There appear to be no serious measures proposed to ensure that all residents of the North can capture real and enduring benefits from so-called 'northern' development.

4.2 THE PATTERNS OF NORTHERN DEVELOPMENT

Since the mid-nineteenth century, when it forcibly came under *de facto* (as distinct from *de jure*) colonial control, the development of northern Australia has followed a pattern that has been characterized as a 'pulse and pause' dynamic.[1] If we follow current policy trends, the future will be the continuing erratic economic development of the region and the separate but continuing marginalisation and impoverishment of its Aboriginal population.

4.2.1 Northern Economic Development History

Economic development in northern Australia historically features intermittent booms punctuated by long periods of economic stasis or even decline. The second half of the nineteenth century saw a dramatic 'pulse' as industries such as mining, pastoralism, and pearling irrupted into northern Australia (as did agriculture, particularly sugar, on the Queensland Coast). Miners, now so important, were also significant in expanding the frontier of the North in the nineteenth century. In the 1950s, pearling disappeared, destroyed by the plastic button.

After the rapid expansion of pastoralism from the 1880s to World War I, this industry has mostly paused. There have been brief revivals – as in the 1960s with the Commonwealth beef road programme and mechanization (road trains, helicopters, and motor bikes) and contractualisation (for bore sinking, fencing, and mustering) replacing Aboriginal labour. More recently the industry has sustained itself by live exports, although it appears some companies are interested in developing a more sophisticated model, based on vertically integrated supply chains. Some observers claim that, with properly managed lands and adequate water supplies the northern cattle industry has the potential to triple on current levels,[2] though it is not clear what the environmental and social risks of such an expansion would be. In the last fifty years tourism has become a major industry across northern Australia, but it has been in relative stasis for some years.

Apart from along the maritime littoral of northern Queensland, agriculture has not driven economic development in the North. The east coast of Queensland has managed intensive development, initially around sugar, and features a density of large towns and cities that portend self-sustaining development. This region exhibits economic dynamics (coal mining aside), that make it distinct from the rest of northern Australia. It was created by agriculture, but this has been in stasis for the last 50 years,[3] development since then being driven by services, tourism, and latterly coal mining.

In the rest of the North large-scale agriculture has not been successful. In the 1950s and 1960s, large schemes, such as Humpty Doo rice, failed and the Ord River project has never repaid its capital. Even its current stage 2 expansion is not strictly economic[4] (Box 4.1). Agricultural development is often the *political* driver of northern development; rarely has it been an *economic* driver. It seems that

BOX 4.1 THE ORD SCHEME – A SPECTACULAR FAILED VISION

HOWARD PEDERSEN

NYAMBA BURU YAWURU LTD

BROOME

As noted in the White Paper on Developing Northern Australia, 'many failed projects and plans litter the north'. Few projects have failed as spectacularly as the Ord River irrigation scheme. The scheme has two dams, a 'diversion dam' close to Kununurra, the town built to administer the project, and Lake Argyle, Australia's largest water reservoir.

The two dams provide water to the Ord River Irrigation Area (ORIA). The first stage of ORIA resulted in 14,000 hectares (ha) of irrigated farmland being developed, a modest outcome given the public investment. For comparison, this represents 1% of the Murray Darling Basin's 1.4 million ha.

The White Paper includes a proposal to expand the Ord Irrigation area by 14,500 ha: expanding the size of the Ord irrigation area offers the potential to achieve scale and establish a variety of commercial crops in the district, which would enhance both the profitability and economic resilience of the region.

The agricultural potential of the Ord River valley has excited European Australians since it was first explored by Alexander Forrest in 1879. Numerous proposals were put forward before and after the Second World War, but it was not until 1959 that the Menzies Federal Government provided funding for the initial Ord diversion dam. Work began on the diversion dam in 1961, and it was officially opened in 1963. Farms were then allocated for this first phase of Ord irrigation between 1962 and 1965.

The main dam on the Ord was approved and federal grants provided two years later in 1967. This decision followed extensive public debate around the economic merit of the main dam, most notably by economist Bruce Davidson in his book *The Northern Myth*. Federal reluctance to further fund an agricultural region dangerously dependent on one crop was derided in Western Australia (WA) as 'chronic procrastination'. Notably, no review or assessment of the first part of the Ord development was conducted before a decision was made to build and fund the main dam and associated infrastructure. The main dam was commissioned and completed in 1972, and just two years later cotton crops were abandoned in the Ord altogether because of pest problems, insecticide costs, and transport and marketing costs.

Despite the lack of financial viability of the scheme, the WA Government borrowed millions more from the Commonwealth to develop more irrigation holdings in 1974. Regardless of the overwhelming commercial and environmental problems associated with the Ord, the WA Government, led by Sir Charles Court, decided to invest more public money to expand the scheme. Sugar replaced cotton as the commercial hope for the scheme but following investment in a sugar mill and planting sugar cane crops it too failed. Proposals to build 'Ord Stage II' grew through the late 1990s until they were eventually funded mainly by the federal government partly in response to the global financial crisis in 2008.

Ord Stage II construction of irrigation infrastructure took until December 2014 and cost the WA Government $334 million, a cost overrun of $114 million according to WA's auditor general. Only one developer has taken up a land allocation in the expanded agricultural area. Hassall & Associates produced a comprehensive cost–benefit analysis of ORIA from 1959 to 1991 and found that by 1991 total public and private costs were returning just 55 cents in the dollar. The public return was much worse with governments only receiving 17 cents in each dollar invested.

(Continued)

BOX 4.1 (Continued) THE ORD SCHEME – A SPECTACULAR FAILED VISION

The calculated cost of private and public cost of the Ord River Scheme does not take into account the social consequences of flooding the traditional owner's *country* and relocating two to three hundred Mirriwung/Gajerrong people in fringe camps and reserves in Kununurra in appalling living conditions. Nor does it take into account the enormous environmental costs and the significant public investment by the state to fight the Mirriwung/Gajerrong claim to native title in protracted legislation over many years.

Whilst the Ord River Scheme is the most spectacular failure of the northern development paradigm, there have been many other unsuccessful attempts at irrigated agriculture through northern Australia. Some of the most notable failed agricultural ventures include attempts to grow cashews and sugar cane in the Daly River and Katherine areas before the second world war, rice growing enterprises covering more than 300,000 ha in sub-coastal plains from the Adelaide River to Arnhem Land in the late 1960s, sorghum development near Humpty Doo in the 1960s, and the disastrous Camballin Irrigation Scheme that commenced in the early 1970s and was finally abandoned in 1983.

What all these schemes had in common was an optimistic vision that abundant water and fertile lands could promote concentrated and sustainable settler populations. An important and consistent political dimension in the promotion of these irrigation schemes has been a vision of transformational economic benefit for Indigenous people; a dream that the Ord River scheme shows is elusive.

this is currently the situation. There is a case for more intensive agriculture in the North but designed sustainably around small locally efficient production and not dependent upon federal investment in environmentally suspect dams for irrigated broad-acre cropping. Horticulture and specialist tropical crops appear to have the most environmentally and economically sustainable prospects.

4.2.2 Recent Economic Development

Since the 1960s, northern Australia's burgeoning resource extraction industries have been linked with the Asian industrialization that started with Japan. The 1960s and 1970s saw economic 'pulses' in the North based on these industries. Because they saw the construction of new towns (mostly in the Pilbara), this initial phase had more impact on long-term northern development than the most recent twenty-first century boom/pulse.[5,6] The 1960s was the first time 'northern' development had impacted on the greater Australian economy.[7]

The core features of Australia's resource extraction economy are the concentration of a growth industry's central management in a growth metropole (Perth) with much of the value-added in the mining industry being created there. Although WA only employs 36% of mining, including oil and gas, industry workers (Queensland (QLD) employs 27% and New South Wales (NSW) 21%), it garners more than 48% of the mining industry's value-added. Perth has become the centre of a wide range of resource extraction services companies engaged in support logistics, mining construction and design services, IT, finance, explosives, and so on. Some of these, such as Leighton Holdings, Orica, Worley Parsons, and Incitec, are now very large companies servicing resource extraction enterprises worldwide. This is arguably the most significant development in the Australian economy since the economic restructuring of the 1980s restored Australian competitiveness and a shift to service industries. It is a non-northern consequence of so-called northern development.

The problem for northern Australia is the poor multipliers into the region of this resource extraction industry.[8] Remote central and northern Australia are probably going to be weakly coupled to the engine of Australian growth over the foreseeable future. Hopes for a northern agricultural revolution on the scale of the recent resource extraction pulse are unrealistic, but the signals in the White paper and elsewhere are that a serious, infrastructure-driven attempt will be made.

This pattern of economic development has features that impact on northern Australian Governments. The resource extraction labour force is largely supplied by fly-in/fly-out (FIFO) workers commuting from large towns not in remote or northern Australia (Queensland has a variation of this drive-in/drive-out workers commuting from coastal cities to the coal mines of the Bowen and Gallilee basins). Such arrangements are a consequence of how mining companies have developed new 'greenfields' projects – that is no longer providing a town to house workers because of the front-end costs and the associated exchange rate and interest rate risks. The provision of housing, education, health, and other services for the workforce is left to the relevant state and territory (or local) governments. The extra costs of running a mine with FIFO labour comes off the Commonwealth's company tax receipts. As a result, FIFO makes eminent sense for mining companies; its only downside (for mining companies) is in exacerbating labour turnover.[9] Governments have to worry about services, maintaining viable towns, and relevant infrastructure and so bear most of the social and fiscal costs of resource extraction. The high cost of labour in the northern resource extraction industries means that mining companies will increasingly automate their operations. Over the next two or three or so decades, mining output growth will not be matched by employment growth, and FIFO will wane. The resource extraction industries will drive northern output growth but not northern development. The services industries will supply most future labour market and population growth and economic development. Commonwealth defence expenditure will be an important component of these developments.

Consequently, future research must turn to the services sector to define growth possibilities. That does not mean that services will remain separate from resource extraction or agriculture. These industries can provide services opportunities, as indicated by the development of logistics and engineering services that accompanied the most recent resource pulse (boom). To grow current agricultural productivity requires a range of agronomic and soil science services (as well as finance, workforce training, and other services). Environmental services work has grown from a small base to now include a wide array of activities including biosecurity, fire, and wildlife management, protected area management and control of pest plants and animals paid for by a mix of public and private purchasers.

The services sector is where future conventional labour and consumer markets will grow. The issue for northern Australia (apart from the Queensland coastal cities and Darwin) is attracting labour to maintain services and economic viability. Overseas migration can be one means.[10] Principally this remains an issue of amenity: ready access to goods and services and at reasonable prices (DRG 2013), which is difficult to supply to remote and small towns. Labour constraints will continue to be a problem for northern Australia. These issues are not solved by the White Paper proposals. That set of policy prescriptions naively assume that population will follow investment.

4.2.3 Current Policy Failures

Key issues in contemporary policy failure in economic and regional development, illustrated by particular reference to the Northern Territory (NT), include

- The concentration of investment, public and private in the largest centres. A particular case is, in the NT, the redirection of federally sourced funds from remote and regional to urban priorities (see tables in Gerritsen).[9,11]
- The asymmetry of connections between sectors in remote economies, with benefits often flowing from activity in the Aboriginal economy to non-Aboriginal people but much less often in the other direction. The tendency for Aboriginal landowners to be treated as largely passive responders to development initiatives from others rather than making their own development plans.
- The dismantling of systems for community engagement in effectively irreversible resource allocation decisions, constraining options. This has, for example, seen the transfer of program delivery to remote communities from local Aboriginal organisations to metropolitan-based NGOs.[12] We deal with this in more detail in the following.

- The treatment of employment in land management as inferior to other forms of work, despite critical needs for active management of northern landscapes and the increasing availability of private funds for such work.
- The private disinvestment of northern Australia. The people of northern Australia do not invest in northern development. For example, the Commonwealth's tax favouritism for negative gearing of housing transfers a large proportion of salary-earning north Australians' investment into housing in other parts of Australia. The superannuation system also transfers investment to southern Australia as retirees return to their places of origin.
- The inefficient labour markets that prevent endogenous growth. Non-Indigenous North Australians, particularly away from the Queensland coastal cities, arrive in the North as 'escalator' migrants (that is, they secure positions they would not in southern cities). They stay for some time and then leave, taking their skills with them. Its demographic patterns are one of the major problems faced by North Australia.[13]

In the following we identify the policy changes needed to mitigate these failures and to encourage both new economic processes (including payment for environmental services) and equitable participation of Aboriginal people in northern development.

Whilst the sorts of conflicts these narratives illustrate are by no means unique to North Australia, we argue that the dominant influences on government policy have been particularly damaging to Aboriginal society. And current policy prescriptions appear likely to continue to compromise Aboriginal roles and participation in northern economic and social development, despite official commitment to 'closing the gap'. No serious measures are proposed to ensure that residents of the North can capture real and enduring benefits from so-called 'northern' development.

There are core problems in northern Australia that merit serious research including

- *The environment*: How to manage land, seas, and natural resources to achieve sustainable economic growth and broadened economic opportunities.
- *The population*: How to retain immigrants and educate them to produce a forward-looking labour market that can respond to new opportunities, especially in services industries.
- *Remote participation*: How to produce an economy favouring the Aboriginal population.
- *Infrastructure*: How to create the infrastructure that makes North Australia both a worthwhile place to live (education, health, and cultural and recreational infrastructure) and economically efficient (physical infrastructure).

These are the issues that researchers and policymakers interested in the economic development of North Australia must consider.

4.3 WHAT IS THE WHITE PAPER/COMMONWEALTH'S PRESCRIPTION FOR THE REGIONAL ECONOMIC DEVELOPMENT OF NORTHERN AUSTRALIA?

In strictly economic terms, successful regional development means the ability to raise the level of regional exports relative to goods and services for local consumption. Essentially the White Paper[14] embodies these traditional Australian regional development strategies. The usual definition of regional development[15] posits a growth in regional output or productivity comprising an interacting combination of factors: population, employment, income, industrial value-added. Under different headings, the chapters in the White Paper follow these prescriptions. It ignores recent theorists, such as Richard Florida, who have added creativity (or at least people in 'creative' occupations) and social milieu – the quality of public health and welfare and environmental quality.

Descriptions of how regional development occurs have two dominant/orthodox approaches:

Development-from-above – development occurs in core/centre regions and trickles out to the periphery/hinterland, the peripheral localities depending upon relatively simple inputs/exports (labour, few products) to their cores.[16] The world is a hierarchy of markets. This is the keystone of the White Paper and ensuing policy development.

Development-from-below – a more proactive practice/theory accepts the core-periphery constraint but argues that local regions must be active in shaping their institutions and lifestyle and new forms of economy.[17] This is the approach we support.

The White Paper clearly favours development-from-above, albeit in a neo-liberal framework. The government invests in helping the private sector to create growth that will 'trickle down' to the rest of the target population. A quick look at what has happened under the White Paper process will illustrate this point. The initial Green Paper specified a target of five million people in northern Australia by 2050.[18] This was Kerrigan-esque ('they're dreaming') and was abandoned in the subsequent White Paper. The fact is that amenity is the major internal population driver in Australia[19] and liveability (i.e. cost-of-living) as an important inhibitor of regional settlement[20] was noted. But no investments in amenity or ways of reducing northern costs of living were included in the White Paper, so this population objective was at best fanciful.

Northern Development policy measures in the 2016/2017 budget were as follows

- $200 million from the National Water Infrastructure Development Fund. This will probably go to fund weirs in Queensland's Fitzroy River and feasibility studies for dams on the Mitchell River, water studies in the Darwin region, and feasibility studies for the Ord Stage 3.[21] The Nullinga Dam is already an electoral commitment.
- Northern Australia Infrastructure Facility (nominal funding $5 billion) to provide concessional financing to complement private investment.[22]
- $75 million for a Townsville-based Northern Development CRC to do industry-driven research.[22]
- $700 million for transport infrastructure ($600 million for roads), spent over four years. This includes rail feasibility studies,[23] the Cape York Project ($200 million), and expenditure on the Outback Way ($42 million).[24]

There are also the usual unfunded and general policy objectives:

- Austrade (Australian Trade and Investment Commission) initiatives to attract overseas investment in 'gateway' trade infrastructure (no funding)[25]
- Unspecified (and unfunded) White Paper Land Reforms

The great bulk of the expenditure actually allocated is within Queensland. A cynic might suggest that the fact that north Queensland has seven of the 9.5 federal electorates in northern Australia (versus 2 for the NT and one half (of Durack) for WA (see Maps 2.15, 2.16), played some role in this distributional outcome.

Nevertheless, the core tenet of the current northern development policy is that the government both invests in infrastructure and subsidises private investment to expand current output and create new markets. The lessons of the Ord have not been learned. It is also assumed that large-scale investment won't have deleterious effects on pre-existing business by distorting prices. However, the Heckscher-Ohlin price equilibrium,[26] core to these assumptions, assumes geographic cost-price equilibrium won't work in the northern context due to a lack of a skilled population, remoteness from markets, and the high cost of input factors. The new northern development will be very narrowly based, as was the growth in mining and hydrocarbons extraction in recent decades.

Irrespective of this argument, we feel the current approach to northern development will fail because it does not deal with three factors: Firstly, the core position and perspectives of the North's Aboriginal population (the true northern Australians, most of the rest are effectively expatriates) who own outright or have native title interests in much of the land (mostly in the NT). Secondly there are very real biophysical impediments to development of the officially preferred type. And thirdly, alternative approaches that embed developments in functioning natural and cultural landscapes instead of entirely replacing them are not considered. We now turn to these issues.

4.4 THE PROBLEM WITH 'TRICKLE DOWN' DEVELOPMENT – PARTICULARLY FOR INDIGENOUS PEOPLE

As we have noted earlier, the northern Australian economy is not simply a smaller-scale version of the broader Australian economy. First, industries which provide most income and employment in southern regions (e.g. trade [both wholesale and retail], manufacturing, property services, and finance), are not those which support communities in remote parts of northern Australia (sometimes retail, but mostly government, mining, and agriculture).[27] This has implications for the way in which the costs and benefits of development are distributed across space. Second, although Aboriginal and Torres Strait Islanders comprise only 4% of the entire population of Australia, they comprise almost 16% of northern residents,[28] and in some parts of northern Australia, more than 90% of the population is Indigenous.[29] Not only are the Indigenous people at significant socioeconomic disadvantage compared to non-Indigenous people,[30,31] there is an asymmetric divide between the Indigenous and non-Indigenous economies.[32] Both of these points have implications for the way in which development plays out across the North.

To address the issue of industry structure first, the smaller and more remote a regional economy is, the smaller will be the regional economic impact of development in that area.[33,34] This is because total impact depends not only on how much is spent initially (the first-round effects), but how any money earned in that first round of expenditure is subsequently re-spent (the second-round effects). Small remote economies have fewer and a narrower range of shops, manufacturing, and other businesses than urban centres, so there are few opportunities for people to 'buy local'. Thus, the second-round impacts of expenditure are typically small (sometimes even zero) in small remote areas since the supply chain of industries extends to other areas. When industries are established or expanded in northern Australia, the largest share of benefits associated with that development can accrue to areas in southern centres, as illustrated by studies of the mining industry in Queensland[35,36] and elsewhere.[37,38] If regions across the North accrue much smaller benefits than those in the South, then residents of the North might feel that development in their local area is not 'true' northern development, but rather use of northern resources for the benefit of businesses and residents in other regions. This is a particularly pertinent problem if development generates negative social or environmental externalities felt by residents of the North, without generating compensatory benefits, a situation to which Indigenous people are most vulnerable.[39]

There are, however, substantial differences in the impact of development (local and regional) of different industries. Capital-intensive industries (such as mining, and some forms of agriculture), have very small employment multipliers[40–42] compared to labour-intensive industries (such as those in the service sector). Simply put, these industries do not require much labour and so employ few people, so second-round benefits of expenditure are likely very small. Similarly, industries with a high percentage of long-distance commuters in their workforce will also generate relatively little local economic benefit (compared to those which use local residents). Money earned by the workers is spent elsewhere (e.g. in Perth, when companies use FIFO staff from there in northern mine-sites), thus the second-round benefits accrue elsewhere.

This problem of low-local benefit for capital intensive industries is exacerbated by the fact that the manufacturing sector (a required provider of equipment for capital intensive industries) is distinctly absent across most of northern Australia. Thus, while capital-intensive industries may generate substantive benefits (both direct and indirect) to regions with a strong manufacturing sector, regions without much manufacturing (like much of northern Australia[43]) do not see those benefits. Thus, even if there are no localized negative social and environmental externalities of development (or if externalities are managed), capital intensive development in Australia's North may still be viewed not as 'true' northern development, but rather as development that generates benefits for economies elsewhere: specifically, economies with a vibrant manufacturing sector in which the FIFO labour force resides and spends its money.

Lack of localized economic benefit is a problem for both Indigenous and non-Indigenous residents of northern Australia. Unless the structure of the northern economy is altered, Indigenous people will see even less benefit from northern development than their non-Indigenous counterparts. This is because there is not one but rather two (Indigenous and non-Indigenous) types of economies across northern Australia, and there is a disjuncture between the two (termed the asymmetric divide[32]).

The non-Indigenous economy operates much like the broader Australian economy. Put simply, businesses provide goods and services to households in exchange for money, and households provide labour to businesses, also in exchange for money. But very few Indigenous people work within non-Indigenous businesses,[44] and very few Indigenous people own businesses that supply goods and services (such as food or accommodations) to non-Indigenous people in northern Australia.[45] They may supply services to, for example, the Australian Government when employed as Rangers. As such, Indigenous people are constrained in their ability to exchange money with non-Indigenous people for labour or for other goods and services. What this means is that when the non-Indigenous sector is stimulated or 'developed', the money generated by the stimulus circulates within the non-Indigenous economy (with much flowing out of the North as taxes, investments, savings, or imports); very little money finds its way into the Indigenous economy. Perversely, when the Indigenous economy is stimulated, much of that money flows to the non-Indigenous economy when, for instance, food and accommodation are purchased from businesses that are owned by non-Indigenous people.

Traditional development options (which might, for example, expand the mining or agricultural sectors) are thus likely to generate more significant increases in employment and income for non-Indigenous people than for Indigenous people.[46,47] Rather than helping to close the gap[48] and thus contributing to the federal government's Indigenous Advancement Strategy,* this type of development may effectively widen the gap between Indigenous and non-Indigenous outcomes. The problem is not one of no benefit for Indigenous people, it is one of relativities. Indigenous people typically have lower incomes than non-Indigenous people; so if development generates relatively small increases in income for Indigenous people, the gap will inevitably widen. The only way to close the gap is to select development options that give a larger rise in income for Indigenous than non-Indigenous people, effectively allowing Indigenous people to 'catch up'.

Total economic impacts are the sum of first and second round impacts. It is tempting to conclude that one could allow Indigenous people to 'catch up' by promoting developments that generate all (or at least significant) first-round benefits for Indigenous people. Nevertheless, the current structure of economies across northern Australia suggests that even that may not be enough. To test the potential significance of this problem, Jarvis and Stoeckl[49] modified three publicly available input-output (I-O) tables[41,42,50] to estimate the regional economic impact of spending on Indigenous Land Management Programs (ILMPs) in the Kimberley, the Northern Territory, and Far North Queensland. Details of the total expenditure on ILMPs within these regions for the year 2014/2015 was provided by Hill et al. (2013),[51] who developed a database of expenditure on all ILMPs across Australia from 2001/2002 onwards. The three different IO models were adjusted to facilitate cross-regional comparisons and to disaggregate financial flows to Indigenous and non-Indigenous households. Jarvis and Stoeckl[49] then used the modified models to assess the potential impact on both Indigenous and non-Indigenous incomes of that ILMP expenditure. Their analysis showed clear differences in the scale of impact on each regional economy and in the financial benefits flowing to Indigenous and non-Indigenous households in each region (Table 4.1). It also highlighted the vitally important role of secondary spending. In Far North Queensland, for example, the secondary

* http://www.dpmc.gov.au/indigenous-affairs/indigenous-advancement-strategy.

Table 4.1 Estimated Financial Benefits from Expenditure on Indigenous Land Management Programs, by Region (2014/2015)

	Kimberley $	Northern Territory $	Far North Queensland $
Initial (direct) expenditure on ILMPs in 2014/2015	10,438,656	46,507,861	22,642,596
Estimated initial incomes paid to Indigenous households included within direct expenditure	3,966,689	17,672,987	8,604,187
Estimated initial incomes paid to non-Indigenous households included within direct expenditure	1,043,866	4,650,786	2,264,260
Total impact on the economy of each region	19,771,637	109,109,786	54,528,465
Indirect ('knock-on') impact	9,332,980	62,601,925	31,885,869
Direct and indirect increase in Indigenous incomes	4,418,764	18,750,869	9,116,141
Direct and indirect increase in non-Indigenous incomes	3,933,131	18,747,151	11,845,551
Indirect increase in Indigenous incomes (resulting from multiplier effects)	452,074	1,077,881	511,955
Indirect increase in non-Indigenous incomes (resulting from multiplier effects)	2,889,266	14,096,365	9,581,292

(indirect) benefits that accrue to non-Indigenous households are so large that their total benefits exceed those accruing to Indigenous people.

Evidently, the large-scale expansion of traditional industries may generate financial benefits for those directly involved in those industries or involved further down the supply chain. But relatively few existing northern residents own or operate such large industries. Many of these industries employ relatively few local residents and relatively few northern residents are involved in businesses that comprise part of the supply chain of these large-scale industries. Thus, unless or until there is a change in the structure of northern economies, few northern residents will see large benefits from such expansions (this is particularly true for Indigenous people – the traditional owners of this land). Instead, the largest share of benefits of such top-down type approaches to development will likely accrue to those outside the region, for the most part in urban centres. That may well be the consequence of the federal government's northern development policy. However, if instead the intent is for development that benefits northern residents, or contributes to the Indigenous Advancement Strategy, then more bottom-up approaches may be required. These will overwhelmingly depend on the management of non-extractive natural resources, to which we now turn.

4.5 HOW DO WE ORGANIZE ALTERNATIVE EMPHASES FOR NORTHERN DEVELOPMENT BASED ON SUSTAINABLE USE OF NATURAL RESOURCES IN THE LIGHT OF THE NORTH NEVER BEING HEAVILY POPULATED?

We have shown that the large-scale developments of the sort pursued through the white paper are either unlikely to be realised or, where they are taken up, will deliver benefits to few of those locals who must adjust to their presence. Here we seek alternatives that are favoured by or will less obviously compromise the biophysical and social assets of the North.

Whilst we approach this task with some optimism that positive change is possible, we also want to avoid utopianism. We base our brief treatment (which will be expanded on in other parts of this volume) on an acknowledgment of present conditions in the North that generate political and social pressures for orthodox development. From this starting point, we seek to show how less orthodox approaches to engage local people in design and delivery of important elements of such developments could also deliver much greater benefits to local, especially Indigenous, people.

We have already touched on many of the issues that demand more from government and industry than shared distaste for regulation – especially in areas of environmental and heritage protection – and preference for direct and indirect, immediate and delayed public subsidy of large-scale developments that struggle to demonstrate net long-term benefit.

To reiterate, the most important are discussed in the following paragraphs.

Aboriginal people own or have native title rights in most of the northern landscape which, outside a few major centres, is occupied by mostly Aboriginal residents, many of whom are of or about to enter working age. Aboriginal expectations (and emerging recognition in law) that their interests in land should enable commercial use of resources is not matched by landholder technical and other capacity to sort real commercial opportunity from the unfavourable or implausible. Paradoxically, given relentless rhetoric about economic inclusion, governments have supported planning to include Aboriginal lands in the conservation estate, but not for local choices in commercial development.

Northern landscapes are relatively poorly understood biophysically. Distributions of natural resources like soils and water required for agriculture are weakly described. There has been no equivalent of continuing publicly-funded 'pre-competitive' surveys of mineral and petroleum prospectively for other non-extractive activity, meaning that, for example, proposals for large-scale agricultural developments are speculative while smaller-scale opportunities may be missed. Risks are exacerbated by intense seasonality and great year-to-year variation in the strength and timing of alternating floods and seasonal droughts. Landscapes often appear structurally intact but have been degraded by long-term effects of grazing for production. Over large areas fires, invasive plants, and feral animals are effectively unmanaged. Chronic pollution through acid drainage from many (in the NT arguably most) closed and operating mines with inadequate bonds for rehabilitation (Box 4.2) will continue indefinitely to damage waters and their capacity to support other use.

No policy responses have been framed to this combination of local Aboriginal interest in development but limited capacity to explore options, generally weak technical knowledge and pervasive and growing impacts on landscape integrity. Architects of the White Paper declined to grapple with such issues. Regional land use planning is absent or desultory, contributing to uneconomic or unsustainable development trajectories. Local people are effectively excluded from decision-making about developments affecting their lands and lives. Rights in resources needed to deploy land rights for commercial activity are denied.

The stumbling of successive NT administrations into over-allocation of groundwater, despite starting at self-government in 1978 with unstressed water systems, highlights the risks if laissez faire approaches are continued. Indeed, some recent (2013–2016) policy decisions appear actively hostile to improved decision making on development options delivering local benefits. Environmental offset schemes favouring Aboriginal providers have been dismantled. Government procurement policies favouring Aboriginal providers are not interpreted to cover land management, conservation, and recreational services, which remain in the hands of mostly non-Indigenous staff and managers of public parks and reserves. In contrast to deregulatory 'hands-off' approaches elsewhere, governments have sought to intervene in the use of Aboriginal freehold lands.

To summarise, Aboriginal landowners and their communities want economic development but on their own terms. They are eager to participate in active management of land to address multiple objectives where these are compatible with cultural obligations. Widespread environmental difficulties require management of ongoing pressures and repair of damaged landscapes, offering great scope for environmentally positive work. However, the capacity to manage existing and emerging problems has been weakened by deliberate policy choices. Predicting and managing negative effects of new development will be particularly challenging because land use planning is absent or weak and the knowledge and interests of local people are disregarded – governments and industry seek to override landholder and community views to push favoured developments into biophysically or culturally unsuitable places. All of this is exacerbated by weak technical and other support available to locals to understand options and make their own decisions about commercial use of their lands.

BOX 4.2 emPOWERING AUSTRALIA'S NORTH

Low Cost, Geographically Available Energy: The Key Factor for Sustainable Economic, Ecological, and Social Prosperity

IAN CHAMBERS

CRAWFORD SCHOOL OF PUBLIC POLICY
AUSTRALIAN NATIONAL UNIVERSITY, CANBERRA

JULIAN CRIBB

AUSTRALIAN NATIONAL CENTRE FOR THE PUBLIC AWARENESS OF SCIENCE
AUSTRALIAN NATIONAL UNIVERSITY, CANBERRA

JEREMY RUSSELL-SMITH

DARWIN CENTRE FOR BUSHFIRE RESEARCH
CHARLES DARWIN UNIVERSITY, CANBERRA

Current situation: A review of the literature and proposals for the development of Australia's North, including the federal government's 2015 'White Paper', demonstrates a surprising lack of focus on available and future energy requirements of the region. This is surprising for a number of reasons.

Firstly, North Australia's abundant mineral resources (e.g. bauxite) often have to be transported vast distances southwards and internationally to where there are sufficient supplies of fossil fuels to transform those resources into 'value-added' products. This not only deprives North Australia of significant opportunities for industry and 'value-added' product and service development, it adds significant costs to Australian goods and services when competing in international markets.

Secondly, dependence on such fossil fuel resources exposes regional enterprises to significant energy market volatility.

Thirdly, while energy distribution networks are currently very underdeveloped across the North, the region possesses some of the richest reserves of clean and renewable energy in the world. With rapidly expanding global requirements for clean and renewable energy, and North Australia's geographical proximity to some of the world's fastest growing markets in South East Asia, development of renewable energy resources offers a significant opportunity for the region's long-term economic development and social sustainability. It is notable that there is no mention of the renewable energy option in the Northern Development White Paper. It is important therefore to further explore this significant opportunity.

The opportunity: The renewable energy opportunity is perhaps most graphically captured in Figure B4.2.1, developed by researchers at the Australian National University. This graphic demonstrates how solar energy capture from just a fraction of the North could potentially supply enough solar energy to meet not only Australia's needs but most of the world's energy requirements also.

Professor Blakers and colleagues describe how a solar energy generation facility could be developed to raise water (e.g. at Lake Argyle in the Kimberley) into elevated storage tanks, to then drive hydro-energy generation. Such hydro-energy could then be distributed through high voltage direct current (HDVC) long-distance energy transmission cable technology as part of a national energy 'superhighway'. An alternative approach to the high cost of construction of hydro storage is to connect the solar and wind energy rich regions in the north with an energy hub in the South East via the high geothermal regions of central Australia

(*Continued*)

BOX 4.2 (Continued) emPOWERING AUSTRALIA'S NORTH

Figure B4.2.1 The vast renewable energy resources of Australia's north and national and international connectivity opportunities. (From Blakers, AW. et al., *GREEN – The International Journal of Sustainable Energy Conversion and Storage* **2**, 189–202, 2012.)

(Figure B4.2.1). Such transmission and connectivity capability are already a feasible option and well established in other regions such as Europe.[2]

Development of a national energy superhighway would allow, firstly, for diversifying Australia's energy security options by interconnectivity of current conventional fossil fuel energy sources generated in the South, with the vast renewable energy resources of the North. Secondly, further connection of the national superhighway with the Association of Southeast Asian Nations (ASEAN) super-grid network (Figure B4.2.2) would allow Australia to become a major regional energy supplier based on abundant renewable energy generated on a tectonically and politically stable platform.

There are a number of benefits from this approach:

- It would allow Australia to become a major player in the clean energy market for the region with the associated economic opportunities this would bring.
- It would provide a strategy for transitioning from Australia's high dependency and current exposure on fossil fuels (especially coal) to alternative and abundant revenue sources from renewable and clean energy.
- It would contribute significantly to Australia's international climate change commitments as well as assisting other countries in the region meet theirs. The immediately accessible Indonesian energy market, forecast to grow by more than 500% by 2050,[3] has committed to a 23% renewable energy target over the same period.[4]

Other regional renewable energy options also exist. For example, the vast areas of salt pans in North Australia could be used to produce alternative algal transport fuel products.[5,6] Key limitations for extensive production of these types of fuels include space requirements and low-cost, readily accessible energy sources. The development of a national energy superhighway

(Continued)

BOX 4.2 (Continued) emPOWERING AUSTRALIA'S NORTH

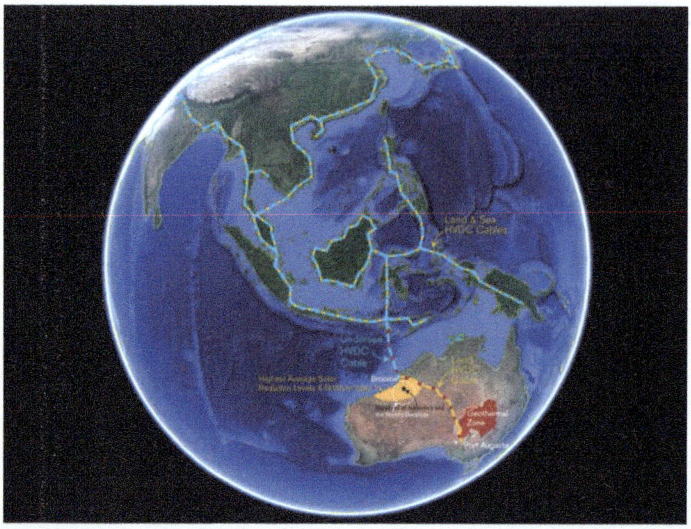

Figure B4.2.2 Connecting Australia to the existing and planned ASEAN super-grid network. (From Blakers, AW. et al., *GREEN – The International Journal of Sustainable Energy Conversion and Storage* **2**, 189–202, 2012.)

overcomes these energy limitations, allowing Australia to potentially meet its own requirements for transport fuels as an alternative to oil. This is a market currently worth AUD 44 billion in Australia[7] and also provides opportunities for significant export earnings.[8]

The rapid development and associated declining costs of renewable energy and transmission technologies, coupled with the international growth of clean energy markets, means that investment costs associated with development of a national energy superhighway and connection to the ASEAN super-grid are now eminently feasible. Estimates are AUD 7.2 billion for connectivity to the ASEAN super-grid[9] and AUD 6 billion for the national energy superhighway based on recent Australian feasibility studies and European contracts.[2] In total, this is approximately a quarter of the cost of the National Broadband Network (AUD 45 billion) or Australia's new submarine fleet (AUD 50 billion), and less than three years of current Australian Government annual subsidy of AUD 5.6 billion to the fossil fuel industry.[10]

emPowering remote communities: Harnessing the vast and well-distributed clean and renewable energy resources of North Australia has significant implications to empower remote communities. Decreasing dependence on expensive and transported fossil fuels for remote community energy needs can allow not only economic but also social and ecological sustainability.[11] The rapid growth of solar combined with battery storage and electric vehicles, can enable not only self-sufficiency of remote communities but also economic and community development. Costs of installation are immediately offset by significant reductions in costs of imported fuels such as diesel, and decreased costs for other infrastructure and community support resources such as water transport and storage.[12] Energy production can also be easily scaled up to support developing economic and community developments – and reduce exposure to external prices of fossil fuel products.

(*Continued*)

BOX 4.2 (Continued) emPOWERING AUSTRALIA'S NORTH

Conclusions: emPowering the North has historically been seen as a challenge for both regional and remote communities. The rapid growth of renewable and clean energy technologies and rapidly declining price points reversing this equation, providing significant economic and sustainable development opportunities for North Australian communities – opportunities that can be harnessed now.

REFERENCES

1. Blakers AW, Luther J, Nadolny N (2012) Asia Pacific Super-Grid – Solar electricity generation, storage, and distribution. *GREEN – The International Journal of Sustainable Energy Conversion and Storage* **2**, 189–202.
2. WNA (World Nuclear Association) (2018) Home/Information Library/Current and Future Generation/Electricity Transmission Systems.
3. BZE (Below Zero Emissions) (2015) Australia as an energy superpower. Retrieved from http://bze.org.au/renewable-energy-superpower. 288 Brunswick St, Fitzroy, Victoria (accessed October 2015).
4. Renew Economy (2017) Pilbara Solar expands eyes. Retrieved from http://reneweconomy.com.au/pilbara-solar-eyes-naif-funding-plan-export-wa-solar-asia-77040/. ASEAN super-grid (TBC).
5. Murphy S (2013) Algae farmers spuik potential for WA biofuel boom. *ABC News*. 23 June 2013.
6. ARENA (2016) First step towards an Australian green-fuel biorefinery *Media Release*, 23 June 2016. Retrieved from arena.gov.au.
7. DOEAE (Department of Environment and Energy) (2016). Australia Petroleum Statistics, October 2016.
8. Garvey P (2017) Coal demand suffers record fall: Global energy revolution well and truly underway. *The Australian Business Review*, 12 December 2017.
9. Vorrath S (2017) Pilbara solar eyes. NAIF funding for plan to export WA solar to Asia'. Retrieved from Renew Economy.com.au/Pilbara-solar-eyes-naif-funding-plan-export-wa-soal-asia-77040.
10. Gordeljevic V (2017) Health and Environment Alliance, Hidden Price Tags.
11. Indigenous Business Association (2017) Renewable energy transition to benefit Indigenous Australians. *Media Release*, 8 June 2017.
12. Frydenburg J, Scullion N (2017) Delivering solar to remote Indigenous communities. *Joint Media Release*, 11 August 2017.

In such convoluted biophysical, social, and policy terrain, mapping a development path that is accessible to the region's Aboriginal people requires a leap of faith: that there will at some point be a real interest in development that respects northern landscapes and their local peoples. But to ground such an attempt it is necessary to acknowledge that, in the short to medium term, infrastructure and public subsidy 'push' will override other considerations. We assume, optimistically, that ideological influences will weaken as evidence of economic, social, and environmental failures mount and alternatives are tested.

4.5.1 Orthodox Development Trajectories

To form a view of environmental services needed and deliverable in northern landscapes, it is useful to make some projections of at least plausible orthodox development directions. A starting point is to consider commitments already made that appear unlikely to be reversed, irrespective of economic, social, or environmental credentials. We do not review here in detail other technically and economically plausible options (e.g. Box 4.2) that were not considered for the white paper or in any other government program of which we are aware.

4.5.1.1 Commitments to Agriculture

Developments for irrigated agriculture are strongly favoured by present federal, state, and territory governments. The Northern Territory, with federal government funding support, is engaged in searches for conjunctions of favourable soils and groundwater availability. In addition, the NT has relentlessly pursued bringing Ord River water to the Spirit Hills (western Keep River) region,[52] despite studies undertaken by CSIRO and others indicating that it is a relatively unfavourable site.[53,54]

Whilst the full extent and detail of agricultural developments that may be driven by new dams across the North remains unclear, it appears safe to assume that substantial areas will be developed. In some locations, broad-scale production of relatively low unit value commodities is likely. In others, including parts of the NT, 'mosaics' of relatively small areas of higher value horticultural products may be favoured, because good soils are patchily distributed. In areas where soils are less favourable, irrigation to improve pasture for more intensive cattle grazing is also likely.

A huge aquaculture development proposed for the eastern side of the Keep River will use large volumes of fresh and seawater.[55] For Stage 1, freshwater will be diverted from wetlands on an NT Site of Conservation Significance (the Legune Wetlands). If it progresses to the ultimate scale proposed (~10,000 ha of production ponds plus large areas of water conditioning and processing ponds), it will require additional local dams or water from the Ord River or both, which may in turn drive development of the biophysically unfavourable Spirit Hill lands, which local Aboriginal people regard as incompatible with the cultural significance of the area.

All such developments will require clearing of native vegetation, causing large pulses of greenhouse gas emissions as remnants are burned or decay. At regional scales biodiversity losses may be substantial where broad-scale clearing is permitted. Sedimentation and pollution of associated waterways will increase.[56–58] Northern rivers may be particularly susceptible to pollution effects because most stop flowing relatively early in the dry season and the concentration of nutrients and other pollutant rise as remnant pools shrink with evaporation and seepage. Embedding such developments in unmodified landscapes with populations of large seed-, fruit- (and prawn-) eating birds as well as fruit bats, will demand control of native pest species (as well as the usual feral animals).

4.5.1.2 Commitment to Extractive Industries

Large areas of the NT, and northern Queensland and Western Australia, including much Aboriginal land are regarded as prospective for shale oil and gas,[59–61] which requires large numbers of regularly spaced wells with horizontal 'fracked' drill holes within favourable strata. Development push comes from existing and approved future investments in infrastructure to handle and export products and, in the Northern Territory, a proposed gas pipeline from near Tennant Creek to Mt Isa. An existing pipeline from the Gulf of Carpentaria to Darwin might service the prospective McArthur River Basin, which runs deep into south-eastern Arnhem Land.

Governments continue to celebrate public subsidies of mining, particularly in 'pre-competitive' data gathering and analyses. Rates of establishment of new mines are likely to have been slowed by recent falls in commodity prices, but a number of new sites remain under active development across northern Australia. Some sites abandoned following the drop-in iron ore prices from their 2011 peak may be reactivated if prices continue to rise, as they have through much of 2016 and early 2017.

4.5.2 Scale of Probable Change

In combination, these politically 'inevitable' developments alone will fundamentally change large areas, generate many millions of tonnes of greenhouse gases, and require active management of waterways and other wetlands and protection of cultural (including sacred) sites. The area and resources likely to be affected in the medium term is difficult to estimate with confidence because it will depend on the determination of governments and industry to "coerce" development into marginal settings within convenient reach of relevant infrastructure. However, the relentless push by successive governments of both major parties to use the Spirit Hills sites suggests that pressure will continue, despite a long history of failure of such interventions.[62,63]

For agricultural land, estimates of areas affected might reasonably be bounded by two sets of CSIRO-supported analyses of suitable conjunctions of water and soils. In 2009 growth in groundwater irrigable land in north Australia was estimated at 20,000–40,000 ha. New surface water dams were treated as undesirable given few topographically suitable sites (for deep storages) and high rates of evaporation.[64] A few years later estimates were much greater: groundwater could support 100,000–150,000 ha of irrigation and dams over 1.0 million ha.[54] A 30-fold increase cannot be attributed to improved information but to the susceptibility of such estimates to underlying assumptions about acceptability or otherwise of environmental and social impacts. As noted earlier, the current federal government has committed funds to many of the dam options identified. Incremental small-scale horticultural development, of the sort seen in the Darwin hinterland (presently producing more than the entire Ord River scheme), is likely to continue around major centres where access to relevant infrastructure and local and export markets is easier. A conservative estimate might put the 'probable' commercial land usage at several hundred thousand ha for northern Australia, and up to 100,000 ha in the Northern Territory.

Areas likely to be affected by unconventional oil and gas extraction are also difficult to estimate. But again, using infrastructure push as a starting point, the capacity of a proposed east-west pipeline[65] that runs near some of the area's most prospective for shale gas may help. Subject to a number of assumptions about production and sourcing of gas, mostly following Broderick et al. (2011),[66] proposed design capacity of 90TJ/day could probably be met by a few hundred wells operating from significantly fewer drill pads. Even if wells are managed impeccably to prevent damage to groundwater, this sort of activity increases (fugitive) emissions of potent greenhouse gases, pollutes the large volumes of fracking water that must then be stored, and needs infrastructure (pipes and service roads) to transport condensate and gases. Wells would be linked across production landscapes by several hundred km of tracks and pipes. The area potentially affected in the relevant regions is large but much less than proposed agricultural expansions.* But managing such large-scale intrusions into previously undisturbed landscapes will require interventions for control of fire and to prevent incursions of invasive plants and animals and to allow people to continue safe use of the intervening *country* .

Areas from which minerals are taken are relatively small compared with agricultural developments, even for bulk commodities like iron ore. But as already noted, the NT and Queensland[67] share a dismal history of poorly-managed northern mines, with acid drainage and related metal pollution affecting connected waterways.[68,69] A preferred regulatory response is to keep these poorly designed, previously inadequately regulated, and weakly supervised mines operating, with their new operators tasked to keep a lid on problems. To stop digging when already in a deep hole and to correct the existing problems doesn't yet appear to be treated as a serious option.

But at some point even institutionalised co-dependency must collapse, with the public ultimately bearing even greater costs. Despite the areas creating the problems being relatively small compared with other development types, the public costs of effective intervention to protect the wider landscape – especially waterways and wetlands that support such iconic northern activities as recreational and guided sports fishing for barramundi – will run into many billions across northern Australia (Box 4.3).

Our purpose here is not to map in full the likely trajectory of orthodox development in the North but to show that in addition to the entrenched and ongoing environmental damage already

* A comprehensive government inquiry into shale gas development in the Northern Territory recently concluded that environmental and social risks could be satisfactorily managed and also presented a number of development scenarios much larger than considered here. (The reference for the inquiry report is Pepper R (2018) Final Report. March 2018. Scientific Inquiry into Hydraulic Fracturing in the Northern Territory, Darwin. 505 pp. https://frackinginquiry.nt.gov.au/inquiry-reports/final-report).

BOX 4.3 MINE SITE REHABILITATION AS AN OPPORTUNITY FOR ABORIGINAL ECONOMIC DEVELOPMENT IN NORTHERN AUSTRALIA

MARK R. GARDENER
RESEARCH INSTITUTE FOR THE ENVIRONMENT AND LIVELIHOODS
CHARLES DARWIN UNIVERSITY, DARWIN

PETER CHRISTOPHERSEN
KAKADU NATIVE PLANTS SUPPLIES
JABIRU

Mine-site rehabilitation potentially offers Aboriginal Australians long-term, local, land-management-based employment opportunities. Before this can be realised, funding models must become longer term and be earmarked for Aboriginal people; major closure risks must be adequately addressed; and government policy must strongly support training and business development.

One of the biggest barriers to success in both mine rehabilitation and Aboriginal economic development is absence of long-term funding. Each of northern Australia's jurisdictions have a different funding model for mine rehabilitation. West Australia's Mining Rehabilitation Fund (MRF) is the most unique; instead of a bond, all operators contribute annually based on severity and size of disturbance and the cost of rehabilitation (e.g. a tailings dam has a much higher cost than a quarry). Operators draw on the fund when rehabilitation commences. The interest earned on contributions is used at the government's discretion for legacy mines they manage. Queensland holds a bond based on the predicted cost of rehabilitation and funds legacy mines as part of the government's budget. The Northern Territory has a hybrid model with both a bond and a 1% annual levy on mining security bonds for work on legacy mines.

Whilst discretionary funds for legacy mines available from the Western Australian and Northern Territory models are currently modest, they are ongoing and continue to grow. As of 30 June 2017, interest from the MRF available for the Abandoned Mines Program was reported to be approximately $3.6 million.[1] Whilst it is not clearly reported in the Northern Territory, based on published mining securities,[2] it is conservatively estimated that more than $18 million has accumulated from the annual levy for work on legacy mines. Though there are a number of demands on these funds, some is put aside to do targeted work to reduce environmental impacts and improve public safety. Procurement processes are used to favour the engagement of Aboriginal and local businesses to do these works, but contractors must meet requirements and be able to offer value for money. Furthermore, if governments (in Queensland and the Northern Territory) weren't so reluctant to claim bonds for the state, considerable extra funds could be raised. Of course, it would be ideal that operators of existing mining tenements preferentially engaged Aboriginal businesses in rehabilitation activities, but this requires a policy of affirmative action. Both the bond and levy systems are distressingly inadequate to fund the required works and costs can be avoided by using financial securities instead of a cash bond, selling to less viable companies, putting the site into care and maintenance or by going broke.[3]

For land management activities to go ahead, major closure risks must first be resolved: the site must be stable, without contamination, and have a substrate suitable to support flora and fauna. This is often technically complex, requiring highly specialised expertise and has high associated capital costs. Until this capacity is developed by Aboriginal businesses, these

(Continued)

BOX 4.3 (Continued) MINE SITE REHABILITATION AS AN OPPORTUNITY FOR ABORIGINAL ECONOMIC DEVELOPMENT IN NORTHERN AUSTRALIA

works need to be complete before handover. Aboriginal businesses currently have capacity to lead land management activities such as seed collection, seed storage, plant propagation, planting, direct seeding, minor earthworks, weed control, fire management, and fencing.

It makes social and economic sense to decentralise mine site rehabilitation. Many mine sites in northern Australia are very remote, and the only permanent residents are in nearby Aboriginal communities with few economic opportunities. Many in the communities, particularly Traditional Owners, have long-term stakes in rehabilitation success. There are several challenges in developing a successful business model; most importantly, government policy must strongly support small business development, governance, technical training, and networking. Furthermore, businesses need continuity of personnel and must explore working arrangements, such as part-time employment, that recognise local social and cultural conditions and mine site requirements such as drug and alcohol policies.

A good example of an Aboriginal-run mine rehabilitation business is Kakadu Native Plants Supplies (KNPS). A family business established in 2005, it supplies services to Energy Resources of Australia (ERA). ERA operates the Ranger uranium mine embedded in Kakadu National Park (KNP) (Figure B4.3.1). KNPS runs a specialist plant nursery providing local

Figure B4.3.1 Ranger mine site in 2012 showing tailings dam on upper RHS and waste rock dumps in upper LHS. A trial rehabilitation site is in the front LHS. The mine operator, Energy Resources of Australia, is in the process of decommissioning the mine and rehabilitating the site.

(Continued)

BOX 4.3 (Continued) MINE SITE REHABILITATION AS AN OPPORTUNITY FOR ABORIGINAL ECONOMIC DEVELOPMENT IN NORTHERN AUSTRALIA

provenance species and rehabilitation services such as fire and weed management. They also service a number of external clients in the region. ERA and its parent company Rio Tinto, are taking seriously their commitment to return the mine to a state that is similar to KNP. They are in the process of burying tailings, backfilling pits and constructing a final landform, which cost hundreds of millions of dollars. The final landform will be revegetated progressively over about ten years, but it is expected to take decades of management before the site can be successfully relinquished.

KNPS is built on the family's traditional ecological knowledge and 25 years of experience in the natural resource management sector. Their business and technical skills have mostly developed by trial and error supplemented with some formal training. Currently, KNPS offer Aboriginal youth opportunities by training up the next generation in preparation for the following 10 years, when the workforce will expand from about three to 50 to produce and plant about one million seedlings. The major risks identified by KNPS are expanding the workforce and a continuity of funds. KNPS has identified a number of Aboriginal ranger groups within the adjacent west Arnhem region to provide extra capacity. Their main tasks would involve seed collection, nursery management and tree planting, with KNPS working closely to provide ongoing training and mentoring. While work for ERA is likely to last until 2026, it will be seasonal in nature. Securing other work in the region will allow for a model that provides real and ongoing jobs once staff are fully trained.

REFERENCES

1. Government of Western Australia (2017) Mining rehabilitation fund yearly report 2017. Accessed 19 December 2017. http://www.dmp.wa.gov.au/Documents/Environment/ENV-MRF_YearlyReport2017.pdf.
2. Northern Territory Government (2017) Mining securities. Accessed 19 December 2017. Retrieved from https://dpir.nt.gov.au/mining-and-energy/mines-and-energy-publications-information-and-statistics/authorised-mining-sites/mining-securities.
3. Lamb D, Erskine PD, Fletcher A (2015) Widening gap between expectations and practice in Australian minesite rehabilitation. *Ecological Management & Restoration* 16, 186–195.

documented, much more change is on its way. The total quantum of existing, emerging and likely future need for better environmental management is more than large enough to accommodate a major increase in active protection and rehabilitation of biophysical and cultural values, which Aboriginal people are best placed to provide.

4.5.3 Opportunities for Environmental Services

Skilled environmental interventions will be essential if the North is to maintain environmental quality and the sociocultural systems that depend directly on healthy landscapes.[70] The crude, ideologically drawn trajectory mapped in the White Paper is likely to generate suboptimal or destructive outcomes, but it also creates some opportunity.

The opportunity is to design for and manage change in ways that go beyond minimising environmental and social costs at new developments, by deploying investments in ways that halt or reverse existing chronic degrading processes and repair already severely damaged sites. Given the scale of proposed public investments already committed and the accompanying private investments, directing a small proportion of funds and skills to environmental services can do much to improve North

Australia's biophysical and social environments. However, redirection of funds will not occur in the absence of supporting policy and law.

4.5.3.1 Policy Change – Demand for Environmental Services

Corrections of present policy incoherence will require a number of changes, including to

- Reinvigorate and refocus environmental offsets policy and law. The Commonwealth, Western Australia, and Queensland all have specific offset policies backed by environmental assessment and other law. The NT Government dismantled environmental offsets policy in 2012.

  Contrary to nationally and internationally recognised best practice accepted by industry,[71,72] the NTEPA formalised abandonment of compensation for residual environmental damage with at least equivalent environmental benefits, and now confounds this distinct obligation with social impacts.[73] The resultant local policy vacuum focuses attention on quirky matters of national environmental significance, that are often poorly known and patchily distributed in northern landscapes, instead of targeting effects that erode the very fabric of the landscape.[74]

  Offsets law and policy should be refocused to support restoration of wider ecological health by better management of chronic issues like fire and poorly managed grazing that severely degrade biological communities, including threatened species.[75,76] Emphasis should also go to fixing mining-related pollution of waterways that not only damage wildlife populations but render fish and other wildlife using those waters unavailable as customary foods.
- Abandon the publicly funded Emissions Reduction Fund (ERF) and replace it with cap-and-trade arrangements designed to actually reduce emissions and allow industry to choose the credits they access to meet obligations.

  The ERF has damaged the reputation of Australia's offsets regimes, particularly through weak safeguards that are not designed to reduce emissions.[77,78] Methods using public funds to reward industry for using techniques that have been common practice for decades, like flaring of 'fugitive' emissions from gas wells to convert methane to less potent CO_2,[79] amount to little more than an unnecessary and unacknowledged subsidy.

  In the long run, the ERF is likely to collapse under the weight of its internal contradictions, including demands on public funds to achieve a meaningful reduction in emissions. In the meantime, land-based offset providers should also consider voluntary markets, including direct relationships with industry. The ConocoPhillips arrangement with the WALFA project provides a working example of what is achievable.[80]
- Improve procurement policies by emphasising services in which the skills of Indigenous people and landowners in particular are well matched to need.

  Present arrangements offering preference to Indigenous providers are effectively restricted to activities in which remotely based people are rarely positioned to compete due to inexperience, limited access to capital for equipment and diseconomies of scale.

  An active program to transfer all publicly funded remote and most regional land management work (defence lands, parks and reserves, weed and feral animal management) to Indigenous providers within five years would provide an immediate boost to Indigenous enterprise and employment, open pathways to service-industry NRM needs (agricultural pest management, pollution management), and ultimately other forms of enterprise and employment.
- Program the transfer of all present public-sector activities for on-site land management to Indigenous people with interests in relevant lands.

  Jobs in natural and cultural heritage management are too often slighted in bureaucratic and political circles as somehow less than real work, despite the difficult conditions in which the work is often done. Yet state, territory, and federal governments continue to offer large numbers of non-Indigenous people permanent employment in such work on parks and reserves, many on Indigenous-owned lands. The latest Kakadu Plan of Management, which runs until nearly a half-century after the first lease was signed, still talks earnestly about making better progress with transfer of jobs to the park's Aboriginal owners and residents. Continued failure to make good on employment goals in lease commitments is a striking example of the hypocrisy inherent in providing modest support for Indigenous people to declare and operate their own protected areas, counting towards national reserve system targets and international commitments, while continuing to employ mostly non-Indigenous people in government-controlled parks on Indigenous land.[81]

It is absurd to urge Aboriginal people to leave ancestral lands to take up jobs for which they are presently ill-equipped, while importing and subsidising the living conditions of others for employment on their lands, with the imports doing work that Aboriginal people have shown that they want and are well-qualified to do.

- Implement a major mine rehabilitation program directing work to Indigenous land managers, particularly those with recognised traditional interests in affected sites

  Estimates of costs of repair of north Australian 'legacy' mines is in the billions, and we appear to be in the process of creating more such obligations.[82] Full details of the individual mines and standards on which these estimates are based are unavailable* but – given failure to provide adequate rehabilitation bonds – it seems safe to assume that they are conservative (see Box 4.3). Repair of these sites offers both much work and a considerable diversity of work, involving skills that are either already held by Aboriginal landowners or can be developed.

  A partial list includes landscape design; wetland construction and management; fire management; wildlife management; weed and feral animal control; plant nursery establishment and operation; harvesting of seed and other propagules; monitoring water quality; monitoring and assessing recovery of biodiversity; erosion prevention and management; earthmoving equipment operation and maintenance; geographic information systems, database design and management, data analysis; and website design and maintenance for presenting data and analysis. Many of these opportunities will be transferable to other sectors, facilitating redeployment of skills as projects shift focus.

- Recognise Aboriginal history of and contemporary rights to trade in resources[83] on traditional lands and restructure resource allocation† regimes accordingly

  Present arrangements treat Aboriginal interests as exclusively customary and exclude them from resource allocations based on landownership or non-exclusive native title interests. States and territories should ensure that Aboriginal interests have decision-making roles in resource allocations.

- To provide a program of support for Aboriginal landowners and their communities to make high-quality, technically informed plans for future use of their lands, including compatible commercial uses

  The intent, set out by the Aboriginal research group NAILSMA[84] and summarised in Chapter 1, is to position Aboriginal landowners to chart approaches to management of their lands to optimise benefits including intangibles like cultural values and meeting traditional obligations to land and the plants and animals dependent on them.

Important features common to members of this incomplete list are that they emphasise known immediate needs; are compatible with each other in that they can be done in tandem on the same or adjoining sites; can be done alongside orthodox commercial opportunity in smaller-scale agriculture (Box 4.4), tourism and local fisheries that landholders have already approved or will be offered in future; and are consistent with case law recognising wider native title rights.

They also help address a particular concern for Indigenous people regarding food security. Agricultural development in northern Australia is unlikely to contribute meaningfully to food security at national and international scales. The most plausible development targets in such high cost settings are high-value irrigated crops for export to meet 'niche' demands from a growing middle class in Asia. As demonstrated elsewhere in this volume, systems for animal production on lands marginal for this purpose are

* The Northern Territory government has recently provided access to information on quanta of mining securities held. (https://dpir.nt.gov.au/mining-and-energy/mines-and-energy-publications-information-and-statistics/authorised-mining-sites/mining-securities).

† Water allocation policy in the Northern Territory has recently been changed to restore Indigenous landholders' access to a portion of the consumptive pool for commercial use.

BOX 4.4 SACRED FOOD ECONOMIC OPPORTUNITY: THE CASE OF *TERMINALIA FERDINANDIANA* AND THE CREATION OF AN ECONOMIC BASE FOR NORTHERN AUSTRALIAN HOMELANDS, REMOTE AND REGIONAL AREAS

PAUL LANE

KIMBERLEY INSTITUTE, BROOME

Over the past three decades the world has discovered some of the ancient, sacred foods of Aboriginal Australia. The scientific and commercial literature on the nutritional value and the medicinal and industrial potential of these foods grows every year. Many of these bush foods, previously regarded as worthless, are now being hailed as 'super foods'. Aboriginal cultural entities celebrate these foods as something more than food – as something that might be called a sacred life essence. Science and Aboriginal traditional knowledge of Australian native foods is coming together to create a unique social and economic opportunity for North Australia.

Leading the interest of the commercial and scientific worlds is *Terminalia ferdinandiana*, a pale olive fruit native to North Australia that grows naturally in tropical savanna woodlands. The fruit, well known to Aboriginal people, has several names: Billy Goat plum, Kakadu plum, salty plum, bush plum (in English); *gubinge* (Bardi people in Kimberley); *mi marrari* (Wadeye); *'ka-bakarra* (Yolngu/North East Arnhem land); *gabiny* (Yawuru people in and around Broome); *kabiny* (Nyul Nyul); and *muranga* (eastern Arnhem Land). Each of the Aboriginal language names leads us to a special food culture that needs to be documented and understood.

Even at this early stage of development, *Terminalia ferdinaniana* has the capacity to create a unique economic and commercial foundation for Aboriginal homelands and remote communities. The ability of the fruit to be stored for up to four years at below –25°C without losing quality or nutritional and biomedical properties means that it can easily be distributed by road and plane to national and international markets. Even in its raw state, communities harvesting *Terminalia ferdinaniana* can currently earn $20,000 for one tonne of fruit over a short harvest period. This is a useful income supplement for pickers in northern Aboriginal communities where the average income is around $341 a week. However, the potential to create more value from the raw food can also lead to an important economic foundation for rural and remote regions, with the Broome and Wadeye communities leading the way forward in innovation and development.

The potential of *Terminalia ferdinaniana* is enormous. It has the highest vitamin C content of any plant in the world – 100 times that of oranges. It has superior water and oil soluble antioxidant properties with five times antioxidant capability of blueberries, giving it a role in potentially preventing and delaying diseases such as Alzheimer's, autoimmune and cardiovascular disease, cancer, and diabetes. It contains vitamin E, lutein (eye health and well-being), folate, zinc, magnesium, and calcium. The high potassium levels can potentially prevent hypertension. Three international health and cosmetic companies have already included *Terminalia ferdinandiana* in beauty products such as cleansers, body lotions, hand creams and lip balms.

Terminalia ferdinandiana is already used commercially as a preservative for the seafood industry (1000 tonnes of farmed prawns processed in 2016/2017). With the proven application in seafood, the red meat processing industry is now showing some interest in developing a similar preservative to extend the shelf life of processed meats by up to eight days (Figures B4.4.1 and B4.4.2).

(*Continued*)

BOX 4.4 (Continued) SACRED FOOD ECONOMIC OPPORTUNITY: THE CASE OF *TERMINALIA FERDINANDIANA* AND THE CREATION OF AN ECONOMIC BASE FOR NORTHERN AUSTRALIAN HOMELANDS, REMOTE AND REGIONAL AREAS

Figure B4.4.1 Seed collection for Gubinge to assess suitable varieties for best production in Enrichment Orchards, involving workers from Mamabulanjin Aboriginal Corporation, Dampier Peninsula.

Figure B4.4.2 The processing of the Kakadu plum (Gubinge) is being undertaken by the Palngun Wurnanget Aboriginal Corporation at the Women's Centre in Wadeye.

anything but secure, on most sites producing negative net incomes propped up by high and growing debt (Chapter 5), while accruing environmental costs that degrade natural and production values.

But Indigenous people are among the most food insecure of Australians. They particularly depend on wild plants and animals when low incomes constrain access to mainstream commercial sources.[85] Outside the larger centres, costs of imported food are high and nutritional value and quality are often low: people who do not access wild foods are often less healthy.[86]

Most forms of development favoured in the White Paper have the potential to reduce availability of customary foods directly, by reducing wild populations or denying access to foraging sites or, more perniciously, by damaging confidence in the health of foods through actual or potential contamination by various toxicants (see Chapter 7). Risks to Indigenous food security must be considered in decisions about development options and the environmental management responses to them.

4.5.4 Capacity to Deliver Environmental Services

The capacity of northern Australian Aboriginal landowners and managers to provide the sorts of services here advocated draws particularly on one of the few successful federal government programs since the removal of the Community Development Employment Projects (CDEP). The extraordinarily successful Working on *Country* (WoC) programme has enabled at least several thousand Indigenous people to gain experience in meaningful jobs that enjoy considerable prestige within communities, while delivering an impressive array of other environmental and socioeconomic benefits.[87–90]

However, the apparent government ambivalence about land management as real work appears to have inhibited follow up to consolidate gains. At the time of writing, there is no commitment to continue the program beyond 2021, and there has been no work with participating communities to examine opportunities in related work on the land, drawing on the skills and experience gained. Government appears to cling so tenaciously to fantasies about trickle-down benefits as to be unwilling to pursue their own acknowledged direct successes, even when they are so well matched to Aboriginal aspirations, commitments to lands, and their present social conditions. Coercion of homeland residents to move to pursue mainstream jobs for which they may be individually poorly suited appears to remain the preferred option.[91]

A better way may be for Aboriginal people to themselves examine opportunities to make wider use of the skills and motivation that their experience in WoC and elsewhere has helped build. That would also depend on government but perhaps in a less ideologically challenging vehicle, by supporting landowners and their communities to plan their own economic development trajectories, including commercial use of their land, by offering the best available technical support and analysis. This would allow local people to adapt land uses to avoid conflict with cultural norms and special sites and, in many situations, position them to better meet obligations to lands. Experience shows that environmental services designed to meet national goals (such as the conservation, recreation, and watershed protection goals met by the national reserve system, that is now heavily dependent on Indigenous Protected Areas: Box 4.5) as well as local aspirations and obligations will be an important part of the mix.

We have shown that there is more than sufficient work in maintaining environmental quality to sustain a sizeable workforce that responsible governments might reasonably support. But it could be argued that this approaches the key question from the wrong direction. Perhaps we should be asking what incentives the nation can offer to keep – in a forever sparsely populated and climatically harsher northern Australia – a workforce large and skilled enough to sustain landscapes in which are embedded small but shifting nodes of high risk orthodox economic activity that disperse their impacts off-site. Landscapes that we know, even in the absence of such disturbance, demand a sustained, localised, knowledgeable human presence and active management to keep their natural and cultural values.

BOX 4.5 VALUING THE NORTH AUSTRALIAN CONSERVATION ESTATE

KAMALJIT K. SANGHA
DARWIN CENTRE FOR BUSHFIRE RESEARCH
CHARLES DARWIN UNIVERSITY, DARWIN

ANDREW EDWARDS
DARWIN CENTRE FOR BUSHFIRE RESEARCH
CHARLES DARWIN UNIVERSITY, DARWIN

JEREMY RUSSELL-SMITH
DARWIN CENTRE FOR BUSHFIRE RESEARCH
CHARLES DARWIN UNIVERSITY, DARWIN

The formal conservation estate occupies 320,000 km^2, over a quarter of the total 1.2m km^2 North Australia region. This includes Indigenous Protected Areas (IPAs), National Parks (NPs), biodiversity hotspots, wilderness areas, conservation reserve/areas, nature refuges and reserves, and officially recognised habitat or species management areas.[1] IPAs contribute the largest area of the conservation estate (154,000 km^2) followed by NPs (109,000 km^2).

Additionally, a further 72,000 km^2 of high conservation value is wetlands and riparian areas (Table B4.5.1). All these areas are appreciated for their direct and indirect services or benefits to society—Ecosystem Services—in terms of their uniqueness, morphological features, diversity, and cultural, social, and economic values.

Table B4.5.1 Area of Conservation Estate in >600 mm Rainfall Region of North Australia Using CAPAD (Collaborative Australian Protected Area Database 2014), and the Related IPA, NP and Other Management Costs

Category	Northern Territory	Queensland	Western Australia	Total
Indigenous Protected Area (IPA) (km^2)[5]	74,214	9,451	70,842	154,507
IPA – annual management costs[6]	*$10,315,746*	*$2,476,162*	*$3,471,258*	*$16,263,166*
National Parks (NP) (km^2)	52,443	42,519	14,089	109,052
NP – annual management costs using state/territory relevant costs[7]	*$45,363,195*	*$80,786,100*	*$3,536,339*	*$129,685,634*
Other areas of conservation significance (biodiversity hotspots, conservation reserve/area, nature refuge/reserve etc.) (km^2)	7,840	29,416	18,592	55,847
Other areas- annual management costs[8]	*$20,642,720*	*$77,452,328*	*$48,952,736*	*$147,047,784*
Wetland area (km^2)	11,936	48,184	2,381	62,501
Wetlands – annual management costs		*NOT COSTED*		
Riparian area (km^2)	3,055	4,746	2,837	10,638
Riparian area – annual management costs		*NOT COSTED*		
Total costs for managing IPAs, NPs and other conservation areas ($/yr)	*$76,321,661*	*$160,714,590*	*$55,960,333*	*$292,996,584*

Source: Collaborative Assessment of Protected Area Database (CAPAD 2014 – http://www.environment.gov.au/land/nrs/science/capad/2014; Commonwealth Government.

(Continued)

BOX 4.5 (Continued) VALUING THE NORTH AUSTRALIAN CONSERVATION ESTATE

Management costs associated with maintaining ES flows from the formal conservation estate range widely, both between jurisdictions, and between different conservation institutional arrangements (e.g. NPs, IPAs, conservation reserve or refuge). Using State/Territory specific average costings associated with managing IPAs in North Australia,[2] the total cost for managing all IPAs is currently around $16.2m/yr. Regional NPs cost another ~$130m/yr to manage (Table B4.5.1). In total, management costs, excluding the costs of managing wetlands or riparian areas, can be estimated to incur ~$292m/yr (Table B4.5.1). While the conservation estate is recognised as contributing very significantly to the regional nature-based tourism economy, estimated to be worth ~$3B annually,[2] investment in IPAs alone also provides very substantial social, cultural, environmental and economic benefits, realizing returns on investment of the order of 3:1.[3]

The Australian and Queensland Governments currently fund 490 ('Working on *Country*') and 65 ('Indigenous Land and Sea Ranger') management positions respectively, predominantly in North Australia. While Full-time Equivalent (FTE) employment figures associated with regional IPAs are not readily available,[4] we estimate that the formal conservation estate currently provides FTE employment for around 600 Indigenous people in North Australia.

REFERENCES

1. See Collaborative Assessment of Protected Area Database (CAPAD 2014- http://www.environment.gov.au/land/nrs/science/capad/2014; Commonwealth Government).
2. Clark E, Abel N, Measham T, Morison J, Rippin L (2009) Chapter 11: Water based tourism and recreation in northern Australia. In *Northern Australia Land and Water Science Review (full report)* (Ed P Stone). Final report to the Northern Australia Land and Water Taskforce, CSIRO, Canberra.
3. Social Ventures Australia (2016) Consolidated report on Indigenous Protected Areas following Social Return on Investment analyses. Report to Department of the Prime Minister & Cabinet. SVA Consulting.
4. Taylor M, Fitzsimons J, Sattler P (2014) *Building Nature's Safety Net 2014: A decade of protected area achievements in Australia*. WWF-Australia, Sydney. Available at www.wwf.org.au/buildingnaturesafetynet2014.
5. Source: Department of Prime Minister & Cabinet, April 2017.
6. Average annual management costs as for the NT = $1.39/ha; WA = $0.49/ha; and Qld = $2.62/ha were estimated from the IPA expenditure data obtained from the DPMC for each state/territory in northern Australia.
7. The annual management cost data (2015–2016) were obtained from the relevant Parks Departments for five parks in WA, and two in the NT. The average costs of $251/km^2 for WA, and $865/km^2 for the NT were used to estimate overall costs. In the absence of any direct data from Queensland, we used the average annual costs for Queensland as $1900/km^2 from the WWF–Australia report (by Taylor et al. 2014).
8. An average cost of managing covenants ($2,633/km^2) from WWF-Australia report (2014) was used.

4.6 CONCLUSION

We have argued that current policy on northern development is misconceived. It is based on theoretical assumptions and political expediencies which, for several reasons, are not applicable to the situation in northern Australia, particularly in its remoter regions. For example, the White

Paper ignores recent research on regional development that indicates the importance of amenity as a regional economy driver. It does not recognize that the historical process of economic growth in northern Australia is erratic, exhibiting a pulse-pause pattern. The White Paper makes assumptions that seem overly optimistic on the supply side, ignoring the history of the Ord Scheme, which shows that a large-scale public investment in a centralized infrastructure project has not driven sustainable regional development let alone achieved an acceptable rate of return on investment. The White Paper shows no understanding of how Aboriginal persons in northern Australia are systemically disadvantaged. Only a small proportion of funds ostensibly directed to Aboriginal welfare and progress actually reaches them.

What is needed is an alternative paradigm to that of the White Paper. We need to have a northern economy based upon sustainable long-term development principles. A sustainable economy based much more on the stewardship, not just the exploitation, of the northern estate. An economy that incorporates and uses the land management skills of the Aboriginal population who are the permanent residents of the North. They do not decamp and take their assets away. We also need to be open to new opportunities, such as managing increased bio-security risks that will follow climate change. In sum, we need to have a northern development policy that reduces the systemic transfer of economic surplus from the North to Australia's South, while leaving rural and remote residents of the North in despair of the future.

ACKNOWLEDGEMENTS

The research underpinning the economic modelling of the impact of spending on Indigenous Land Management Funding was supported through funding from the Australian Government's National Environmental Science Programme. We would like to thank Diane Jarvis for allowing us to reproduce results from her modelling of those impacts within our section on *trickle-down* development.

REFERENCES

1. Gerritsen R (2018) Pulse and Pause: Researching the economic future of northern and remote Australia. In *Rethinking Northern Australian Development*. Canberra (Eds Harwood S, Gerritsen R, Prideaux B, Brewer T, Rosenman L, Dale A), in press. ANU e-Press, Canberra, Australia.
2. Deane, P (2014) *Molehill to Mountain: Agriculture in Northern Australia*. ANZ Bank Report, Melbourne, Australia.
3. Cameron D (2005) Closer settlement in Queensland: The rise and decline of the agrarian dream, 1860s–1960s. In *Struggle Country: The Rural Ideal in Twentieth Century Australia* (Eds Davison G, Brodie M), pp. 6.1–6.21. Monash University Press, Melbourne, Australia.
4. WA Auditor General (2016) *Ord-East Kimberley Development. Report 20: September 2016*. Western Australian Auditor General, Perth, Australia.
5. Connolly E, Orsmond D (2011) The mining industry: From bust to boom. In *Proceedings of The Australian Economy in the 2000s* (Eds Gerard H and Kearns J), pp. 111–156. Reserve Bank of Australia, Sydney.
6. Edwards J (2014) *Beyond the Boom. A Lowy Institute Paper*. Penguin Books, Melbourne, Australia.
7. McLean IW (2013) *Why Australia Prospered: The Shifting Sources of Economic Growth*, p. 312. Princeton University Press, Princeton, NJ.
8. Bashar OHM (2015) The trickle-down effect of the mining boom in Australia: Fact or myth? *Economic Record* **91** (Special Issue), 94–108.
9. Gerritsen R (2010) A post-colonial model for north Australian political economy: The case of the Northern Territory. In *North Australian Political Economy: Issues and Agendas* (Ed Gerritsen R). Charles Darwin University Press, Darwin, NT.
10. Taylor A, Bell L, Gerritsen R (2014) Benefits of skilled migration programs for regional Australia: Perspectives from the Northern Territory. *Journal of Economic and Social Policy* **16**(1), Article 3.

11. Gerritsen R (2010) The post-colonial state as an impediment to Aboriginal development: An Arnhem Land example. In *North Australian Political Economy: Issues and Agendas* (Ed Gerritsen R). Charles Darwin University Press, Darwin, NT.
12. Colyer G (2014) The best of intentions. Mainstreaming the not-for-profit sector and indigenous Australians. PhD Thesis, Macquarie University, Sydney, Australia.
13. Taylor A, Payer H, Brokensha H (2015) The demography of developing Northern Australia. *Northern Institute Research Brief*, Technical Report, doi:10.13140/RG.2.1.2900.7207.
14. Commonwealth of Australia (2015). *Our North, Our Future: White Paper on Developing Northern Australia.* Commonwealth of Australia, Canberra, Australia.
15. Nelson AC (1993) Theories of regional development. In *Theories of Local Economic Development: Perspectives across the disciplines* (Eds Bingham RD, Mier R). Sage Publications, London, UK.
16. Jacobs J (1984) *Cities and the Wealth of Nations.* Random House, New York.
17. Vella K, Snipe N, Dale A, Taylor B (2015) Not learning from the past: Adaptive governance challenges for Australian natural resource management. *Geographical Research* **53**(4): 379–392.
18. Australian Government 2014. *Green Paper on Developing Northern Australia.* Australian Government, Canberra, Australia.
19. BITR (2014) Bureau of infrastructure, transport and regional economics. *The Evolution of Australian Towns.* Report 136. Department of Infrastructure and Regional Development, Canberra, Australia.
20. DRG (Department of Regional Development) (2013) *Regional price index 2013.* Department of Regional Development, Government of Western Australia, Perth, Australia.
21. Nash F (Senator) (2016) Investing in regional growth. *Budget Statement by the Minister for Regional Development 3 May 2016 (on* page *16), p.166.* Parliament of Australia, Canberra, Australia.
22. Nash F (Senator) (2016) Investing in regional growth. *Budget Statement by the Minister for Regional Development 3 May 2016 (on* page *107), p.166.* Parliament of Australia, Canberra, Australia.
23. Nash F (Senator) (2016). Investing in regional growth. *Budget Statement by the Minister for Regional Development 3 May 2016 (on* page *127), p.166.* Parliament of Australia, Canberra, Australia.
24. Nash F (Senator) (2016). Investing in regional growth. *Budget Statement by the Minister for Regional Development 3 May 2016 (on* page *127–129), p.166.* Parliament of Australia, Canberra, Australia.
25. Nash F (Senator) (2016). Investing in regional growth. *Budget Statement by the Minister for Regional Development 3 May 2016 (on* page *69), p.166.* Parliament of Australia, Canberra, Australia.
26. Ohlin, B (1933) *Interregional and International Trade.* Harvard University Press, Cambridge.
27. Stoeckl N, Stanley O (2007) Key industries in Australia's tropical Savanna. *Australasian Journal of Regional Studies* **13**(3): 255–286.
28. Carson D, Taylor A, Campbell S (2009) *Demographic Trends and Likely Futures for Australia's Tropical Rivers.* Charles Darwin University, Darwin, NT. Retrieved from http://track.org.au/sites/default/files/managed/file-attach/biblio/TRaCK_3.1_CDU_Report_final.pdf.
29. Larson S, Alexandridis K (2009) *Socio-Economic Profiling of Tropical Tivers.* CSIRO Land & Water Australia, Canberra, Australia.
30. Hunter B (1999) *Three Nations, Not One: Indigenous and Other Australian Ppoverty. CAEPR Working Paper No. 1.* Centre for Aboriginal Economic Policy Research, The Australian National University, Canberra, Australia. Retrieved from http://caepr.anu.edu.au/Publications/WP/1999WP1.php.
31. Taylor A, Larson S, Stoeckl N, Carson D (2011) The haves and have nots in Australia's Tropical North – New perspectives on a persisting problem. *Geographical Research* **49**(1): 13–22. doi:10.1111/j.1745-5871.2010.00648.x.
32. Stoeckl N (2010) Bridging the asymmetric divide: Background to, and strategies for bridging the divide between Indigenous and non-Indigenous economies in northern Australia. In *Northern Australia Political Economy* (Ed Gerritsen R). Charles Darwin University, Darwin, NT.
33. Allen Consulting Group (2011) *Assessment of the economic and employment outcomes of the Working on Country program.* Retrieved from https://www.environment.gov.au/indigenous/workingoncountry/publications/woc-economics.html.
34. Stoeckl N (2007) Using surveys of business expenditure to draw inferences about the size of regional multipliers: A case-study of tourism in Northern Australia. *Regional Studies* **41**(7): 917–931. doi:10.1080/00343400601142803.
35. Rolfe J, Miles B, Lockie S, Ivanova G (2007) Lessons from the social and economic impacts of the mining boom in the Bowen Basin 2004–2006. *Australasian Journal of Regional Studies* **13**(2): 134–153.

36. Rolfe J, Gregg D, Ivanova G, Lawrence R, Rynne D (2011) The economic contribution of the resources sector by regional areas in Queensland. *Economic Analysis and Policy* **41**(1): 15–36.
37. Fleming DA, Measham TG (2014) Local job multipliers of mining. *Resources Policy* **41**(1): 9–15. doi:10.1016/j.resourpol.2014.02.005.
38. Fleming DA, Measham TG, Paredes D (2015) Understanding the resource curse (or blessing) across national and regional scales: Theory, empirical challenges and an application. *Australian Journal of Agricultural and Resource Economics* **59**(4): 624–639. doi:10.1111/1467-8489.12118.
39. Stoeckl N, Jackson S, Pantus F, Finn M, Kennard MJ, Pusey BJ (2013) An integrated assessment of financial, hydrological, ecological and social impacts of 'development' on Indigenous and non-Indigenous people in northern Australia. *Biological Conservation* **159**: 214–221. doi:10.1016/j.biocon.2012.12.007.
40. Ivanova, G. (2014) The mining industry in Queensland, Australia: Some regional development issues. *Resources Policy* **39**(1): 101–114. doi:10.1016/j.resourpol.2014.01.005.
41. Johnson PL (2001) *An Input-Output Table for the Kimberley Region of Western Australia*. Economic Research Centre, Department of Economics, University of Western Australia, Perth, Australia.
42. Murti S, Northern Territory Office of Resource Development (2001) *Input-output multipliers for the Northern Territory 1997–1998*. The Office of Resource Development, Darwin, NT.
43. Stoeckl N, Stanley O (2009) Maximising the benefits of development in Australia's far North. *Australasian Journal of Regional Studies* **15**(3): 255–280.
44. Hunter, B (2014) *Indigenous employment and businesses: Whose business is it to employ Indigenous workers? CAEPR Working Paper No. 95/2014*. Centre for Aboriginal Economic Policy Research, The Australian National University, Canberra, Australia. Retrieved from http://caepr.anu.edu.au/sites/default/files/Publications/WP/CAEPR_Working_Paper_95.pdf.
45. Stoeckl N, Esparon M, Stanley O, Farr M, Delisle A, Altai Z (2011) *Socio-Economic Activity and Water Use in Australia's Tropical Rivers: A Case Study in the Mitchell and Daly River Catchments: Final report for The Tropical Rivers and Coastal Knowledge Research Consortium*. Charles Darwin University, Darwin, NT.
46. Stoeckl N, Esparon M, Farr M, Delisle A, Stanley O (2013a) Distributional and consumptive water demand impacts of different types of economic growth in two northern Australian river catchments. *Australasian Journal of Regional Studies* **19**(3): 396–432.
47. Stoeckl N, Esparon M, Farr M, Delisle A, Stanley O (2013b) The great asymmetric divide: An empirical investigation of the link between indigenous and non-indigenous economic systems in Northern Australia. *Papers in Regional Science* **93**(4): 783–801. doi:10.1111/pirs.12028.
48. Kalb G, Le T, Hunter B, Leung F (2014) Identifying important factors for closing the gap in labour force status between indigenous and non-indigenous Australians. *Economic Record* **90**(291): 536–550. doi:10.1111/1475-4932.12142.
49. Jarvis D, Stoeckl N (in review) Enumerating the financial benefits of ILMPs and the distribution of those benefits within the broader community. A report prepared for the Department of Environment and Energy, to provide an update on research undertaken as part of the Northern Australia Environmental Resources Hub of the National Environmental Science Program, Project 5.3.
50. Office of the Government Statistician Queensland Government (2004) *Queensland Regional Input-Output Tables 1996–97 34 Industries*. Retrieved from http://www.qgso.qld.gov.au/products/reports/qld-reg-io-tables/index.php.
51. Hill R, Pert P, Davies J, Robinson C, Walsh F, Falco-Mammone F (2013) *Indigenous Land Management in Australia: Extent, Scope, Diversity, Barriers and Success Factors*. CSIRO Ecosystem Services, Cairns, Australia.
52. Northern Territory Government, Northern Territory Land Corporation (2016b) *Ord Stage 3 Expansion: Request for Proposal*, p. 6. April 2016.
53. Warchivker I, Lange V, Price N, Douglas-Hill C, Andrén L (2012) *Victoria River District Economic Development Opportunities. STAGE 1: Scoping Study*. Report for Northern Land Council, p. 76. Top End Farm for Centrefarm Aboriginal Horticulture, Alice Springs, Australia.
54. Petheram C, Gallant J, Wilson P, Stone P, Eades G, Rogers L, Read A et al. (2014) *Northern Rivers and Dams: A Preliminary Assessment of Surface Water Storage Potential for northern Australia*. Technical Report. CSIRO Land and Water Flagship Technical Report, p. 79. CSIRO Land and Water Flagship, CSIRO, Canberra, Australia.

55. Seafarms Group Limited (2016) *Project sea Dragon Stage 1 Legune Grow Out Facility*. Draft Environmental Impact Statement. Volume 1 Project Overview. Chapter 3 Project Description, p. 105. Seafarms Group PL, Perth, Australia.
56. Brodie JE, Mitchell AW (2005) Nutrients in Australian tropical rivers: Changes with agricultural development and implications for receiving environments. *Marine and Freshwater Research* **56**: 279–302.
57. Harris G (2001) Biogeochemistry of nitrogen and phosphorus in Australian catchments, rivers and estuaries: Effects of land use and flow regulation and comparisons with global patterns. *Marine and Freshwater Research* **52**: 139–149.
58. Harris G (2001b) *A Nutrient Dynamics Model for Australian Waterways: Land Use, Catchment Biogeochemistry and Water Quality in Australian Rivers, Lakes and Estuaries*. Australia: State of the Environment Second Technical Paper Series No. 2 (Inland Waters), p. 90. CSIRO Land & Water, Environment Australia, Canberra.
59. CSIRO (2012) *Australia's shale gas resources*. CSIRO, Canberra, Australia.
60. Geoscience Australia, ABARE (2014) *Australian Energy Resource Assessment*. Geoscience Australia and Australian Bureau of Agricultural and Resource Economics (ABARE), p. 336. Canberra, Australia.
61. Northern Territory Government (2016c) *Northern Territory Acreage Release 2016 Re-release Round [oil and gas]*. Department of Mines and Energy, Darwin, NT.
62. Woinarski J, Dawson F (2002) Limitless lands and limited knowledge: coping with uncertainty and ignorance in northern Australia. In *Ecology, Uncertainty and Policy: Managing Ecosystems for Sustainability* (Eds Handmer J, Norton T, Dovers S), pp. 83–115. Prentice-Hall, Harlow, UK.
63. Cook, G (2009) Chapter 6: Historical perspectives on land use development in northern Australia: With emphasis on the Northern Territory. In *Northern Australia Land and Water Science Review: Full Report*, pp. 6.1–6.30. CSIRO Sustainable Agriculture Flagship and Northern Australia Land and Water Taskforce, Darwin, NT.
64. Ross J, Ah Mat R, Baffsky D, Blanch S, Crombie D, Gardiner E, Hill K et al. (2009) Sustainable Development in Northern Australia. A report to Government from the Northern Australia Land and Water Taskforce, p. 38. Department of Infrastructure, Transport, Regional Development and Local Government, Canberra, Australia.
65. Northern Territory Government (2015a) North East Gas Interconnector: Fact Sheet 1. Private sector and government in partnership.
66. Broderick J, Anderson K, Wood R, Gilbert P, Sharmina M, Footitt A, Glynn S, Nicholls F (2011) *Shale Gas: An Updated Assessment of Environmental and Climate Change Impacts*. Tyndall Centre for Climate Change Research, University of Manchester, Manchester, UK.
67. Queensland Audit Office (2013) *Environmental Regulation of the Resources and Waste Industries*. Government of Queensland, Brisbane, Australia.
68. NTEPA (2013) Section 14–Investigation Report Mt Todd Gold Mine Uncontrolled Discharge, p. 30. Northern Territory Environment Protection Authority, Darwin, NT.
69. NTEPA (2014) Redbank Copper Mine: Environmental Quality Report, p. 57. Northern Territory Environment Protection Authority, Darwin, NT.
70. Whitehead PJ, Woinarski J, Williams R, Jacklyn P, Fell R (2000) Defining healthy savanna landscapes: A north Australian perspective, p. 23. *Tropical Savannas CRC Discussion Paper*, Darwin, NT.
71. MCA (2014) Senate Committee Hearing into Environmental Offsets: Opening Statement 6 May 2014. Minerals Council of Australia, Canberra, Australia. Accessed at http://www.minerals.org.au/news/senate_committee_review_environmental_ offsets in November 2016.
72. ICMM (2006) Good practice guidance for mining and biodiversity. International Council on Mining and Metals. Accessed at http://www.icmm.com/publications/pdfs/13.pdf.
73. NTEPA (2013) Guidelines on environmental offsets and associated approval conditions. Version 2.0, p. 6. Northern Territory Environment Protection Authority, Darwin, NT.
74. Whitehead PJ, Woinarski J, Franklin D, Price O (2002) Landscape ecology, wildlife management and conservation in northern Australia: Linking policy, practice and capability in regional planning. In *Landscape ecology and resource management: Linking theory with practice* (Eds Bissonette JA, Storch I), pp. 227–259. Island Press, Ithaca NY.
75. Woinarski J, Milne D, Wanganeen G (2001) Changes in mammal populations in relatively intact landscapes of Kakadu National Park, Northern Territory, Australia. *Austral Ecology* **26**: 360–370.

76. Franklin D, Whitehead PJ, Pardon G, Matthews J, McMahon P, McIntyre D (2005) Geographic patterns and correlates of the decline of granivorous birds in northern Australia. *Wildlife Research* **32**: 399–408.
77. Wood T, Blowers D, Moran G (2016) *Climate Phoenix: A Sustainable Australian Climate Policy.* Grattan Institute, Melbourne, Australia.
78. Burke PJ (2017) Undermined by adverse selection: Australia's direct action abatement subsidies. *Economic Papers* **35**, 216–229.
79. CoA (Commonwealth of Australia) (2015) *Carbon Credits (Carbon Farming Initiative—Oil and Gas Fugitives) Methodology Determination 2015.* p. 29. Australian Government, Canberra, Australia.
80. Russell-Smith J, Cook G, Cooke P, Edwards A, Lendrum M, Meyer C, Whitehead P, Yibarbuk D (2013) Managing fire regimes in north Australian savannas: Applying customary Aboriginal burning approaches to contemporary problems. *Frontiers in Ecology and Environment* **11**, E55–E63.
81. NAILSMA (2014) Beyond respect: A central role for Indigenous people in Australian land and sea management. Paper drafted for World Parks Congress, Sydney, November 2014. *North Australian Indigenous Land and Sea Management Alliance*, Darwin, NT, p. 28.
82. Northern Land Council (2017) Draft Environmental Impact Statement for the McArthur River Mine Overburden Management Project: Northern Land Council Submission to the Northern Territory Environment Protection Authority. *Northern Land Council,* Darwin, NT, p. 33.
83. Mansfield J (2016) Rrumburriya Borroloola Claim Group v Northern Territory of Australia NTD 6014 of 2000, NTD 6003 of 2003. 30 June 2016, p. 111. Federal Court of Australia, Canberra, Australia.
84. NAILSMA (North Australian Indigenous Land and Sea Management Alliance) (2013) An Indigenous prospectus for northern development: Setting the agenda. *Knowledge Issues Series*, 20/2013 Policy Paper, p. 78. NAILSMA, Darwin, NT.
85. Ferguson M, Brown C, Georga C, Miles E, Wilson A, Brimblecombe J (2017) Traditional food availability and consumption in remote Aboriginal communities in the Northern Territory, Australia. *Australian and New Zealand Journal of Public Health* **41**, 294–298.
86. Bussey C (2013) Food security and traditional foods in remote Aboriginal communities: A review of the literature. *Australian Indigenous Health Bulletin* **13**(2): 9.
87. Burgess C, Johnston F, Bowman D, Whitehead PJ (2005) Healthy land: Healthy people? Exploring the health benefits of indigenous land management. *Australian New Zealand Journal of Public Health* **29**: 117–122.
88. Garnett S, Sithole B, Whitehead PJ, Burgess C, Johnston F, Lea T (2009) Healthy country, healthy people: Policy implications of links between Indigenous human health and environmental condition in tropical Australia. *Australian Journal of Public Administration* **68**: 53–66.
89. Social Ventures Australia (SVA) (2016a) *Consolidated report on Indigenous Protected Areas following Social Return on Investment analyses for the department of the Prime Minister & Cabinet.* p. 50. SVA Consulting.
90. SVA (2016b) *Social Return on Investment Analysis of the Warddeken Indigenous Protected Area and Associated Indigenous Ranger Programme.* Full Report, p. 59.
91. Jordan K, Fowlkes L (Eds.) (2016) Job creation and income support in remote indigenous Australia: moving forward with a better system. *Centre for Aboriginal Economic Policy Research, Australian National University,* Canberra, Australia, p. 34.

CHAPTER 5

Towards a Sustainable, Diversified Land Sector Economy for North Australia

Jeremy Russell-Smith, Kamaljit K. Sangha, Robert Costanza, Ida Kubiszewski, and Andrew Edwards

CONTENTS

5.1 INTRODUCTION

Previous chapters have set out a compelling case for developing sustainable economies across the North that are inclusive of the wellbeing and aspirations of Indigenous people. Despite being very significant components of the North's population, especially in remoter areas, and increasingly 'land rich' through ongoing acquisitions and Native Title determinations, it is widely acknowledged that Indigenous people remain severely economically and socially disadvantaged. In this chapter we address the challenge of developing a culturally, environmentally, socially, and economically sustainable regional land sector economy, focusing on alternatives to the current spatially dominant land use sector of the North, the pastoral industry.

By some estimates, beef cattle production activities occur over as much as 90% of North Australia, particularly involving extensive, as opposed to intensive grain-fed, production systems.[1] Pastoral lands include not only pastoral leases but also swathes of *country* involving Indigenously-owned Land Trust areas, such as Arnhem Land that supports large numbers of feral water buffalo, cattle, and horses. Available data indicate that there are around four million cattle in our focal area of 1.2 million km^2 of the North Australia study region (Chapter 2; Map 2.3), with a 'northern Australian' (including all of Queensland) herd of around 13.7 million, and a national herd of around 25 million.[2]

The pastoral industry in our focal area generates approximately $400 million gross value from about 650 pastoral enterprises with about 2800 direct employees, of whom, based on available 2016 census data, about 390 are Indigenous.[3–5]

Despite the vast geographic spread of this industry, all is not economically, socially, or environmentally well. For a start, a landmark North Australian industry situation report covering the period 2001–2012 found that most pastoral enterprises were neither economically viable nor sustainable.[4] Average operating profitability across the northern pastoral industry over the 12-year assessment period was just 0.2%, and 2.6% for the top 25% of business performers; average return on assets was <1%.[4] Such low levels of profitability reflect conditions across the industry of typically low fertility soils, seasonal access restrictions, limited infrastructure, high labour and input costs, limited financial management skills, and distant and volatile markets.[4,6–12] If there is one mitigating bright spot, it is the longer-term trend of exponentially increasing land values, averaging around 6% annually over 20 years for a median-priced northern pastoral property[13] – albeit subject to market fluctuations such as occurred with the 2011 collapse of the Indonesian live-export market (Box 5.1).

While the industry has enjoyed high cattle and property prices in recent seasons, over the long-term they have experienced poor financial returns, fluctuating land values, and significant debt levels extenuated by prolonged drought conditions in some regions, contributing to major

BOX 5.1 VALUING PASTORAL PROPERTIES IN NORTHERN AUSTRALIA

FRANK PEACOCKE

RURAL VALUATIONS, HERRON TODD WHITE (NORTHERN TERRITORY) PTY LTD, DARWIN

1. NT/KIMBERLEY PASTORAL MARKET OVERVIEW (SEE FIGURES B5.1.1)

2001–mid-2009 NT Pastoral Property Market Boom

Beef cattle properties in northern Australia in general experienced rapid value increases over a period of about eight years to around mid/late-2009. This rapid growth created a snowballing effect in cattle property demand. It appears that the market reached its peak in the Northern Territory (NT)/Kimberley in 2008/2009 (Queensland a bit earlier in 2007). Factors influencing the buoyant market conditions up until that stage included

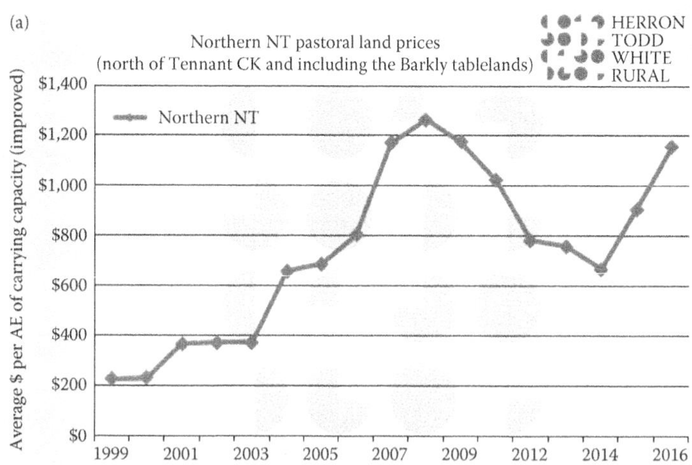

(Continued)

BOX 5.1 (Continued) VALUING PASTORAL PROPERTIES IN NORTHERN AUSTRALIA

(b)

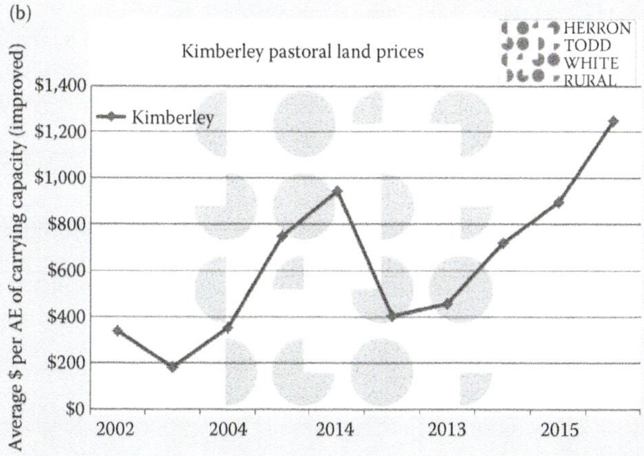

(c)

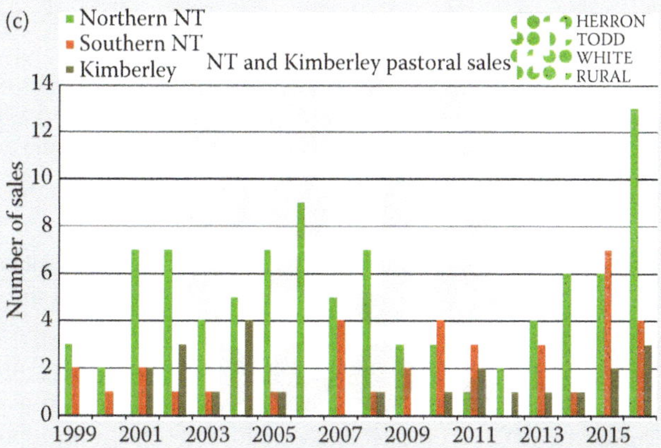

Figure B5.1.1 Pastoral property prices represented as average $per Adult Equivalent (AE) of carrying capacity (improved) over time. (a) Price changes in the NT from 1999 to 2016. (From HTW [Northern Territory] sales analysis based on HTW's own long-term carrying capacity assessments.) (b) Price changes in Kimberley from 2002 to 2016. (From HTW [Northern Territory] sales analysis based on HTW's own long-term carrying capacity assessments.) and (c) Number of sales from 1999 to 2016. (From HTW [Northern Territory] analysis of settled sales.)

- Generally strong cattle prices (particularly Indonesian live export through Darwin) and favourable seasonal conditions boosting confidence in the industry. It must be remembered that the Indonesian live export market has historically provided a second trade option for Barkly Tableland cattle stations for out of spec cattle (culls).
- Low interest rates and bank lending policies encouraging equity borrowing to fund purchases on long term interest only loans.
- Strong demand and a shortage of listings.

Mid-2009–2013: NT pastoral property market downturn

Beyond 2009 and through to the end of 2012 there were a number of features that negatively affected the values of pastoral enterprises in the northern NT region, especially

(*Continued*)

BOX 5.1 (Continued) VALUING PASTORAL PROPERTIES IN NORTHERN AUSTRALIA

- The Indonesian Government's step towards the goal of self-sufficiency in 2010, which hanged the import specifications to a cap on cattle live weights of 350 kgs and no more females.
- The June 2011 suspension of the live cattle trade to Indonesia, which had the combined effect of significantly reducing the size of the market to Indonesia and therefore the gross revenues achievable from Kimberley/Pilbara/NT Top End cattle stations.
- The suspension's impact on gross revenue decline in markets for the base product (live export cattle) meant that previously manageable debt levels held by many pastoralists became unsustainable at prevailing levels and this appears to have forced a number of northern properties onto the market.

Late 2013 can be viewed as representing the most recent low point in the pastoral property market in the Kimberley region, and 2014 in the Northern Territory Top End.

2013 to Present

Since late 2013 a combination of strengthening live export cattle prices, continuing low interest rates, and an influx of overseas investor capital into the northern pastoral market has seen value levels strengthen. Of most importance to Top End cattle properties is the live cattle export market, which has strengthened significantly since late 2013. Although as of June 2016 live cattle shipments from Australia (630,547 head) were down 12% year-on-year, it has still been the third highest on record. And while exports to Indonesia were down 14% in that year to June, exports to all other major markets were up 19% year-on year. However, keeping up with the cattle supply to south-east Asian markets is anticipated by exporters to be a problem for the rest of 2017 as pastoralists focus more on herd rebuilding.

Other positive drivers during this period have included the opening of Australian Agricultural Company's (AACo) new abattoirs at Livingstone, south of Darwin, and the NT Government's announced changes to the Pastoral Land Act, whereby pastoralists are given greater freedom to potentially utilise their property's natural attributes and diversify the potential income streams (for example, develop a groundwater resource that lies under arable *country* for the purposes of horticulture; or clear an area of appropriate land type for forestry). It is proposed that the diversification permit would run for 30 years and would be registered on title and so would run with the land in the event of a sale. Also, in 2015 the NT Government formalised (subject to consent) the use of pastoral lease property for non-pastoral uses. Although some pastoral leases have carried out activities such as tourism for some years, a formal process now exists to legitimise these activities. It is expected that the most common non-pastoral uses would be for tourism-related purposes and ancillary purposes such as haymaking for sale. It is therefore expected that the number of pastoral leases which will be able to benefit is relatively small.

2. VALUATION METHOD: PASTORAL LEASES AND CARRYING CAPACITY

In a nutshell, the use of the $/AE (Adult Equivalent) approach in Herron Todd White's (HTW) sales analysis of a pastoral lease interprets economic worth by relating land values to long-term, sustainable productivity as determined by district standards (i.e. the assessed long-term sustainable carrying capacity [CC] reflects a buyer's, vendor's, or valuer's opinion of a property's *highest and best use*). By multiplication of a property's current carrying capacity by the $/AE rate resulting from the analysis of a group of comparable property sales (i.e. their sale price excluding livestock, plant and equipment divided by current produces a value which HTW contends is a fair indicator of what the market is likely to pay.

(Continued)

BOX 5.1 (Continued) VALUING PASTORAL PROPERTIES IN NORTHERN AUSTRALIA

CC assessments are generally undertaken using either the utilisation method or historic method. Importantly, we note that in applying the methodology and assumptions below the valuer is also maintaining consistency with how he analyses property sales.

Utilisation Method

We consider this method the best method to gain consistency in carrying capacity (CC) assessment across the northern pastoral estate. The utilisation methodology is always based on the most recent pasture research. The long-term CC of the property is estimated using methodology outlined in Meat and Livestock Australia's Edge Network Grazing Land Management (GLM) package. The experience gained from various grazing trials on the Barkly Tablelands, Victoria River District, Katherine Region is also incorporated here.

The method requires that areas of the different *country* types (land systems or land units) within the cadastral boundaries of a pastoral lease to be mapped and calculated. This is done electronically using a GIS program such as ArcGIS. A typical land system map with 3 km grazing radii on permanent waters follows (Figure B5.1.2):

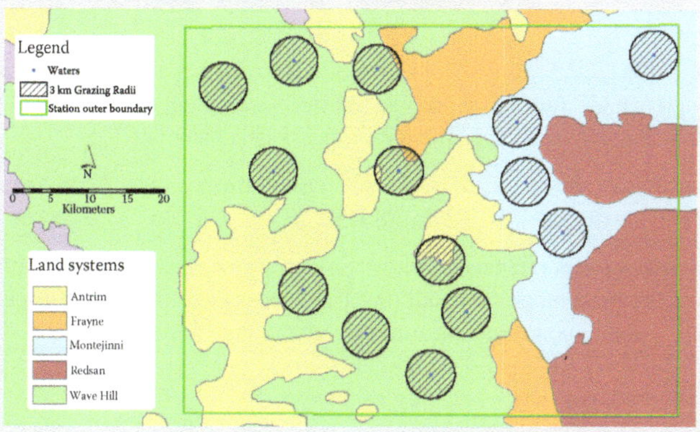

Figure B5.1.2 Land systems and water resources mapping to estimate CC of a pastoral property.

For each land system:

- The area (ha or km^2) within 3 km grazing radii of permanent water is calculated.
- The estimated expected annual growth/supply of pasture is calculated, taking into account land type, climate, and land condition. In the example in Table B5.1.2 we have adopted a land condition rating of B out of a range of A to D in accordance with the assessment parameters in Table B5.1.1.
- Apply a safe annual utilisation rate, taking into account the forage demand of the grazing animal:

$$\text{Long-term stocking rate (ha/AE/yr or AE/sqkm/yr)} = \frac{\text{Annual pasture growth (kg dry matter/ha)} \times \text{Utilisation (\%)}}{\text{Forage Demand per AE (kg/yr)}}$$

where annual (median) pasture growth is estimated using the GRASP model developed by Queensland Department of Primary Industries and calibrated to local conditions. Utilisation rates are taken from the GLM package, and forage demand is calculated as the amount of forage that an AE (again, a 450 kg dry animal) consuming a percentage of live-weight on average each day over a year (around 2,920 kg based on 8 kg per day or around 1.8%–2% of bodyweight).

(Continued)

BOX 5.1 (Continued) VALUING PASTORAL PROPERTIES IN NORTHERN AUSTRALIA

Table B5.1.1 Assessment of Pastoral Land Condition (A, B, C and D) Using Soil, Pasture Types, Weed and Woodland Attributes to Evaluate Pastoral Properties

Land	Soil	Pasture	Weed	Woodland
A All of these features	No erosion and good surface condition	Good coverage of 3P grasses[a], little bare ground (<30%) in most years	Few weeds and no significant infestations	No signs of woodland thickening
B (At least one or more of these features)	Some signs of previous erosion and some current signs of erosion	Some decline in the presence of 3P grasses and/or bare ground (more than 30%, but less than 50% in most years)	Small infestations of weeds	Some thickening in the density of woodland
C (One or more of these features)	Obvious signs of past erosion and/or current susceptibility to erosion	General decline in the presence of 3P species and/or bare ground (>50% in most years)	Obvious presence of weeds	General thickening in the density of woody plants
D (One or more of these features)	Severe erosion, scalding or compaction resulting in a hostile environment for plant growth	General lack of and perennial grasses or forbs	Large weed infestations covering significant areas	Thickets of woody plants that cover significant

Source: NT Government.
[a] 3P grasses = Palatable, Productive, Perennial.

- Once the long-term CC for each land type has been estimated, this is applied to the different land systems areas within 3 km grazing radii previously calculated.

Table B5.1.2 provides an example of the CC calculation on a typical area of pastoral land in the Victoria River District. Note, in Table B5.1.2:

- *Supp Util %* means the percentage of pasture utilised assuming mineral supplementation.
- *Access* means the percentage of actual physical access the animal has for pasture given the topography (e.g. steep banks on a pastured riverbank may not be fully accessible). In the following example, everything is fully accessible.
- *Intake* means an AE consumes 2,920 kg pa on average.

Historic Method

Estimates of sustainable long-term CC made using the historical approach utilises the experience and historic long-term stocking evidence of many long-term graziers and pastoralists who have managed *country* without causing significant reduction in land condition over long periods. These CC estimates are usually applied to the areas of *country* within 5 km of permanent water and are generally utilised in districts where pasture growth and utilisation modelling has not been well researched (mainly south of Tennant Creek, in the more remote districts of the NT, and in the Kimberley).

DEFINITIONS

Market value: The estimated amount for which an asset or liability should exchange on the valuation date between a willing buyer and a willing seller in an arm's length transaction, after proper marketing, where the parties had each acted knowledgeably, prudently, and without compulsion.

Highest and best use: The use of an asset that maximises its potential and that is physically possible, legally permissible and financially feasible.

Adult equivalent: An adult equivalent (AE) is defined as a 450 km non-lactating beast. This is a long term, year-in-year-out sustainable average level of productivity through fluctuating annual seasonal conditions.

(Continued)

BOX 5.1 (Continued) VALUING PASTORAL PROPERTIES IN NORTHERN AUSTRALIA

Table B5.1.2 An Example of CC Calculations of Pastoral Land in the Victoria River District of the NT

Land System	Class Description	Land System Area within 3 km (ha)	Yield (B cond)	Supp Util %	Unsupp Util %	Access	Intake (kg pa)	AE/km²	AE
Wave Hill	Plains and rises on basalt, agglomerate and tuff, some dolerite; clayey soils with surface stone	26,489	2,425	20	15	100	2,920	16.6	4,400
Frayne	Plains and rises on basalt, agglomerate, tuff, some dolerite; clayey soils with surface stone	394	2,200	15	10	100	2,920	11.3	44
Antrim	Low hills and hills on basalt, agglomerate and tuff, some dolerite; mostly rock rock outcrop with surface stone and pockets of clayey soils	1,391	1,900	15	10	100	2,920	9.8	136
Montejinni	Plains, rises and plateaux on weathered and unweathered Cambrian limestone, dolomite, chalcedony, shale, sandstone and siltstone with associated sand	11,309	1,200	10	8	100	2,920	4.1	465
TOTAL		39,582							5,045

financial stresses on many pastoral businesses.[12,14] At the family level, such stresses have contributed to serious emotional health issues – including high rates of suicide in north Australian rural and remote communities.[15] The ongoing impacts of dispossession on Aboriginal people and communities are well documented (Chapter 3).

The industry also generates significant long-term environmental liabilities, which are not accounted for in current market pricing, including: land degradation[9,11,16–22]; gully erosion,[23–25] and associated down-stream siltation effects on maritime systems[26–28]; impacts on freshwater resources and quality, including unrestricted access to and exploitation of ecologically fragile springs and soaks[29–31]; impacts on native species biodiversity, including granivorous birds and small mammals[7,32–35]; and ruminant greenhouse gas emissions, which are estimated to contribute 4% of national emissions.[36] However, it needs to be recognised that, in many northern regions, significant weed and feral animal management issues are common to all types of land use.[11,19]

Collectively, these issues point to a pressing need for assessing the value proposition of the northern pastoral industry, and to explore novel diversified approaches for developing sustainable land use practices and broader community benefits. In this chapter we focus on the development of diversified economic opportunities that currently, and in the near future, can be derived through enhanced and sustainable management of natural ecosystems and the services these provide.

The chapter begins with a general description of what we mean by 'ecosystem services', derived from natural systems, followed by illustration of the value of such services nationally under different land use scenarios. The second part of the chapter describes and assesses the current economic and environmental condition of the pastoral industry in our focal region, and associated impacts on ecosystem services, based on available documented sources. A more localised geographic assessment is provided for the Northern Territory's gulf region as a regional case study in Chapter 7. The final section of the paper addresses opportunities for developing a diversified regional land sector economy based on sustainable ecosystem services.

Wherever possible, our intention is to provide quantitative monetized valuations of both the financial condition of the industry as well as associated cultural, social, and environmental benefits and costs. However, as might be expected, many core attributes cannot readily be expressed in monetary units (e.g. spiritual values), or insufficient data currently exist (e.g. valuing threatened species or dry season above-ground water sources). Nonetheless, using these assembled data, we show that people and communities occupying the great majority of the current North Australia pastoral region will benefit substantially from the development of a diversified ecosystem services-based land sector economy.

5.2 ECOSYSTEM SERVICES

Ecosystem services can be defined as the ecological characteristics, functions, or processes that directly or indirectly contribute to human well-being – the benefits people derive from functioning ecosystems.[37,38] Ecosystem processes and functions may contribute to ecosystem services, but they are not synonymous. Ecosystem processes and functions describe biophysical relationships and exist regardless of whether or not humans benefit.[39] Ecosystem services, on the other hand, only exist if they contribute to human well-being, and they cannot be defined independently. The following categorization of ecosystem services has been used by the Millennium Ecosystem Assessment.[38]

1. *Provisioning services*: Services that offer 'provisioning' benefits to people such as bush tucker, medicine, timber, and fibre.
2. *Regulating services*: Services that regulate different aspects of the integrated system, such as flood control, storm protection, water regulation, human disease regulation, water purification, air quality maintenance, pollination, pest control, and climate control. These services are generally not marketed but have clear value to society.
3. *Cultural services*: Services that offer recreational, aesthetic, religious, scientific, cultural identity, sense of place, or other 'cultural' benefits. For example, customary lands provide sites for Indigenous people to perform ceremonies and transmit knowledge to future generations.
4. *Supporting services*: Services that maintain basic ecosystem processes and functions such as soil formation, primary productivity, biogeochemistry, and provisioning of habitat. These services affect human well-being *indirectly* by maintaining processes necessary for provisioning, regulating, and cultural services. For example, net primary production (NPP) is an ecosystem function that supports carbon sequestration and removal from the atmosphere, which provides the benefit of climate regulation. Some would argue that these 'supporting' services should rightly be defined as ecosystem 'functions' because they may not yet have interacted with the other three forms of capital to create benefits. We agree with this in principle but recognize that supporting services/functions may sometimes be used as proxies for services in the other categories.

This categorization suggests a very broad definition of services, limited only by the requirement to contribute to human well-being. Even without any subsequent valuation, explicitly listing the services derived from an ecosystem can help ensure appropriate recognition of the full range of potential impacts of a given policy option. This can help make the analysis of ecological systems more transparent and can help inform decision makers of the relative merits of different options before them.[40]

Examples of these services include the maintenance of the composition of the atmosphere; amelioration and stability of climate, flood controls, and drinking water supply; waste assimilation; recycling of nutrients; generation of soils; pollination of crops; provision of food; maintenance of species and a vast genetic library; and also maintenance of the scenery of the landscape, recreational sites, and aesthetic and amenity values.[37,41–45] Biodiversity at genetic, species, population, and ecosystem levels all contribute in maintaining these functions and services.[46] An environmentally literate society would probably accept the assertion that most, if not all, ecosystem functions are in the long term beneficial to society[47] and should not be damaged or traded off lightly.

Many ecosystem services are public goods. This means they are non-excludable, and multiple users can simultaneously benefit from using them. This creates circumstances where individual choices are not the most appropriate approach to valuation. Furthermore, ecosystem services (being public goods) are generally not traded in markets. We therefore need to develop other methods to assess their value.

5.2.1 Valuation of Ecosystem Services

Valuation is about assessing trade-offs towards achieving a goal.[48] All decisions that involve trade-offs involve valuation, either implicitly or explicitly.[40] When assessing trade-offs, one must be clear about the goal. Ecosystem services are defined as the benefits people derive from ecosystems – the support of sustainable human well-being that ecosystems provide.[37,38] The value of ecosystem services is therefore the relative contribution of ecosystems to that goal. There are multiple ways to assess this contribution, some of which are based on individual's perceptions of the benefits they derive. But the support of sustainable human well-being is a much larger goal,[41] and individuals' perceptions are limited and often biased.[49] Therefore, we also need to include methods to assess benefits to individuals that are not well perceived, benefits to whole communities, and benefits to sustainability.[49] This is an ongoing challenge in ecosystem services valuation, but even some of the existing valuation methods like avoided and replacement cost estimates are not dependent on individual perceptions of value. For example, estimating the storm protection value of coastal wetlands requires information on historical damage, storm tracks and probability, wetland area and location, built infrastructure location, population distribution, and so on.[51]

It is also important to note that ecosystems cannot provide any benefits to people without the presence of people (human capital), their communities (social capital), and their built environment (built capital). This interaction is shown in Figure 5.1. Ecosystem services do not flow directly from natural capital to human well-being – it is only through interaction with the other three forms of capital that natural capital can provide benefits. This is also the conceptual valuation framework for the recent UK National Ecosystem Assessment (http://uknea.unep-wcmc.org) and the Intergovernmental Platform on Biodiversity and Ecosystem Services (IPBES; http://www.ipbes.net). The challenge in ecosystem services valuation is to assess the relative contribution of the natural capital stock in this interaction and to balance our assets to enhance sustainable human well-being. The relative contribution of ecosystem services can be expressed in multiple units – in essence any of the contributors to the production of benefits can be used as the 'denominator' and other contributors expressed in terms of it. Since built capital in the economy, expressed in monetary units, is one of the required contributors, and most people understand values expressed in monetary units, this is often a convenient denominator for expressing the relative contributions of the other forms of capital, including natural capital. But other units are certainly possible (i.e. land, energy, time, etc.) – the choice is largely about which units communicate best to different audiences in a given decision-making context.

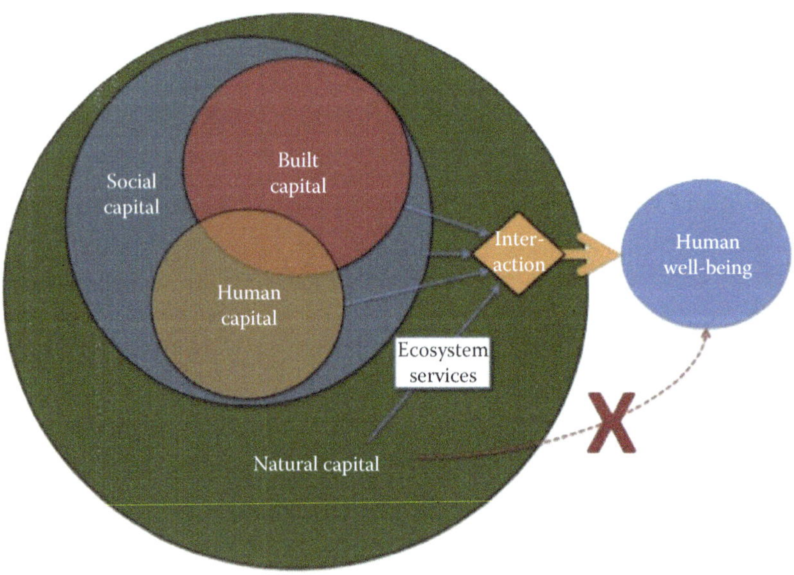

Figure 5.1 Interaction between built, social, human, and natural capital required to produce human well-being. Built and human capital (the economy) are embedded in society which is embedded in the rest of nature. Ecosystem services are the relative contribution of natural capital to human well-being; they do not flow directly. It is therefore essential to adopt a broad, trans-disciplinary perspective in order to address ecosystem services. (From Costanza, R. et al., *Glob. Environ. Change*, 26, 152–158, 2014.)

5.2.2 Ecosystem Services of North Australian Savannas

North Australian Savannas have not been extensively studied in terms of their ecosystem services contributions. Table 5.1 shows a summary of the estimates that have been assembled as part of the Ecosystem Services Value Database (ESVD) (http://www.fsd.nl/esp/80763/5/0/50) for 17 ecosystem services for various ecosystems as of 2012. For woodlands, 21 estimates have been included with a mean value of $1,600/ha/year and a range from $1,300 to 2200/ha/year. For grasslands, 32 estimates have been included with a mean value of $2,900/ha/year, but a broader range of from $125 to 6,000/ha/year. Of these only one study was performed in North Australia, but 20 were performed in similar ecosystems in Africa, the Americas, and Asia. But based on this, we can estimate that North Australian savannas contribute about $1,000–2,000/ha/yr.

5.2.3 Possible Futures under Different Scenarios

As part of a recent global study,[49] the value of ecosystem services was estimated for four future scenarios (Figures 5.2 and 5.3) to the year 2050 built around the four great transition initiative archetypes: (1) market forces (MF); (2) fortress world (FW); (3) policy reform (PR); and (4) great transition (GT).[52–54] Two of these scenarios (market forces and great transition) are similar to those described elsewhere in this chapter. The market forces scenario is well captured in the government white paper on northern Australia development,[55] despite continued public subsidies of some industries. The great transition scenario is consistent with our vision for sustainable and desirable development.

Detailed great transition initiative (GTI) scenarios exist for both the global system and several regions. Brief narrative descriptions of each scenario, extracted directly from the GTI website (www.greattransition.org/explore/scenarios), are reproduced here:

Market forces (MF): The market forces scenario is a story of a market-driven world in the twenty-first century in which demographic, economic, environmental, and technological trends unfold without major surprises relative to unfolding trends. Continuity, globalization, and convergence are key characteristics of world development – institutions gradually adjust without major ruptures, international economic integration proceeds apace and the socioeconomic patterns of poor regions converge slowly towards the development

Table 5.1 Summary of the Number of Estimates, Mean, Standard Deviation, Median, Minimum and Maximum Values from the ES Value Database. Values Are in International $/ha/yr

	No. of Estimates	Total of Service Mean Values	Total of St. Dev. of Means	Total of Median Values	Total of Minimum Values	Total of Maximum Values
Open oceans	14	491	762	135	85	1,664
Coral reefs	94	352,915	668,639	197,900	36,794	2,129,122
Coastal systems	28	28,917	5045	26,760	26,167	42,063
Coastal wetlands	139	193,845	384,192	12,163	300	887,828
Inland wetlands	168	25,682	36,585	16,534	3018	104,924
Rivers and lakes	15	4267	2771	3938	1446	7757
Tropical forest	96	5264	6526	2355	1581	20,851
Temperate forest	58	3013	5437	1127	278	16,406
Woodlands	21	1588	317	1522	1373	2188
Grasslands	32	2871	3860	2698	124	5930

Source: de van der Ploeg S, de Groot R (2010) The TEEB Valuation Database – A searchable database of 1310 estimates of monetary values of ecosystem services. Foundation for Sustainable Development, Wageningen, the Netherlands.

Figure 5.2 Four development scenarios about the future of ES.[52] (From Kubiszewski, I. et al., *Ecosys. Ser.*, 26, 289-301, 2017.)

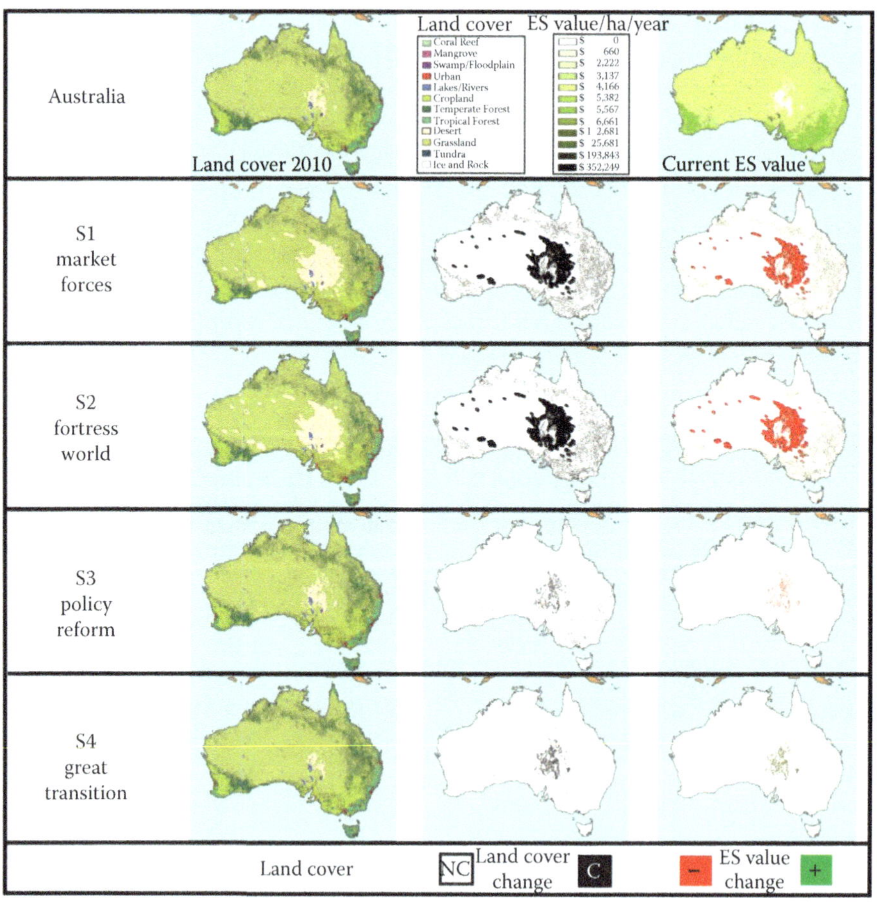

Figure 5.3 Land cover and ecosystem service values for Australia in 2010, and for four future scenarios to 2050.[52] (From Kubiszewski, I. et al., *Ecosyst. Ser.*, 26, 289-301, 2017.)

model of the rich regions. Despite economic growth, extreme income disparity between rich and poor countries, and between the rich and poor within countries, remains a critical social trend. Environmental transformation and degradation are a progressively more significant factor in global affairs.

Policy reform (PR): The policy reform scenario envisions the emergence of strong political will for taking harmonized and rapid action to ensure a successful transition to a more equitable and environmentally resilient future. Policy Reform is designed to achieve a set of future sustainability goals. The analytical task is to identify plausible development pathways for reaching that end-point. Thus, the policy reform scenario explores the requirements for simultaneously achieving social and environmental sustainability goals under high economic growth conditions similar to those of market forces.

Fortress world (FW): The fortress world scenario is a variant of a broader class of Barbarization scenarios, in the hierarchy of the global scenario group.[52] Fortress world features an authoritarian response to the threat of social breakdown brought about by growing inequality. Ensconced in protected enclaves, elites safeguard their privilege by controlling an impoverished majority and managing critical natural resources, while outside the fortress there is repression, environmental destruction, and misery.

Great transition (GT): The great transition scenario explores visionary solutions to the sustainability challenge, including new socioeconomic arrangements and fundamental changes in values. This scenario depicts a transition to a society that preserves natural systems, provides high levels of welfare through material sufficiency and equitable distribution, and enjoys a strong sense of local solidarity.

Each of these scenarios has been produced and vetted by a large network of scholars and have been used as archetypes for a range of other scenario planning studies.[52–54,56] They incorporate a range of world views and policies, and the impacts of these on the entire, integrated system, including population, energy use, equity, environmental change, and climate change. The GTI scenarios also include impacts on land use and management. The interactive web tool Futures in Motion, on the GTI website, was used to derive estimates of land use change (urban, cropland, forest, grassland, desert), population, GDP, and other variables such as inequality and GDP for these four future scenarios to the year 2050 (see www.tellus.org/results/results_World.html).

Based on earlier methodological approaches, Figure 5.3 shows maps of land cover for each biome for the base map and the four scenarios, changes in the land cover between 2011 and each of the four scenarios in 2050 (shown as those pixels that changed or did not change), and the change in ecosystem services value from the 2011 values to each of the four scenarios within that *country* or region.[56] Regions in the vicinity of deserts will experience a greater desertification rate than those regions further away from existing desert. This assessment indicates that the great transition scenario, similar to the scenario we describe later in this chapter, significantly restores ecosystem services in North Australia.

5.3 ASSESSING THE ECONOMIC CONDITION AND ENVIRONMENTAL IMPACT OF THE NORTH AUSTRALIA PASTORAL INDUSTRY

The focal area for this assessment comprises 1.2 million km,[2] including all savannas receiving at least 600 mm and as much as ~2000 mm rainfall/year (Map 2.4). Over this vast region rainfall is highly seasonal, with 90% falling in 4–5 wet season months (generally November–March) (Map 2.4). Soils are mostly infertile,[57] dominated by rudosols, kandosols, and tenosols over 69% of the region, and smaller areas of fertile vertosols covering 15% (Map 2.6).

As noted earlier, extensive beef cattle pastoralism comprises by far the dominant land use in this region (Map 2.7). Pastoral lands in North Australia are recognised generally as supporting much lower levels of pastoral production than in south-eastern Australia, mainly due to low fertility soils and poor-quality pastures, seasonal access restrictions, limited infrastructure (e.g. roads, fencing water points), relatively high labour and input costs, and distant and volatile markets.[4,5,11,36]

The region includes large pastoral properties varying in median size from 100,000 to 200,000 ha in the Top End (NT), Cape York (QLD), and Kimberley (WA), and from 15,000 to 70,000 ha in the rest of QLD. Most properties are under either freehold, or long-term perpetual (99 year) pastoral lease, tenure arrangements depending upon respective state or territory legislation. Annual rental lease payments are low (0.75–1.5% of the land value, depending upon 'term' or 'perpetual lease' arrangements). Average stocking rates vary from <1 to 15 head/km^2, depending upon soil potential, pasture species and infrastructure.[4] The industry's northern beef situational analysis report for the period, 2001–2012, determined that the average gross margin for a median sized beef property in different regions varied from $200,000 to $950,000 per year, with relatively greater returns from Barkly, Kimberley, and QLD Mitchell grass regions.[4]

Pasture types over the study region are dominated by spinifex (*Triodia, Plectrachne* spp.) 19.8%; annual sorghum (*Sorghum* spp.) 16.1%; *Aristida-Bothriochloa* 13.1%; *Shizachyrium* 9.4%; and black spear grass (*H. contortus*) 8.8% (Table 5.2, Figure 5.4a). In terms of pasture capability, 81% of the total area is considered as having low potential, with 4% having moderate and 15% high pasture capability.[58] Following validation of these observations with contemporary understanding of regional pastoral scientists, two revised pasture capability ratings are included here: Queensland black spear grass communities are re-rated from high to moderate; and ribbon grass communities in

Table 5.2 Pasture Species and Area under High, Moderate or Low Pasture Production Potential, Where Total Area = 1.2 million km^2

Pasture Communities Categorised under High, Moderate, and Low Pasture Potential	Area (km^2)	Proportion of Study Area (%)
High	**36,670**	3
Bluegrass-browntop (*Dichanthium fecundum – Eulalia fulva*)	24,831	
Ricegrass (*Xerochloa* sp.)	7,563	
Shortgrass grassland (*Triodia* spp., *Eragrostis* sp.)	4,277	
Moderate	**455,673**	39
Aristida-Bothriochloa spp.	68,377	
Black speargrass (*Heteropogon contortus*)	104,096	
Blady grass (*Imperata cylindrica*)	1,658	
Bluebush/saltbush (*Maireana astrotricha/Atriplex vesicaria*)	59	
Bluegrass-browntop (*Dichanthium fecundum*)	26,627	
Heathland pastures (*Schoenus sparteus*, *Digitaria* sp. etc.)	4,045	
Mitchell grass (*Astrebla* spp.)	32,067	
Plume sorghum (*Sorghum plumosum*)	10,330	
Ribbongrass (*Chrysopogon fallax*)	103,238	
Ricegrass (*Xerochloa* sp.)	131	
Saltwater couch (*Sporobolus virginicus*)	3,184	
Schizachyrium spp.	6,737	
Shortgrass grassland (various species)	12,401	
Spinifex (*Triodia* and *Plectrachne* spp.)	82,206	
Wanderrie grass (*Eriachne* spp.)	516	
Low	**691,596**	58
Annual sorghum (*Sorghum intrans, S. stipoideum, S. australiense*)	190,307	
Aristida-Bothriochloa spp.	85,751	
Blady grass (*Imperata cylindrica* spp.)	11,243	
Heathland pastures	15,262	
Rainforest derived pastures	2,329	
Saltwater couch (*Sporobolus virginicus*)	30,556	
Schizachyrium spp.	103,708	
Spinifex (*Triodia* and *Plectrachne* spp.)	234,071	
Wanderrie grass (*Eriachne* spp.)	18,370	

Source: Adapted from a Regional Report.[58]

WA/NT are re-rated from low to moderate. Accordingly, it is considered that 58% of the focal area is of low, 39% moderate, and only 3% has high pastoral potential (Table 5.2; Figure 5.4b).

We assessed the carrying capacity of regional pastures by integrating data from a number of published sources,[4,5,59,60] and expert opinion (Figure 5.4c). Almost 70% of the total area is considered to support <4 head/km^2, 20% supports 5–9 head/km^2, and just 10% supports ≥10 head/km^2 (Figure 5.4c).

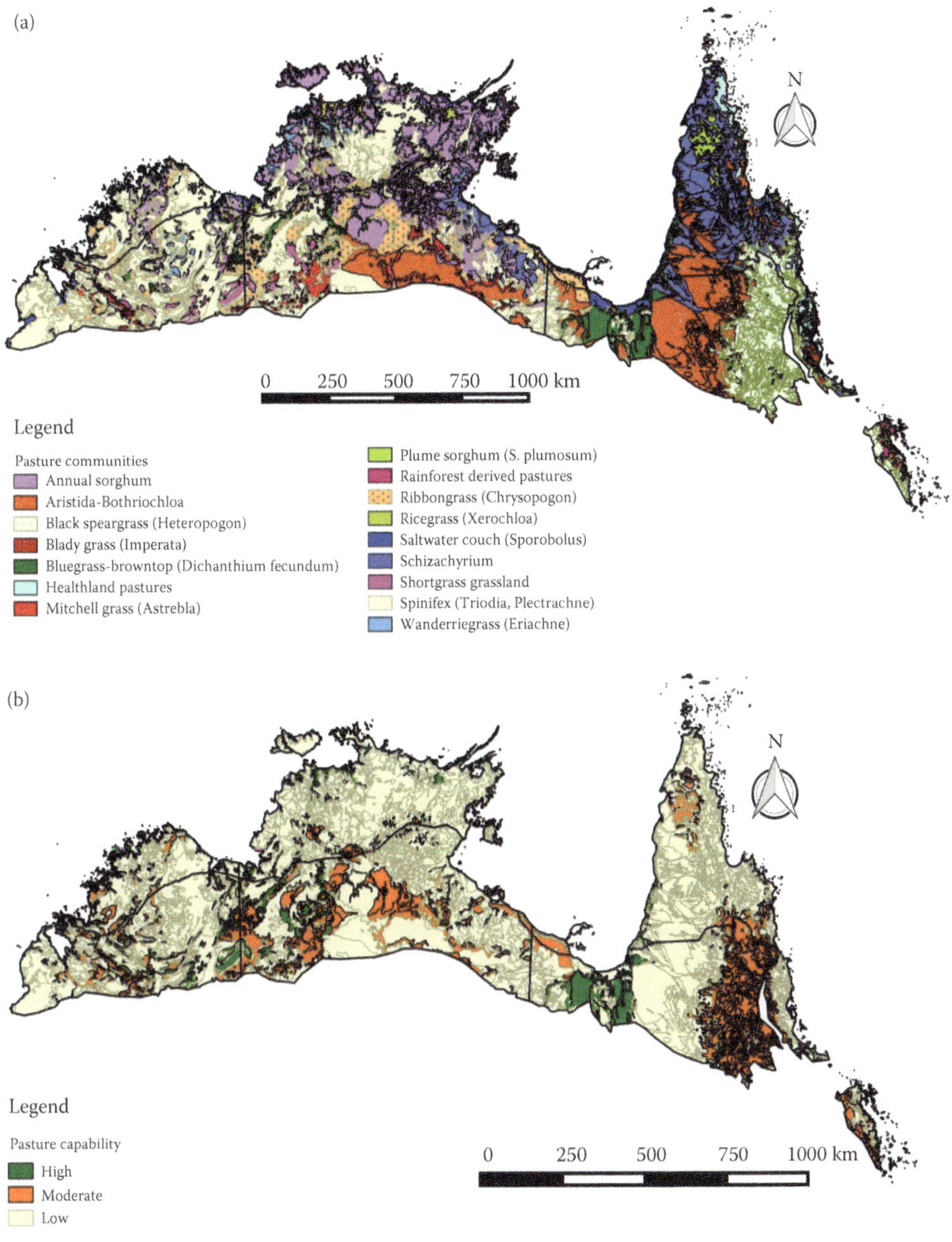

Figure 5.4 Pasture attributes for the study region: (a) pasture communities, (b) pasture capability (the image has been updated with minor modifications for categorising ribbongrass and black spear grass as moderate, following an expert opinion). *(Continued)*

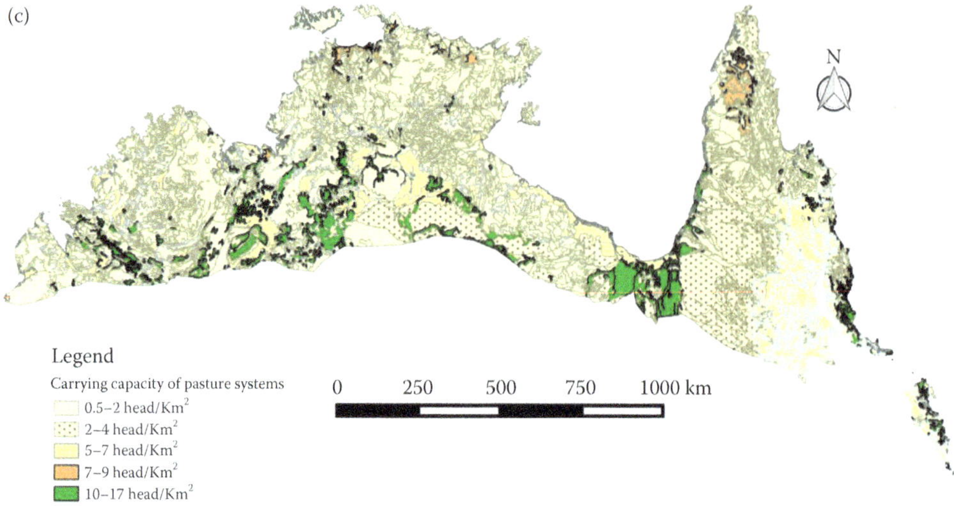

Figure 5.4 (Continued) Pasture attributes for the study region: (c) pasture carrying capacity including current expert opinion. (From Tothill, J. C. and Gillies, C., *The Pasture Lands of northern Australia: their condition, productivity, and sustainability*. Tropical Grassland Society of Australia, on behalf of Meat Research Corporation, Brisbane, Australia, 1992.)

Locations supporting high carrying capacity in our focal region occur mainly in lower rainfall parts of the QLD Gulf, central NT, and Fitzroy plains in WA (Figure 5.4c). In general, this assessment conforms with a recent pastoral industry productivity assessment identifying areas of low, moderate, and high productivity in NT and QLD.[5]

5.3.1 Economic Assessment

We conducted a detailed assessment of the economic condition of a typical regional beef cattle producing enterprise integrating available published data mostly from Australian Bureau of Agricultural and Resource Economics and Sciences (ABARES) farm survey data[60] (http://www.agriculture.gov.au/abares/surveys) and regional sources.[4,5] For this assessment (except where noted below), we used eight regions as identified by Bray et al. (2015) based on their productivity potential, amalgamated from 18 Australian Bureau of Statistics SA2 sub-regions. All values cited are given in 2016 AUS$.

Financial returns from cattle production were assessed in terms of annual earnings before interest and taxes (EBIT) following the northern beef situational analysis report.[4] For each region, we applied long-term (2001–2012) EBIT estimates for a median-sized property operating as a 'typical' pastoral business. Considering the North Australia focal region as a whole, as defined in that report, a 'typical' pastoral business is representative of roughly 70% of all regional pastoral enterprises, with an average herd size of 3200 adult equivalents, and collectively contributing 42% of total beef production in North Australia. Our assessment thus does not include a relatively smaller number of larger, generally corporately owned properties.

Our EBIT financial analysis shows that most northern sub-regions return <$100,000/year/property except for the Barkly and central-west QLD regions (Figure 5.5a). For our focal area, EBIT returns were just $20,000–$50,000/year/median-sized property in north Queensland and the Kimberley, and $50,000–$99,000/year for a median-sized property in the Top End of the NT. The average annual long-term (2001–2012) EBIT for a typical pastoral business in northern Australia is $136,000/year, or $6.16 per adult equivalent (AE), indicating very low financial returns for a majority of pastoral businesses.[4] In reviewing these results, it was noted, however, that 'the calculated EBIT figure includes an imputed value on owner wages. In reality this wage is rarely paid so the businesses may achieve a small cash surplus, even though it is reported as a loss'.*

However, 'exceptional performers' comprise about 27% of total beef production in northern Australia.[4] Financial performance of those pastoral businesses is better (an average EBIT of $348,000/year, or $61.96/AE) than the typical businesses reported here, mainly due to better herd management (especially maintaining stocking rates relative to seasonal carrying capacity), herd productivity, lower operating costs, and better financial management.[4]

Apart from low levels of financial return, most northern pastoral properties are operating with high levels of long-term debt (Figure 5.5b). Using available ABARES debt liability data reported in the northern beef situational analysis report,[4] we estimated earnings after interest but before taxes (EAIBT) by calculating annual interest repayments per median-sized property at a conservative 5% annual rate of interest. When taking debt into account, the EAIBT assessment indicates that almost 80% of pastoral properties in the northern region have been operating at loss, except for high productivity areas in QLD and the NT Barkly (Figure 5.5c). In our focal area, almost all median-sized pastoral businesses have been operating at an average annual loss of $5,000–$21,000 per property. A long-term negative EAIBT for a typical business is also highlighted in the situational analysis report[4] (Figure 5.6).

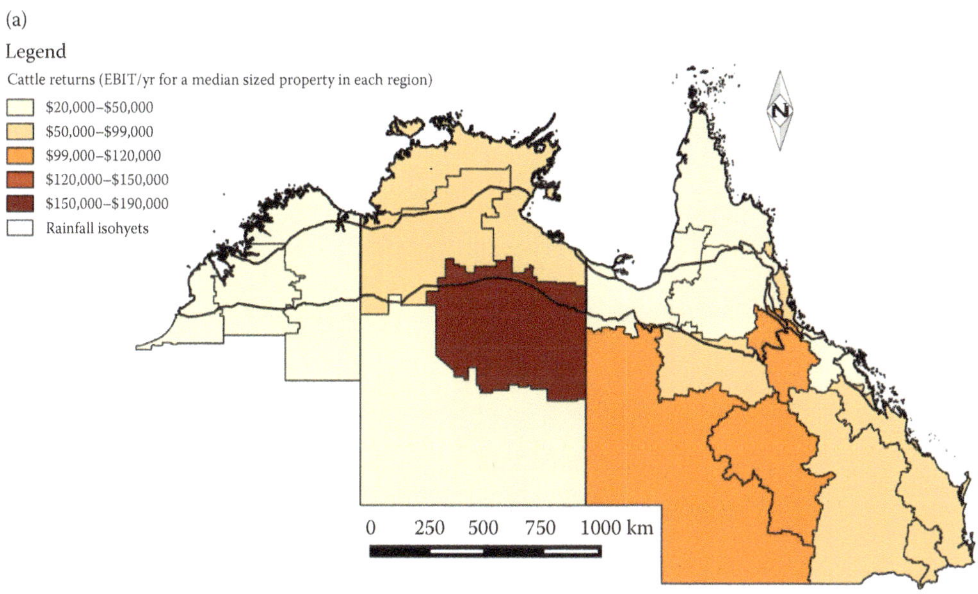

Figure 5.5 Annual long-term (2001–2012) economic returns from a typical, median-sized pastoral business in the north: (a) Earnings before interest and tax (EBIT)/year. (From McLean, I. et al., *The Northern Beef Report – 2013 Northern Beef Situation Analysis*. Meat & Livestock Australia Ltd., North Sydney, NSW, 2059, 2014; Bray, S. et al., 'Desktop research project to provide data on liveweight and liveweight gain in the beef cattle sector in Queensland and the Northern Territory'. *Final report, 2015*. Department of Agriculture and Fisheries, State of Queensland, Rockhampton, Australia, 2015 based upon ABARES (Australian Bureau of Agricultural and Resource Economics and Sciences), Customised data extracted from AAGIS dataset. ABARES. http://www.agriculture.gov.au/abares/surveys frequently accessed from May–December 2016, 2013.) *(Continued)*

* Ian McLean, personal communication, June 2016.

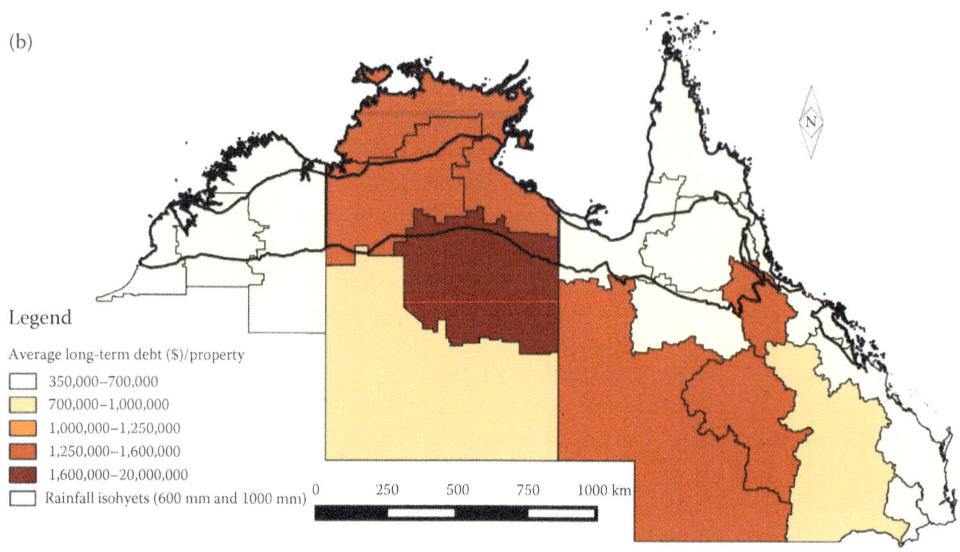

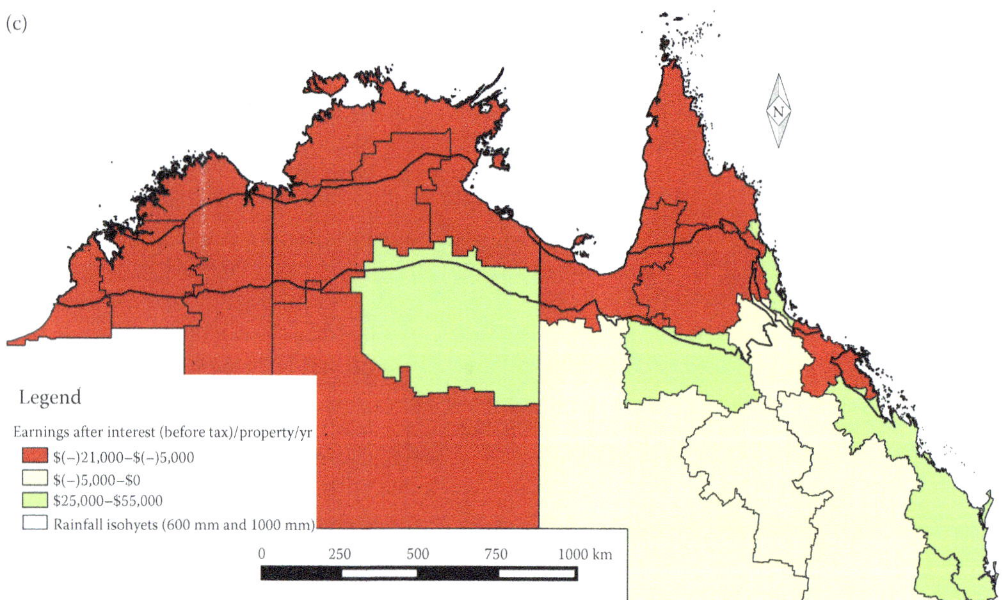

Figure 5.5 (Continued) Annual long-term (2001–2012) economic returns from a typical, median-sized pastoral business in the north: (b) long-term debt (2001–2012). (From ABARE 2001–2012 data [McLean, I. et al., *The Northern Beef Report – 2013 Northern Beef Situation Analysis*. Meat & Livestock Australia Ltd., North Sydney, NSW, 2059, 2014]) and (c) earnings after interest but before tax (EAIBT)/yr. (From McLean, I. et al., *The Northern Beef Report – 2013 Northern Beef Situation Analysis*. Meat & Livestock Australia Ltd., North Sydney, NSW, 2059, 2014; Bray, S. et al., 'Desktop research project to provide data on liveweight and liveweight gain in the beef cattle sector in Queensland and the Northern Territory'. *Final report, 2015*. Department of Agriculture and Fisheries, State of Queensland, Rockhampton, Australia, 2015; ABARES, Regional farm debt: Northern Queensland gulf, south west Queensland, and north west New South Wales. ABARES, Canberra, Australia, December. CC BY 3.0, 2014.)

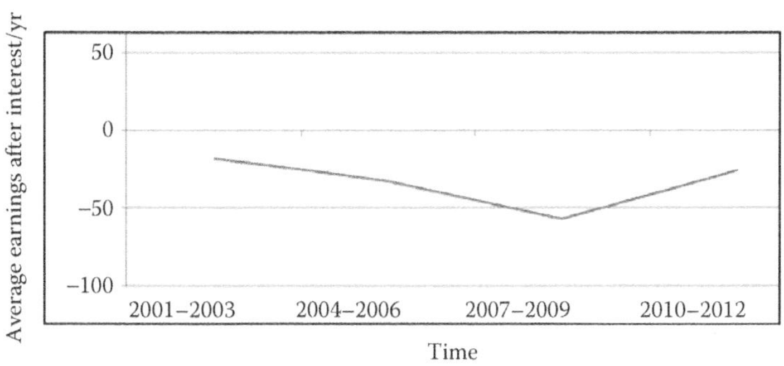

Figure 5.6 'Average annual long-term earnings after interest (EAIBT) for a typical pastoral business in northern Australia'. (From McLean, I. et al., *The Northern Beef Report – 2013 Northern Beef Situation Analysis*. Meat & Livestock Australia, North Sydney, NSW, 2059, 2014.)

Levels of farm business debt across the north are substantial, with 44% of businesses owing more than $0.5 million in the QLD-Gulf region, and 37% owing more than $1 million in the NT.[61] These amounts are much higher than the rest of beef producing regions in Australia. Since 2008, northern pastoralists have taken on obligations to service higher debt levels while farm income has been declining; the ratio of interest payments to total cash receipts for serving debt is 20%, well above the national average (8%).[61] Others have highlighted that many northern beef businesses are in dire financial situation given high debt levels and low productivity.[5,12] Clearly, a majority of the northern pastoral businesses are operating much below the break-even level, with <1% return on assets over the period 2001–2012.[4]

The recurring pattern of long-term poor financial returns from typical, as opposed to 'exceptionally performing' northern Australian pastoral enterprises raises the obvious question why people might want to invest time, effort, and financial resources in this industry. Apart from being an active, if challenging, life-style choice for many participants,[9,11,12,62] a part answer might be that, despite significant inter-annual variability, over the long term property prices have continued to escalate[13] (Box 5.1). However, even where the debt to equity ratio may be relatively small, rising land values in themselves do not provide an annual return sufficient to service associated debt. In economically marginal situations 'it reduces profitability and can force new entrants (who pay the high prices) to run their *country* harder in an effort to achieve a return, thus compromising the environment'.* Rising land values also clearly cannot substitute for losses on communally held Aboriginal lands that will never be sold.

A recent industry report concluded: 'at the individual beef business scale, productivity growth and returns on investment in the northern Australian beef industry is generally static or declining and, together with high debt levels and increasing input costs, many northern grazing businesses are in a dire financial situation'.[5]

5.3.2 Ecological and Environmental Assessment

While recognizing that well-managed pastoral enterprises can provide many positive ecosystem services (e.g. reducing weed, feral animal, fire impacts), heavy grazing by cattle (and associated livestock and feral animals such as horses and Asian water buffalo) incurs well-known impacts on savanna systems which typically are not accounted for in conventional financial assessments, nor taken into account in establishing the real long-term costs (and benefits) of the pastoral industry and its products. Such environmental impacts can be exacerbated by financial imperatives of poorly performing and managed pastoral enterprises to overstock in an endeavour to meet crippling debt.[4]

The character and magnitude of various of these landscape-scale impacts is immense, contributing to land degradation resulting in reduced landscape function and consequent lost pastoral

* Ian McLean, personal communication, June 2016.

production, through losses of topsoil, soil organic matter, nutrients, and surface water retention, and, in severe cases, surface deflation and associated gully erosion; downstream effects of sedimentation on coastal and marine ecosystems; impacts on the quality and quantity of ecologically precious perennial and seasonal surface water resources; greenhouse gas emissions directly associated with cattle rumination and dung; spreading of weeds and facilitating undesirable woody thickening; and significant impacts on vulnerable flora, fauna, and critical habitats (Table 5.3). Effects are ubiquitous but vary in intensity across the landscape.

To illustrate the magnitude of such impacts, in this assessment we consider the environmental costs associated with aspects of pasture condition and greenhouse gas emissions from cattle given ready availability of accepted environmental accounting procedures (Tables 5.4 and 5.5). Costing of other environmental impacts listed in Table 5.3, while feasible to do, has either not

Table 5.3 Effects of Unsustainable Grazing Land Use on Soils (Including Landscape Functions), Biodiversity, GHG Emissions, Woody Thickening, Weeds, and Water Resources

Ecological Effects	Study Areas	References
Loss of Soil, Landscape Health, and Landscape Function		
Effects of over-grazing on land can result in very significant soil erosion, gully erosion, top soil run-off, sediment loss, decline in soil fertility, and loss of landscape functions and processes which further results in loss of future production potential of land.	Northern Australia, including: NT regions – (Daly basin, Arnhem Land, Gulf); QLD – Northern Gulf, Burdekin catchment, GBR catchments; WA – Kimberley.	6, 11, 16, 18, 23, 24, 25, 27, 63–68
Biodiversity Impacts		
Effects of grazing typically result in significant impacts on flora and fauna, including declines in bird and mammal assemblages and diversity, and associated plant and habitat resources	Northern Australia	22, 32–35, 76, 97, 102, 109
GHG Emissions and Climate Change		
GHG from cattle contributing 2–4% of national emissions, estimated at 10.5–25 kg CO_2-e/kg of Live Weight sold, or 1.78–1.83 tCO_2-e/yr per adult equivalent in northern Australia	Australia wide and northern Australia	336, 68, 77, 102, – also see the Australian GHG Emissions Information System (AGEIS) calculator @ (http://ageis.climatechange.gov.au)
Woody Thickening		
Over-grazing, often combined with reduced fire frequency, can result in woodland thickening (involving both native and exotic species) with significant impacts on pasture growth and pastoral productivity. Conversely, C sequestration is promoted through woody thickening	Northern Australia	75, 77–85, 125, 126, 132
Weeds		
Heavy grazing often results in weed infestations. Benefits associated with introduced pasture species (including grasses) typically are far outweighed by long-term environmental costs	Northern Australia	86–93
Impacts on Freshwater and Perennial Water Resources		
Grazing can have significant impacts on quality and quantity of freshwater water resources, downstream ecological communities, and the cultural needs of local indigenous communities	The NT Gulf, GBR, and its catchments on the eastern coast of Queensland	16, 24, 25–28, 30, 64, 94, 95, 113, 118

Table 5.4 Cattle Numbers, Recommended Carrying Capacity, and Long-Term Stocking Rates in North Australia Sub-Regions

Sub-Regions[a]	Proportion of Sub-Region in North Australia (%)	Proportion of Sub-Region with Pastoral Land Use (%)	Pasture Land Use Area ($km^2 \times 1000$)	Beef Cattle Numbers[b] (× 1000)	Recommended Average Carrying Capacity (CC)[c] (hd/km^2)	Long-term Average Stocking Rate (SR), 2001–2012[d] (hd/km^2)	SR/CC	Proportion of degraded land (B-Deteriorating/C-Degraded)[e] (%)
NT–DTE (M)	100	45[f]	91	187	2	1	1	0.6/0.3
NT–VRD (M)	95	80	158	719	6	6	1	0.5/0
Qld–Gulf (M–H)	97	95	199	1,224	4	7	1.75	0.42/0.13
Qld–CYP (L)	100	85	92	125	4	4	1	0.39/0.15
Qld–Coast (L+H)	65	95	44	573	4	17	4.86	0.4/0.15
WA–East Kimberley (L)	65	75	197	992	2	3	1.5	0.35/0.1
WA–Central Kimberley (L)	72	80	76	344	2	3	1.5	0.19/0.07

[a] L, M, M–H, H denotes low, moderate or moderate-high, or high areas of pasture production potential, respectively.

[b] Bray S, et al. 'Desktop research project to provide data on liveweight and liveweight gain in the beef cattle sector in Queensland and the Northern Territory'. Final report, 2015. Department of Agriculture and Fisheries, State of Queensland, Rockhampton, Australia, 2015.

[c] Tothill JC, and Gillies C (1992). *The pasture lands of northern Australia: their condition, productivity and sustainability*. Published by the Tropical Grassland Society of Australia Inc. on behalf of Meat Research Corporation, Brisbane, Australia and expert opinion; note that these average values mask significant productivity variability.

[d] McLean I, et al. 'The Northern beef report – 2013 northern beef situation analysis'. Meat & Livestock Australia, North Sydney, NSW, 2059, 2014; note that these average values mask significant productivity variability.

[e] Tothill JC, and Gillies C (1992). *The pasture lands of northern Australia: their condition, productivity, and sustainability'*. Published by the Tropical Grassland Society of Australia, on behalf of Meat Research Corporation, Brisbane, Australia.[58]

[f] Includes only area under operational cattle enterprises; additional area, especially Arnhem Land, supports large feral stock numbers including very environmentally destructive Asian water buffalo.

Table 5.5 Net Sustainable Economic Benefits (EBIT Less Costs Associated with GHG Emissions from Cattle and Degraded lands) in North Australia Sub-Regions

Sub-Regions	Total Potential Returns from Pasture Enterprise (EBIT) ($/yr × 1000)	Cost of GHG (@$10/ tCO_2–e)/yr) ($/yr × 1000)	Loss of Production from Degraded Land (Accounting 25% for B (Deteriorating) and 50% for C (Degraded) Land Condition) ($/yr × 1000)	Total Costs – GHG Emissions and Loss of Production from Degraded Land ($/yr × 1000)	Net Sustainable Benefits (excl. Cost of Loss of ES from Degraded Land) ($/yr × 1000)	Loss of ES Due to Land Degradation in 'B' and 'C' Land Condition (Based on Tothill and Gillies 1992)[a] ($/yr × 1000)
NT–DTE (M)	1,567	–1,239	–47	–1,286	281	–85,134
NT–VRD (M)	10,972	–4,814	–137	–4,951	6,021	–61,608
Qld–Gulf (M–H)	20,019	–8,132	–3,403	–11,535	8,484	–1,056,591
Qld–CYP (L)	3,152	–1,768	–544	–2,311	841	–497,110
Qld–Coast (L+H)	8,171	–2,999	–1,430	–4,429	3,742	–240,443
WA–East Kimberley (L)	5,143	–3,282	–707	–3,989	1,154	–846,199
WA–Kimberley (L)	2,833	–1,526	–234	–1,760	1,074	–194,654
Total costs/benefits ($)/yr × 1000)	**51,857**	**–23,759**	**–6,502**	**–30,261**	**21,596**	**–2,981,740**

[a] Land degradation estimate assumed to be $156/ha/yr for lands in C condition, and $78/ha/yr for lands in B, following ELD initiative (2015). (From The Economics of Land Degradation (ELD) initiative, *The Value of Land: Prosperous Lands and Positive Rewards through Sustainable Land Management*, ELD Secretariat, Bonn, Germany, 2015. Retrieved from www.eld-initiative.org.).

yet been undertaken for North Australia savanna conditions or poses significant challenges – for example how to put monetary values on biodiversity components and ecologically critical surface and groundwater resources (see Boxes 5.2 and 5.3).

BOX 5.2 WATER, WETLANDS AND ECOSYSTEM SERVICES

JENNY DAVIS

SCHOOL OF ENVIRONMENT

CHARLES DARWIN UNIVERSITY, DARWIN

Wetland ecosystems (including lakes, rivers, marshes, estuaries, and near-shore coastal regions to a depth of 6 meters at low tide) contribute a large range of ecosystem services to human well-being.[1] Provisioning services include supplying water directly for human consumption, as well as water for the production of food, fibre, and fuel. Regulating services provided by wetlands include water purification, climate regulation, flood regulation, and coastal protection. Cultural services include the recreational and tourism opportunities associated with inland freshwaters, estuarine, and near-shore marine regions.[1] Although it is universally accepted that the provision of water is essential to human well-being, it is often not recognised that basic water infrastructure is provided by rivers, lakes, soil, plants, and trees. Investing in and maintaining the land around surface water sources and protecting groundwater resources (particularly aquifer recharge zones) will create a more water-secure future for cities and communities, in addition to generating other benefits, including conservation and protection of biodiversity.[2]

One of the most striking examples of the value of maintaining ecosystem services to support the provision of water is that of the New York City water supply and the restoration of the Catskill/Delaware watersheds.[3] In the late 1990s the quality of drinking water in New York City had declined below acceptable US Environmental Protection Agency (EPA) standards. The cause was found to be watershed (catchment) pollution from various developments, including runoff from agricultural lands and impervious urban surfaces, and discharges from wastewater treatment plants. Pollutants had reduced the ability of the watersheds to provide the ecosystem service of water purification.[3] Restoration of the watershed was estimated to cost $1 to $1.5 billion, while construction of a water filtration plant would cost $6–$8 billion with a further $300 million per year to run the plant. The city of New York saved billions of dollars by choosing the restoration option. The watershed protection program implemented to preserve and restore natural filtration services was clearly a more cost-effective means of providing good quality water than building a water treatment plant.

The major categories of source water protection activities for cities[2] can also be usefully applied to the wetland ecosystems of northern Australia (Table B5.2.1). It is also recognised that the value of source water protection goes well beyond the issue of providing water security for communities. The benefits of investing in water source protection include reduction of our carbon footprint, conserving biodiversity and maintaining critical ecological processes, climate change mitigation, and adaptation and building healthier and more resilient communities.[2]

Protecting and restoring the vegetated fringing zones of tropical wetlands and the riparian zones of tropical rivers is also of critical importance in maintaining the ecosystem services provided by the freshwater ecosystems of Australia's north. These land-water interfaces

(Continued)

BOX 5.2 (Continued) WATER, WETLANDS AND ECOSYSTEM SERVICES

Table B5.2.1 Major Categories of Source Water Protection

Source Water Protection Activity	Description
	*Targeted land protection*: Protecting targeted ecosystems, such as savannas, grasslands, or wetlands
	*Revegetation*: Restoring natural savannas, grasslands, or other habitat through planting (direct seeding) or by enabling natural regeneration
	*Riparian restoration*: Restoring natural habitat that is at the interface between land and water along the banks of a river or stream. These strips are sometimes referred to as riparian buffers.
	*Agricultural and pastoral best management practices (BMPs)*: Changing agricultural/pastoral land management to achieve multiple positive environmental outcomes
	*Fire risk management*: Applying prescribed burning early in the season to reduce grass-fuel and reduce the risk of wildfires
	*Wetland restoration and creation*: Re-establishing the hydrology, plants, and soils of former or degraded wetlands that have been drained or modified
	*Road management*: Deploying a range of avoidance and mitigation techniques that aim to reduce the environmental impacts of roads, and other development activities

Source: Abell, R. et al. (2017). *Beyond the Source: The environmental, economic, and community benefits of source water protection*, Arlington, VA: The Nature Conservancy.

influence and regulate a range of processes, including nutrient and sediment fluxes, water temperatures (shading), and organic matter (leaf litter) inputs. They also influence terrestrial biodiversity through the provision of habitats and food for many faunal species.[4] Land-water interfaces are ecotones, which share elements of both adjoining aquatic and terrestrial ecosystems and are vulnerable to threats from both directions They are especially prone to invasion by exotic species. Invasive plants can reduce plant diversity by smothering and destroying natural vegetation and reduce the availability of riparian habitats for birds and other fauna. Grazing and trampling of fringing and riparian vegetation by cattle and other introduced herbivores also degrades these habitats and reduces water quality through the associated influx of sediments and nutrients.

In the wet-dry tropics perennial river pools play an important role in sustaining populations of aquatic species. The perennial pools that persist within dry river channels after

(Continued)

BOX 5.2 (Continued) WATER, WETLANDS AND ECOSYSTEM SERVICES

monsoonal rains cease, provide important ecological refuges for many species that require permanent water.[4] These pools enable aquatic species to persist until wet season rains, and associated river flows, return. Ensuring that perennial pools are not depleted by over-extraction of groundwater or surface water resources is a major management issue. Accordingly, the determination of environmentally sustainable levels of water extraction is currently a key area of research focus in northern Australia.

REFERENCES

1. Millennium Ecosystem Assessment (MEA) (2005). *Ecosystems and Human Well-Being: Wetlands and Water Synthesis.* World Resources Institute, Washington, DC.
2. Abell R. et al. (2017). *Beyond the source: The environmental, economic, and community benefits of source water protection.* Arlington, VA: The Nature Conservancy.
3. http://ice.ucdavis.edu/node/133.
4. Boulton AJ, Brock MA, Robson BJ, et al. (2014). *Australian freshwater ecology.* UK: Wiley-Blackwell, 300.

BOX 5.3 A SOCIAL, ECONOMIC, AND ENVIRONMENTAL ASSESSMENT OF WATER RESOURCE DEVELOPMENT IN NORTHERN AUSTRALIA

NEVILLE CROSSMAN

SCHOOL OF BIOLOGICAL SCIENCES
UNIVERSITY OF ADELAIDE, ADELAIDE

Developing water resources to increase the value of agriculture is a long-held ambition for northern Australia. Agriculture in northern Australia is characterised by relatively low value beef production with scattered examples of irrigation development to produce higher value crops and feedstock for cattle (e.g. the Ord in WA and the Burdekin in QLD). The June 2015 Australian Government White Paper, *Developing Northern Australia*, outlines a series of federal government investments and reforms to encourage land and water resource development in the north. Any development will come with environmental and social trade-offs from the changes to terrestrial and freshwater ecosystems, and assessing these trade-offs is essential for determining the full range of benefits and costs from water resource development.

In 2013 the Commonwealth Scientific and Industrial Research Organisation (CSIRO) completed an assessment of water resource development opportunities in two catchments of North Queensland, the Flinders and Gilbert rivers, to add to the information base for developing northern Australia. The Flinders and Gilbert Agricultural Resource Assessment identified the soil and water resources available for new irrigated agriculture, the economic viability of irrigated agriculture and documented the ecological systems, industries, infrastructure and values associated with irrigation development. The Flinders and Gilbert Assessment estimated the utility of irrigation development based on the economic, environmental and social costs and benefits of potential irrigation development. The triple bottom line (TBL) assessment includes monetary indicators describing environment and economy as well as broader social considerations.

(Continued)

BOX 5.3 (Continued) A SOCIAL, ECONOMIC AND ENVIRONMENTAL ASSESSMENT OF WATER RESOURCE DEVELOPMENT IN NORTHERN AUSTRALIA

The TBL assessment used a Bayesian decision network (BDN) to estimate the utility of irrigation development in the Flinders and Gilbert catchments. The BDN has many properties that make it useful for performing a TBL assessment, in particular their ease of construction; their ability to handle quantitative and qualitative data types; their preservation of system knowledge and; their ease of use in aiding decision making. The ecosystem services framework was used to structure the BDN and estimate the utility of water resource development.

The TBL assessment:

1. Captured expert knowledge estimating potential impacts to aquatic and terrestrial ecosystems from water resource development.
2. Surveyed regional stakeholders on their values associated with ecosystem services, their priorities for managing ecosystem services, and their preferences for water resource development more broadly (i.e. social values).
3. Reviewed and summarised the unit economic values of ecosystem services supplied by land and water ecosystems in the region.
4. Built a BDN to integrate results from one to three above to estimate the utility of water resource development based on social, environmental and economic benefits and costs.
5. Investigated scenarios that could increase the utility of water resource development.
6. Discussed the institutional and natural resource management options that minimise tradeoffs between irrigation development and ecosystem service impacts.

From the perspective of the TBL, the total utility of water resource development is negative in both the Flinders and Gilbert catchments. The explanation for the negative utility is that the negative impact to ecosystem service values is not balanced by the positive benefits of water resource development. Water resource development, if fully realised, will result in large areas of land use change, impacting on the supply of ecosystem services from the terrestrial land system. The benefit to ecosystem services supplied by the land system will be largely neutral to negative following development.

Diversion of water for irrigation and the subsequent alterations to hydrology and freshwater and riverine ecosystems, while not necessary large overall, will impact on the crucial dry-season pool refugia, as well as potentially impacting on riparian zones and estuarine environments. Results of the ecosystem services social surveys demonstrate that the 'food production' and 'habitat for species' ecosystem services were highly valued by respondents. The estuarine (prawns and fish) and in-stream fish food-based ecosystem services and the freshwater and terrestrial habitat ecosystem services are important components of the BDN model and the medium-to-high impact to them from development, plus their high social and economic value, contribute to the negative utility.

The negative utility is further explained by the low economic value at farm scale of water resource development as demonstrated in the rigorous economic analysis of irrigation costs and benefits completed for the Flinders and Gilbert Assessment, and the high frequency of locations of relatively low land suitability for many irrigated crops. The irrigated agriculture potential estimated in the BDN model is therefore most likely to be low and the benefit for food production (crops and beef) is therefore neutral.

The overall utility of water resource development in both the Flinders and Gilbert catchments could enter positive territory in a number of scenarios: (1) water resource development

(Continued)

BOX 5.3 (Continued) A SOCIAL, ECONOMIC AND ENVIRONMENTAL ASSESSMENT OF WATER RESOURCE DEVELOPMENT IN NORTHERN AUSTRALIA

Table B5.3.1 Utility Value of Water Resource Development in the Flinders and Gilbert Catchments under Different Scenarios

	Flinders Catchment	Gilbert Catchment
Baseline	−296	−378
Scenario 1	2	−54
Scenario 2	53	32
Scenario 3	314	314

Note: Baseline: Plausible scenarios of water resource development; Scenario 1: Water resource development with high economic benefits but ecosystem impacts as per baseline; Scenario 2: No ecosystem impact from irrigation development, but economic benefits as per the baseline; Scenario 3: The economic benefits are high and no ecosystem impact.

is highly sensitive to the environment and there are no or very low environmental impacts, or (2) given an absence of environmental impacts, much higher net economic returns to irrigators eventuate, possibly through higher commodity prices, lower capital costs of water resource development or some combination of both (Table B5.3.1).

REFERENCE

1. Crossman ND, Bark RH (2013). 'Socioeconomics: triple bottom line accounting'. Technical report to the Australian Government from the CSIRO Flinders and Gilbert Agricultural Resource Assessment, part of the North Queensland Irrigated Agriculture Strategy. CSIRO Water for a Healthy *Country* and Sustainable Agriculture flagships, Australia.

For pasture condition, the total area of degraded pastoral land was assessed using data from a regional study published in 1992,[58] classifying northern pastoral lands as being in A (good), B (deteriorating), and C (degraded) states. Although somewhat dated, expert advice suggests that this condition assessment is still broadly representative – if conservative. For example, whereas the report[58] considered only small areas of the NT's VRD pastoral district to be in a degraded state (Table 5.4), more recent assessments document some areas of improvement, and continuing degradation in localised areas of the VRD[96] particularly in association with critical riparian habitat.[97]

Given established relationships that pastures in B and C condition result in reduced productivity,[4,6,16,98,99] we first estimated losses in pasture production and consequent cattle returns for both B and C areas (EBIT/year). As suggested by pastoral scientists,[5,99] we assumed that pastoral lands in B condition reduced pastoral productivity by 25% and by 50% when in C condition. Overall, losses in cattle production from pasture degradation were $1,000–$26,725 per median-sized property/year, with an average of $10,500/property/year. It is notable that land condition in our focal region is generally much better than for the broader northern Australia pastoral estate.[4,17,58]

Costs associated with greenhouse gas (GHG) emissions from cattle were determined using regional cattle numbers,[4] and region-specific cattle GHG emissions ranging from 0.023 to 0.311 tonnes of $CO_{2\text{-equivalent}}$/year/ha (t. CO_2-e/year/ha).[100] The average annual costs for most properties in QLD and the NT ranged between $100,000 and $150,000/median-sized property, and $173,000/property in the Western Australia (WA) Kimberley region.

Taking into account environmental costs associated with lost pastoral production from pasture degradation, and GHG emissions, we can derive a conservative, indicative estimate of the net economic returns of typical pastoral properties in our focal region (Figure 5.7). For the period 2001–2012,

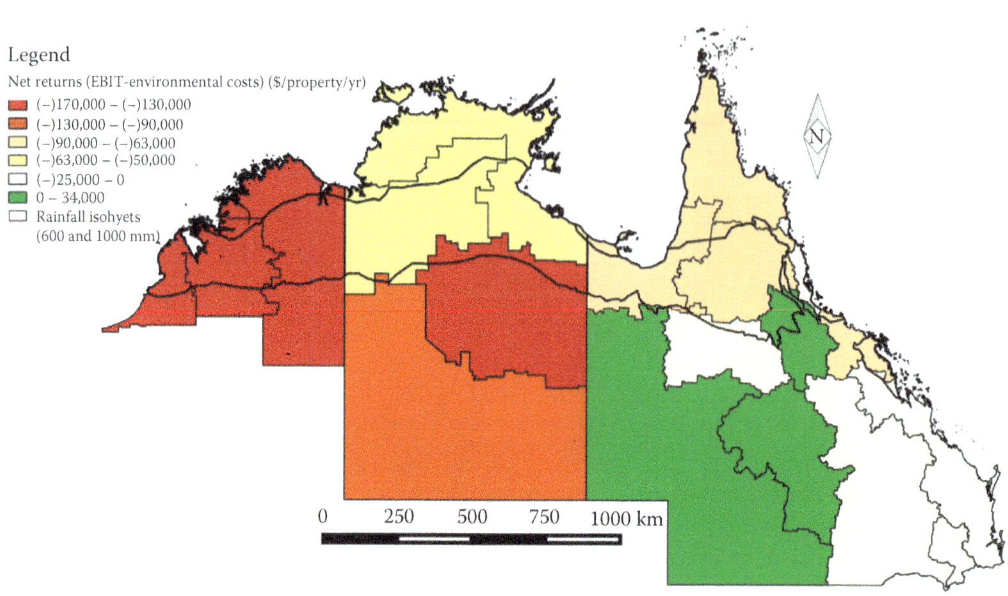

Figure 5.7 'Net annual returns (EBIT) per median-sized North Australia property, adjusted for costs of losses in production due to land degradation and GHG emissions'. (From McLean, I. et al., *The Northern Beef Report – 2013 Northern Beef Situation Analysis*. Meat & Livestock Australia Ltd., North Sydney, NSW, 2059, 2014; Bray, S. et al., 'Desktop research project to provide data on liveweight and liveweight gain in the beef cattle sector in Queensland and the Northern Territory'. *Final report, 2015*. Department of Agriculture and Fisheries, State of Queensland, Rockhampton, Australia, 2015; Eady, S. J. et al., 'Down scaling to regional assessment of greenhouse gas emissions to enable consistency in accounting for emissions reduction projects and national inventory accounts for northern beef production in Australia', *Rangeland J.*, 38, 219–28, 2016 and others.)

net returns for median-sized properties were highly negative both for EBIT (and especially EAIBT), for example: –$70,000/year for Top End NT, –$89,000/year for Cape York, –$133,000 for Kimberley (WA).

Applying a similar approach to that described earlier, we present a complementary assessment of net pastoral enterprise sustainability for seven sub-regions (principally following ABARES statistical boundaries) within our focal region, based on identified areas of low, moderate, and moderate-high carrying capacity (Figure 5.4c). Stocking rates (SR) in five sub-regions were observed to exceed recommended carrying capacity (CC; Table 5.4). Considering only areas of B and C land condition,[56] net benefits (EBIT less environmental costs) for our entire North Australia focal region based on available data for the period 2001–2012 are estimated to amount to just $21 million/year (Table 5.5), before any consideration of interest repayments (EABIT). Although there is no simple robust measure linking overstocking (high SR relative to CC) with land condition class, we can anticipate that associated environmental costs will continue to accumulate with time.

Land degradation also results in significant loss of ecosystem services (Tables 5.3 and 5.5), in addition to those we have costed conservatively above. A recent international assessment of the loss of ecosystem services from degraded lands estimated an average annual indicative cost of $156/ha/year.[101] When applied at a full rate to degraded (land condition C) lands, and a 50% discount to partly degraded (land condition B) lands, in our focal region, annual losses of associated ecosystem services amount to $3 billion/year (Table 5.5).

By any measure, under the current market forces scenario, the North Australia pastoral industry is generally unsustainable. However, as demonstrated by 'exceptionally performing' pastoral enterprises, and especially those which can and have implemented adaptive, seasonally conservative herd management practices,[102–105] enhanced sustainability is attainable.

5.4 TOWARDS THE DEVELOPMENT OF A SUSTAINABLE LAND SECTOR ECONOMY IN NORTH AUSTRALIA

The previous analyses reinforce a number of recent informed assessments concerning the poor financial and environmental performance and sustainability of much of the northern beef industry.[5,11,12,106] These conditions are reinforced further by the most recent authoritative industry assessment for the years 2004–2016, which only came available after the analyses presented here were completed.* Policy imperatives which flow from this assessment include: firstly, recognition that fundamental land sector change is required (in stark contrast with the 'business-as-usual' mantra of current north Australia development policy[52]); and consequently, recognition that more culturally, economically, and environmentally sustainable regional options need to be supported and developed. In practical terms, a key regional sustainable development challenge is to embrace a great transition pathway to a more diversified ecosystem services-based land sector economy.

A first issue to appreciate is that North Australia, far from being a landscape endowed with extensive high pastoral potential (unlike the Barkly and Queensland's Mitchell grasslands), instead supports very significant cultural (Chapter 2), biodiversity conservation,[107] and global carbon stock values[108] which contribute significantly to the socioeconomic well-being of local and regional communities. The North is dominated by lands of international conservation significance[107] (Figure 5.8a): 76% of 0.5°cells ($n = 573$) encompass lands where at least half the cell area is either managed currently for conservation purposes (e.g. National Parks, Indigenous Protected Areas (IPAs), savanna burning project areas), or comprise high conservation value riparian areas (Box 5.2) and biodiversity-rich, topographically rough, hilly terrain[109–111]).

The relatively little structurally modified (by global standards) North Australia landscape resource supports a thriving nature-based tourism industry, anticipated to be worth $2.8 billion in 2018.[112] That industry is based particularly on maintaining intact freshwater systems and associated recreational fishing opportunities.[112,113] Maintaining the integrity and quality of flow regimes is vital also for the productivity of the northern commercial fishing industry,[112] the multibillion dollar tourism industry based on the Great Barrier Reef resource,[27,114–117] and very significant cultural, livelihood, and economic dependencies and aspirations of Indigenous people.[118,119]

In the North Australia focal region, the formal conservation estate (e.g. National Parks, IPAs) occupies 400,000 km^2, and savanna burning projects separately 331,000 km^2 (as at January 2017) (Figure 5.8 a, b), with a combined net area of 595,030 km^2, or 50% of the total region. Both these estates deliver many economic, social, and cultural values to local people. For example, it is estimated that at least 650 Indigenous FTE (full-time equivalent) positions currently are available (mostly through Commonwealth commitments to Working on *Country* and IPA funding programs) to service the North Australia conservation land sector economy (Box 4.5); contrasting with about 390 Indigenous positions currently available through the pastoral industry.* Market-based ecosystem services enterprises offer significant opportunities for growing Indigenous employment opportunities, and concomitantly reducing public expenditures,[120–123] in a diversified North Australia land sector economy (Chapter 4). As addressed also in the following, a key challenge is to reframe current government Indigenous conservation and employment program expenditures as investments for building local Indigenous land sector-based enterprises.

The rapid expansion in market-based savanna burning GHG emissions abatement projects since the introduction of the Australian Government's Carbon Farming Initiative in 2012, and its current iteration as the publicly funded emissions reduction fund, points to the potential both for ongoing expansion in savanna burning projects and the ecosystem services sector generally. Notably, the distribution of current (Figure 5.8b) and prospective (Figure 5.8c) savanna burning projects overlaps markedly with areas identified as having high conservation significance (Figure 5.8a) but has limited overlap with areas identified as supporting (relatively) high pastoral carrying capacity (Figure 5.4c). Savanna burning projects

* See Holmes P, McLean I, Banks, R (2017). *The Australian Beef Report.* (Bush AgriBusiness Pty Ltd: Brisbane.)

can be undertaken profitably also on pastoral properties with eligible fuel types (Figure 5.8d), especially in more fire-prone, pastorally marginal situations,[124] whereas there are genuine concerns about the application of prescribed, relatively low intensity, early dry season fires on pastorally productive grasslands which can, in combination with grazing, result in undesirable woody invasion and thickening.[125–127]

Using the Savanna Burning Abatement Tool (SavBAT),[128] we estimate that an annual average of 7.5 million tonnes (Mt) CO_2-e of accountable GHGs were emitted from savanna fires over the decadal period 2000–2009. Assuming that well-implemented savanna burning projects typically can reduce GHG emissions by at least one third,[129,130] collectively such projects could realize $20–30 million/year based on a currently conservative average carbon price of $10–15 t. CO_2-e. Over the next few years it is likely that additional carbon sequestration opportunities will become available for savanna burning projects through development of methodologies accounting for storage of carbon in woody debris and in living trees. Initial estimates suggest that, for every carbon credit (1 t. CO_2-e) generated by savanna burning emissions abatement projects, complementary sequestration methods could provide an additional three credits for woody debris sequestration,[131] and at least another five credits for sequestration in living trees.[132]

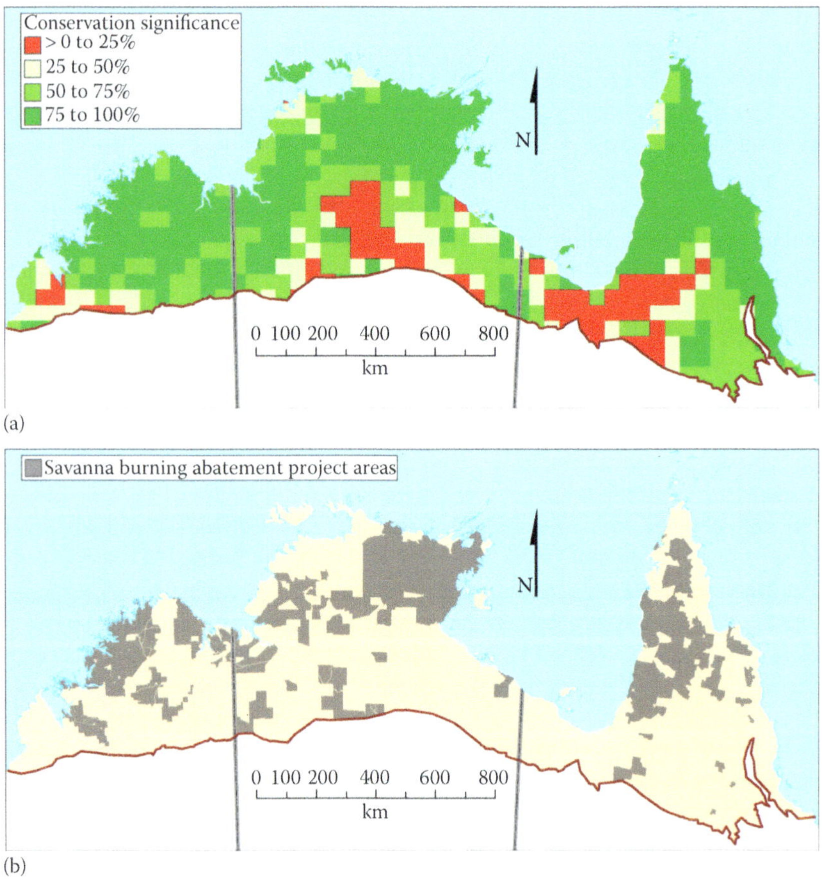

Figure 5.8 Towards a more sustainable land sector economy: (a) Conservation significance, mapped as proportion of 0.5° cells comprising all lands managed for conservation purposes, buffered riparian areas, high topographic roughness – refer Chapter 2 for details; (b) registered 79 savanna burning GHG emissions abatement projects (total area 331,000 km²), as at January 2017. (http://www.cleanenergyregulator.gov.au/ERF/project-and-contracts-registers/carbon-abatement-contract-register.)

(Continued)

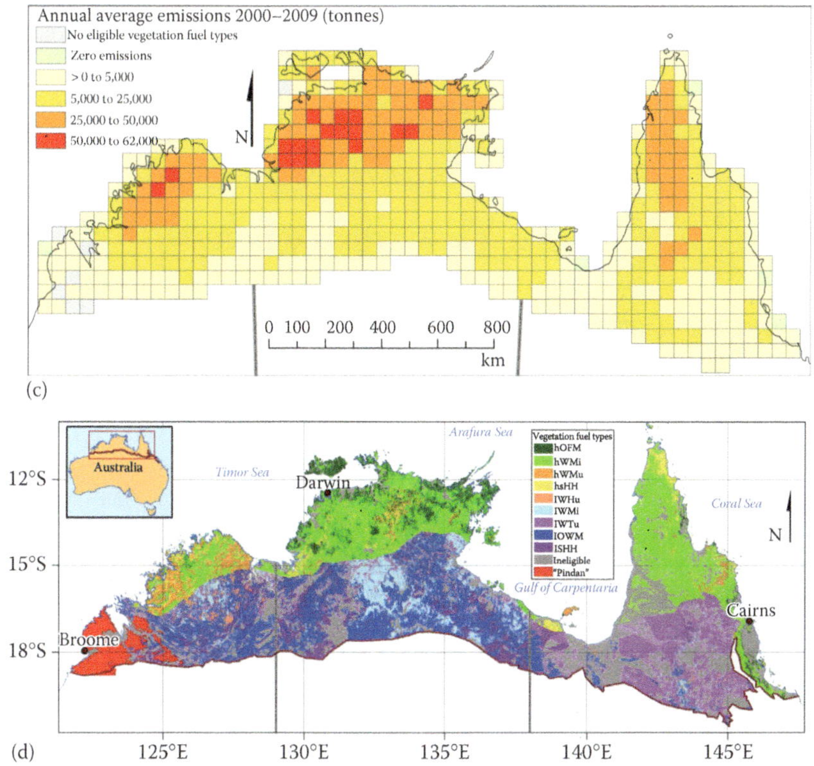

Figure 5.8 (Continued) Towards a more sustainable land sector economy: (c) mean accountable greenhouse gas (GHG) emissions (methane, nitrous oxide) from savanna fires, 2000–2009, summarized per 0.5°cells – based on methodology as at 10 April 2015 (http://www.comlaw.gov.au/Details/F2015L00344.), and following calculation procedures as given in Russell-Smith et al. (2015: Chapter 10[129]); (d) vegetation fuel types: hOFM – open forest with mixed grass, hWMi – Woodland with mixed grass, hWHu/IWHu – Woodland with hummock grass, hSHH – Shrubland (heath) with hummock grass, IWMi – Woodland with mixed tussock/hummock grass, IWTu – Woodland with tussock grass, IOWM – Open Woodland with mixed grass, and ISHH – Shrubland with hummock grass. (From https://data.qld.gov.au/dataset/the-vegetation-of-the-australian-tropical-savannas.)

One evident risk to this scenario is the potential spread of highly flammable, high fuel-load exotic pasture grasses, which are relatively difficult to manage and contain (Box 5.4).

The emergence of the savanna burning industry highlights opportunities for developing further innovative ecosystem services markets and activities based on North Australia's natural and cultural assets. For the pastoral industry, already we are seeing a rapid expansion of carbon market opportunities, including reducing livestock emissions; increasing efficiency of fertiliser use; enhancing carbon in agricultural soils; and sequestering carbon through revegetation and reforestation.[133] It has been suggested also that carbon markets could assist with developing more sustainable grazing practices through incentivising both the restoration and maintenance of productive pastures*. These sorts of affirmative ideas sit at the heart of progressive thinking in the northern pastoral industry (for example, auditable grazing best management practice standards as established by the Fitzroy Basin Association's market chain accreditation scheme[134]). However, the market potential of such systems still has to be developed.

Given the international conservation significance of North Australia, practical incentives are needed for sustainable land management practices on public (typically pastoral leasehold) and

* Steven Bray, personal communication, 2016.

BOX 5.4 HIGH BIOMASS GRASSES THREATEN THE SAVANNA LANDSCAPE

SAMANTHA SETTERFIELD
SCHOOL OF BIOLOGICAL SCIENCES
UNIVERSITY OF WESTERN AUSTRALIA, PERTH

NATALIE ROSSITER-RACHOR
RESEARCH INSTITUTE FOR THE ENVIRONMENT & LIVELIHOODS
CHARLES DARWIN UNIVERSITY, DARWIN

The biodiversity and function of Australia's tropical savannas and wetlands are under threat from a range of high biomass, invasive tropical grasses. Significant areas of invasion occur across northern Australia, particularly in the Top End region of the Northern Territory and Cape York Peninsula, Queensland. Species of particularly high threat are gamba grass (*Andropogon gayanus*), annual mission grass (*Cenchrus pedicellatum*), perennial mission grass (*Cenchrus polystachion*), grader grass (*Themeda quadrivalvis*), and Guinea grass (*Megathyrsus maximus*). These species were introduced primarily for assessment and/or use for pastoral production. They were selected for their persistence under harsh conditions and their higher growth rates and nutritional value compared to native grasses. However, the characteristics were selected to try and find successful pasture plants resulted in making them successful weeds.

Of the suite of invasive grasses, gamba grass poses the greatest threat to northern Australia's savannas. Two varieties of *A. gayanus* were introduced in the 1930 and '40s, subsequently cultivated into the cultivar 'kent' with commercial seed supply available in 1983.[1] Large-scale plantings commenced in the 1980s, but the rate of invasion has been explosive,[2] and it now covers 15,000 km^2 of the NT, with the potential to invade 380,000 km^2 in the NT alone in addition to other large areas in Queensland and Western Australia.[1] The rapid invasion and the significant threat for conservation, aboriginal, pastoral, mining, and defence land users is reflected by its status as one of Australia's weeds of national significance and a key threatening process under the Commonwealth's Environment Protection and Biodiversity Conservation Act.

Gamba grass invasion is considered very high risk across northern Australia because of the resultant changes to fuel beds and fire behaviour.[3] Invasion results in a tall (~4 m), dense fuel bed of up to 25–30 t ha^{-2} that replaces the shorter (<0.5 m) native grass fuel bed (up to 6 t ha^{-2}) savanna communities.[3–5] As a result, fire intensity (the product of the available heat of combustion per unit of ground area and the forward spread of the fire, measured in kilo or megawatts per metre) increases significantly, from typically 1–3 MW m^{-2} in native grass fires to 16 MW m^{-2} in gamba fuelled fires in the early dry season (Figures B5.4.1 through B5.4.3).[5]

The flame height is greater in the invaded sites resulting in more fire damage to the tree and shrub layers.[5,6] If invaded sites are not burnt in one fire season, the fire fuel load increases resulting in more intense, high combustion fires in the following year.[3,5] Plant species diversity is substantially reduced after only a few gamba fuelled fires, and at some sites where we have undertaken repeated vegetation surveys over the past decade, we have followed the vegetation decline from healthy savanna to having only a few unhealthy woody plants remaining (Figure B5.4.4). This represents a major reduction in above-ground savanna carbon store, and therefore a financial risk to landholders engaging in savanna burning projects.[7] This is particularly important given that 75% of the eligible area for savanna burning is spatially coincident with the high suitability range for gamba grass.

(*Continued*)

BOX 5.4 (Continued) HIGH BIOMASS GRASSES THREATEN THE SAVANNA LANDSCAPE

Figure B5.4.1 Dense stands of gamba grass in the Northern Territory. The grass now poses a serious risk to savanna vegetation as it transforms key elements of the ecosystem. (Courtesy of Samantha Setterfield.)

Figure B5.4.2 Dense stands of gamba grass in the Northern Territory. The grass now poses a serious risk to savanna vegetation as it transforms key elements of the ecosystem. (Courtesy of Michael Douglas.)

(Continued)

BOX 5.4 (Continued) HIGH BIOMASS GRASSES THREATEN THE SAVANNA LANDSCAPE

Figure B5.4.3 Gamba grass invasion leads to much hotter fires in the savannas. (Courtesy of Samantha Setterfield.)

Figure B5.4.4 Gamba grass invasion can lead to dramatic declines in tree canopy. (Courtesy of Samantha Setterfield.)

High intensity gamba grass fires have substantial impact on fire management practices and costs. Historically fires were low intensity native grass fires, which occurred in sparsely populated areas, and which could be managed using minimal firefighting equipment. Gamba grass invasion has resulted in hotter fires occurring in residential areas, requiring helicopters and water bombing planes to effectively protect people's lives and properties. Fire management costs have increased

(Continued)

BOX 5.4 (Continued) HIGH BIOMASS GRASSES THREATEN THE SAVANNA LANDSCAPE

none-fold in 10 years in a region of the NT due primarily to gamba grass invasion.[3] Further economic impacts of gamba grass fires are evident each year with damage to dwellings and infrastructure, and the cost of weed management activities to the landholders and government.

Gamba grass presents a long-term threat to Australia's savannas and its biodiversity. Carbon abatement programmes are also at risk. However, many of the infestations on properties are currently small and not financially onerous to eradicate. In addition, current gamba infestations total only 2% of the current potential range. Therefore, immediate containment of the existing gamba infestations would be a significant step in mitigating the risk posed by gamba grass to Australia's iconic savanna region.

REFERENCES

1. Northern Territory Government (2010). 'Weed management plan for *Andropogon gayanus* (gamba grass)'. Northern territory Government, Darwin.
2. Petty AM, Setterfield SA, Ferdinands KB, Barrow P (2012). 'Inferring habitat suitability and spread patterns from large-scale distributions of an exotic invasive pasture grass in north Australia'. *Journal of Applied Ecology* 49, 742–52.
3. Setterfield SA, Rossiter-Rachor NA, Douglas MM, et al. (2013). 'Adding fuel to the fire: the impacts of non-native grass invasion on fire management at a regional scale'. *PLoS One* 8, e59144. doi:10.1371/journal.pone.0059144.
4. Rossiter NA, Setterfield SA, Douglas MM, Hutley LB (2003). 'Testing the grass-fire cycle: alien grass invasion in the tropical savannas of northern Australia'. *Diversity and Distributions* 9, 169–76.
5. Setterfield SA, Rossiter-Rachor NA, Hutley LB, Douglas MM, Williams RJ (2010). 'Turning up the heat: the impacts of *Andropogon gayanus* (gamba grass) invasion on fire behaviour in northern Australian savannas'. *Diversity and Distributions* 16, 854–61.
6. Brooks KJ, Setterfield SA, Douglas MM (2010). 'Exotic grass invasions: applying a conceptual framework to the dynamics of degradation and restoration in Australia's tropical savannas'. *Restoration Ecology* 18, 188–97.
7. Adams VA, Setterfield SA (2013). 'Estimating the financial risks of *Androgon gayanus* to greenhouse gas abatement projects in northern Australia'. *Environmental Research Letters* 8, 025018.

Indigenous lands, especially the maintenance and restoration of critical aquatic systems.[29] While regulatory requirements and approaches evidently need to be better and more forcefully applied by relevant administering institutions, equally, incentivized approaches need to be encouraged (for example, subsidies for riparian fencing and alternative watering point infrastructure; active promotion of market-based sustainable pastoral enterprise accreditation schemes). A land stewardship case for enhancing biodiversity conservation outcomes has been suggested through the application of carbon market-based fire and grazing management.[135]

For regional Indigenous communities and organisations, land sector enterprise opportunities include a variety of contractual services besides the undertaking of current government-funded land and sea natural resource management activities – for example, assuming prime responsibility for remote community emergency management, from prevention to response capabilities[136]; cultural site management; infrastructure maintenance; road works management; degraded land and mine site rehabilitation (Box 4.3); quarantine and surveillance activities; commercial harvest activities; culture-based ecotourism; and industry-funded environmental offset projects that are compatible with community values and needs.[137] In short, any activity that helps develop independent business capability, social capital (Box 5.5), and an extensive

BOX 5.5 SOCIAL CAPITAL AND THE CREATION OF AN INNOVATIVE ENVIRONMENTAL AND CULTURAL ENTERPRISE IN ARNHEM LAND

PETER COOKE

PETER.COOKE@OZEMAIL.COM.AU; NIGHTCLIFFE, NORTHERN TERRITORY

It takes more than one kind of capital to establish an innovative and successful not-for-profit enterprise on Aboriginal land. The genesis of the West Arnhem Land Fire Abatement (WALFA) project demonstrates this point.

For all enterprises, start-up financial capital is needed to pay the bills until an adequate income stream flows. But for WALFA, the first and most critical step was creating social capital to underpin a vigorous collaboration between Aboriginal rangers, communities of traditional owners and governments, industry, and the science community.

It took almost ten years of project development to reach the point in 2006 where a deal brokered by the Northern Territory Government with the operators of the Darwin Liquefied Natural Gas (DLNG) plant provided funds to allow WALFA to go fully operational. In 2006 the DNLG plant owners, energy giant Conoco Phillips, signed up to provide 22 years of funding contingent on WALFA abating a minimum of 100,000 tonnes of CO_2-e emissions annually from an area of 28,000 km^2 in western and central Arnhem Land.

Funding started with \$1.3 million per year (subject to annual indexing) supporting fire abatement work by five Aboriginal ranger groups – Manwurrk (later Warddeken), Djelk, Adjumarlarl, Mimal, and Jawoyn. Since 2013, WALFA has expanded to include most of Arnhem Land under an Aboriginal-owned umbrella entity. With additional income generated from access to the Australian Government's Emissions Reduction Fund, this entity confidently expects a minimum turnover for the 2016 calendar year of ~\$10 million.

But without implementation of a considered and strategic approach to raising social capital, this social and environmental success would not have happened.

Social capital is defined by theoreticians in various ways but the concise description from an Organisation for Economic Co-operation and Development (OECD) publication will suffice in telling the WALFA story: 'Social capital provides the glue which facilitates co-operation, exchange, and innovation'.[1]

In 1997, the fire regime prevailing in western and central Arnhem Land was both a serious local problem and a contributor to the global problems of increasing greenhouse gas (GHG) emissions. During times of assimilation policies, most Aboriginal land managing groups, who for millennia had maintained fire regimes typified by many small and cool fires in the early dry season, had been drawn into government settlements and missions and further afield away from their clan estates. Without this very effective management approach, fire regimes changed to being dominated by late dry season fires, often massive in extent, fierce in intensity and increasingly frequent. Managed fire became feral fire.

Increasing late dry season fires were pushing a fire sensitive cohort within plant communities towards local extinctions and creating big changes in the composition of those plant communities, with negative flow-on effects to animal species.

Much of this was happening out of sight of human observation in a very depopulated, large landscape. Following plant surveys relying on helicopters to access remote, rugged, and trackless areas in the 1980s, alarm bells began to toll amongst conservation scientists. A leadership of emerging Aboriginal ranger groups employing 'two-toolbox' management focused on reinvigorating

(Continued)

BOX 5.5 (Continued) SOCIAL CAPITAL AND THE CREATION OF AN INNOVATIVE ENVIRONMENTAL AND CULTURAL ENTERPRISE IN ARNHEM LAND

customary management and indigenous knowledge but open to using non-Aboriginal management tools, also began to notice the effects of bad fire (for example, declining emu numbers).

These two groups began to talk together and build shared concerns through the cross-sectoral interactions of the Cooperative Research Centre for Tropical Savannas. The CRC's focus on the objective of sustainably healthy landscapes resonated with the aspirations of the ranger group leadership, and the recognition within the CRC of the Aboriginal domain as a sector to be treated equally with mainstream sectors of pastoralism, mining, conservation and tourism began to increase dignity and respect in cross-cultural conversations.

The individuals involved comprised a core group looking for solutions to the commonly perceived problems. But each faced cultural problems in extending engagement sufficiently broadly in Aboriginal and non-Aboriginal society to create the critical collaborative mass needed for effective action.

Bringing a collaboration home to Arnhem Land posed a cultural credibility problem for Ranger leaders. As in many colonised societies around the world, being seen to be 'acting like a white man', or joining a non-Aboriginal agenda, was regarded as conduct unbecoming. In central Arnhem Land the effects of a government-sponsored forestry project that tried to suppress Aboriginal traditional burning was still remembered somewhat bitterly. In south-central Arnhem Land roadside signs erected by the Bushfires Council threatened $1000 fines or six months gaol for fire lighting. The ill-conceived forestry project policies and the manifest non-Aboriginal view that fire was an enemy, expressed in authoritarian signage, were in fact significant drivers of the rise of the catastrophic late dry season fire regimes.

Aboriginal fire management was widely viewed by non-Aborigines as anarchic pyromania, contrasting with the Aboriginal view expressed by Aboriginal Ranger leader Dean Yibarbuk, who said: 'The secret of fire in our traditional knowledge is that it is a thing that brings the land alive again. When we do burning the whole land comes alive again – it is reborn'.[2]

Thus in 1997 the inchoate cross-cultural collaboration focused on bringing back good fire management faced a severe constraint of being overdrawn at the social capital bank.

With funding from the Natural Heritage Trust, the ranger leadership and their non-Aboriginal friends set out to engage with the most respected elders of the region – not necessarily the community councillors or public figures, but ceremonial and cultural leaders mostly of a generation who were born in the bush, swaddled in paperbark, and who grew up in times before fire management was broken down by colonisation.

The middle-aged Ranger leaders brokered friendships and intellectual relationships between Aboriginal and non-Aboriginal 'experts'. They assisted Elders to understand what the scientists' fire history maps were saying – confirming the truth that unfriendly fire was killing *country*. Understanding and trust was being built effectively at large regional meetings over years.

Black and white worked together in the 'big laboratory' of Arnhem Land to create the science of fire, fuel loads, and vegetation communities that led to the accepted savanna burning methodologies (Figure B5.5.1).

Science colleagues worked to create an understanding amongst politicians, bureaucrats and broader science community that Aboriginal customary burning was good for *country* and could tame runaway greenhouse gas emissions. The strength of this collaboration and the evident social

(Continued)

BOX 5.5 (Continued) SOCIAL CAPITAL AND THE CREATION OF AN INNOVATIVE ENVIRONMENTAL AND CULTURAL ENTERPRISE IN ARNHEM LAND

Figure B5.5.1 Fire management planning in WALFA with the aid of satellite-derived fire maps.

Figure B5.5.2 Annual Arnhem Land fire planning meeting 2016, Barrapunta.

capital attendant with it brought on board the federal and Northern Territory environment departments. With their help the DNLG/Conoco Phillips deal was struck in 2006, enabling WALFA to go fully operational. WALFA quickly began to demonstrate emissions abatements well in excess of the 100,000 tonne CO_2-e minimum requirement. WALFA's success has enabled expansion of further viable emissions projects through Arnhem Land, the collaboratively developed science informs emissions accounting nationally and internationally, and importantly real jobs have been created that make for healthy landscapes and healthy people (Figure B5.5.2).

(Continued)

BOX 5.5 (Continued) SOCIAL CAPITAL AND THE CREATION OF AN INNOVATIVE ENVIRONMENTAL AND CULTURAL ENTERPRISE IN ARNHEM LAND

Developing Aboriginal governance capacity has put corporate control firmly in Aboriginal hands; the directors of the umbrella public company ALFA NT Ltd are all elected Aboriginal men and women, as are the directors of the corporations in which the ranger groups are embedded.

CONCLUDING REMARKS

The WALFA story is not a 'model' for development – all development situations must be addressed according to their particular circumstances. But WALFA's success does point to what can be achieved by incorporating the creation and maintenance of social capital as a key element in participatory processes to develop Aboriginal enterprises that resonate with a dynamic Aboriginal culture. Some important elements of social capital,[3] and evident from the WALFA story, are highlighted in the following.

From WALFA 'new networks' have been created – across communities and across cultures and knowledge systems. Many people have been motivated to participate proactively in these networks. Ranger groups' use of indigenous and non-Indigenous 'toolboxes' to manage land has become a new norm. A default Indigenous reluctance to collaborate with the dominant colonising culture has been diminished when benefits of doing so can be demonstrated and trust has been established. Aboriginal land management systems, which have been rightly termed an 'enduring commons', have been renewed, strengthened, and modified to respond to changing social and environmental circumstance.

But more remains to be done. The WALFA experience also points to the need for more attention to be given to the Northern Territory's Aboriginal Land Rights Act (ALRA) so that it can be made more appropriately responsive to and supportive of community-based enterprises like WALFA. Community-driven not-for-profit enterprises are not well served by the formal systems for approvals under the ALRA. Many Aboriginal people and their advocates believe that land council processes favour non-Indigenous enterprise proposals over local not-for-profit proposals. They believe the bar for securing landowner approvals is set higher and made more onerous for local initiatives than for outside interests. Projects suffer long waiting times to be processed.

Without stepping outside the requirements of the ALRA, there is much that can be done by land councils to be a more supportive force for local Aboriginal development initiatives. This is a critical next step in adding value to and expanding the benefits of sustainable enterprise development underpinned by a valuing of social capital.

REFERENCES

1. OECD (2001). The New Economy: beyond the hype/final report on the OECD project/meeting of the OECD council at ministerial level. United Nations, Geneva.
2. Yibarbuk DM (1998). Notes on traditional use of fire on upper Cadell River. In *Burning Issues: emerging environmental issues for Indigenous peoples in northern Australia* (Ed. M Langton), pp. 1–6. Centre for Indigenous Natural and Cultural Management, Northern Territory University, Darwin.
3. Bullen P, Onyx J (1998). Overview of a study 'measuring social capital in five communities in New South Wales: a practitioners guide'. CACOM Working Paper Series (No 41).

range of broader health, education, well-being benefits that, in turn, can substantially reduce the reliance of remote communities on state expenditures.[120–123] In this regard, it is self-evident that current government resource management funding arrangements need to be expanded from sole or primary dependence on public 'green welfare' funding to help build social capital and business capability in local Indigenous organisations. As discussed at length in other chapters, such capability building requires necessary development of typically challenging, culturally appropriate, robust governance arrangements.

So where to for the North Australian pastoral industry, and, given the context of this book, Indigenous interests in land? First, it is evident that there are examples of profitable, adaptive northern pastoral enterprises,[4,102] including those under Indigenous ownership (Box 5.6). In general, however, the industry faces significant sustainability, economic viability, and credibility challenges. Given the poor pasture capability and associated business conditions confronting most of the North, to be

BOX 5.6 INDIGENOUS PASTORAL CASE STUDIES, NORTH QUEENSLAND

RICKY ARCHER

NORTH AUSTRALIAN LAND & SEA MANAGEMENT ALLIANCE LTD. DARWIN

DELTA DOWNS STATION

Delta Downs in a successful Aboriginal owned and managed cattle property in the Gulf of Carpentaria. It is situated on the traditional lands of the Kurtijar people. The property is managed along with two adjoining properties, Maggieville Station and Karumba Downs. The total area of these properties is approximately 390,000 ha, running a herd of 40,000–45,000 Brahman cattle.

Governance: The Kurtijar people have developed a unique governance framework which enables the business to achieve its outcomes (Figure B5.6.1). The framework consists of three components as outlined in the following.

NORMANBY STATION

The Balnggarrawarra Rangers are a newly established Indigenous Ranger group based at Normanby Station, west of Cooktown. Purchased in 1995, Normanby Station is a long-standing pastoral lease held by Normanby Aboriginal Corporation. The cattle station was purchased for the traditional owners of the area with the intention of increasing economic development opportunities and employment.

Fast forward 22 years: the cattle property has not provided sufficient economic return to meet the aspirations of traditional owners. An opportunity was then identified to establish an Indigenous Ranger group, which would better meet aspirational targets and support core values of countrymen (Figure B5.6.2).

The Balnggarrawarra Rangers were established in 2016 and are funded under the Queensland Government's Indigenous Land & Sea Management Program. Through strategic partnerships, Balnggarrawarra manage invasive species, fire, water, and cultural heritage. As a result of these activities they have been able to improve infrastructure on *country*, and planning to redevelop the cattle business in a bid towards self-sustainability. This also enables future environmental and cultural targets to be achieved.

(Continued)

BOX 5.6 (Continued) INDIGENOUS PASTORAL CASE STUDIES, NORTH QUEENSLAND

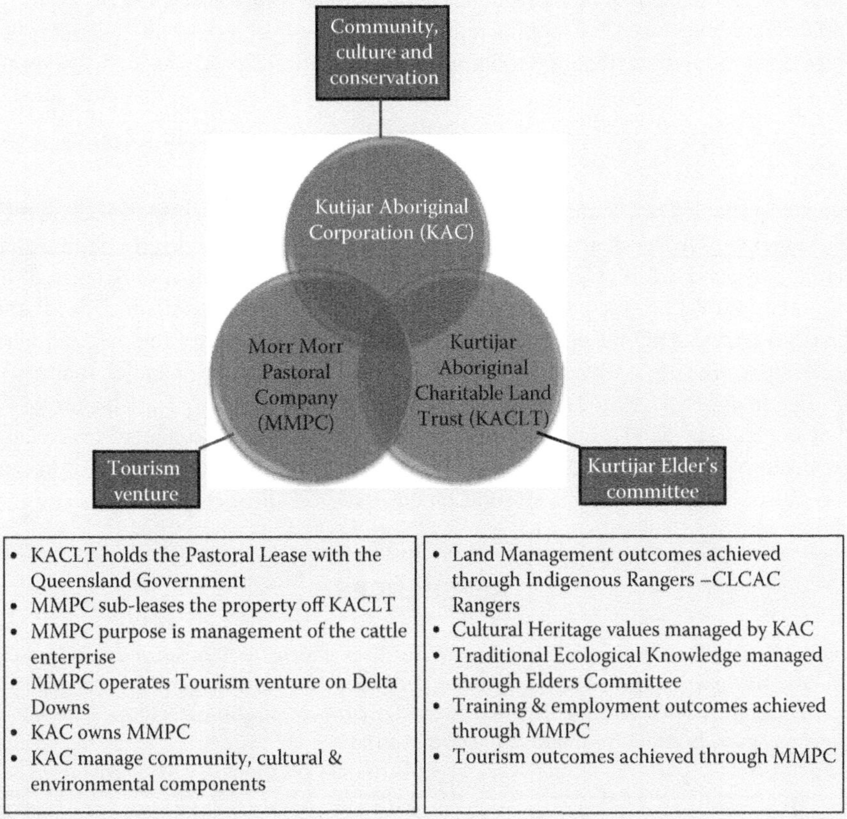

Figure B5.6.1 Concept by Joseph Rainbow, traditional owner and Morr Morr Pastoral Company Board member.

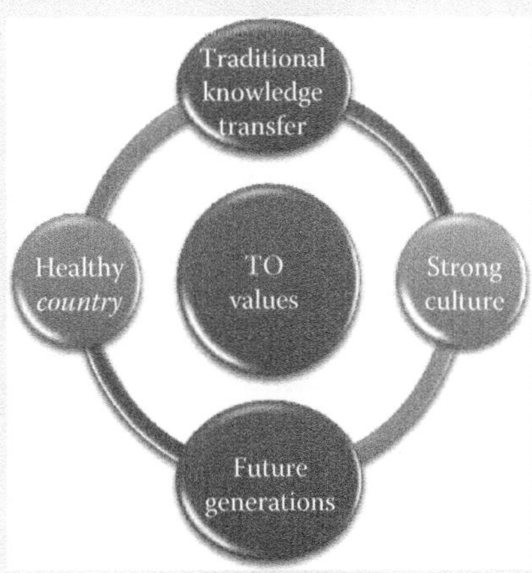

Figure B5.6.2 Concept by Vincent Harrigan, Coordinator, Balnggarrawarra Rangers.

viable and sustainable the industry evidently needs to either substantially improve core business efficiency and productivity where financial resources permit, or adapt and transition to embrace diversified opportunities including, where appropriate, carbon, biodiversity conservation, and culture- and nature-based tourism markets. For Indigenous people, sustainability also includes ensuring cultural responsibilities and obligations for looking after *country*. These are not small undertakings and able and resolute assistance will be required to inform these transitional processes in the years ahead.

ACKNOWLEDGEMENTS

We sincerely thank Dr Dionne Walsh, NT Department of Primary Industries and Fisheries, and Dr Steven Bray, QLD Department of Agriculture and Fisheries, for providing expert advice and research datasets; Dr John Carter and Grant Stone, QLD Department of Science, Information Technology and Innovation, for providing shape files of pasture datasets from Tothill and Gillies (1992); Javi Navarro, CSIRO for providing the shape files of productivity regions; and Ian McLean (Director, Bush Agribusiness Pty. Ltd.) for offering guidance during financial analyses. Without their help, the analyses presented in this chapter were not achievable. The chapter was expertly reviewed by Ian McLean and Dr Dionne Walsh (see earlier), and Professor Peter Whitehead (Darwin Centre for Bushfire Research and NAILSMA). Any errors that remain, including of interpretation, are ours. We also thank all authors who contributed boxes (5.1 through 5.6).

REFERENCES

1. NALaWTF (Northern Australia Land and Water Task Force) (2009) *Sustainable Development of Northern Australia*. A report to government from the Northern Australia Land and Water Task Force. December 2009. Australian Government Department of Infrastructure, Transport, Regional Development and Local Government, Canberra, Australia.
2. Meat and Livestock Australia (2017) Fast facts: Australia's beef industry. Retrieved from http://www.mla.com.au/search/?q=fast+facts.
3. ABS (Australian Bureau of Statistics) (2016) *Census 2016*. ABS, Canberra. Retrieved from http://www.abs.gov.au/census, accessed from September 2016–May 2017.
4. McLean I, Holmes P, Counsell D (2014) *The Northern Beef Report – 2013 Northern Beef Situation Analysis*. Meat & Livestock Australia, North Sydney, Australia, 2059.
5. Bray S, Walsh D, Hoffmann M, Henry B, Eady S, Collier C, Pettit C, Navarro J, Corbet D (2015) *Desktop Research Project to Provide Data on Liveweight and Liveweight Gain in the Beef Cattle Sector in Queensland and the Northern Territory. Final Report, 2015*. Department of Agriculture and Fisheries, State of Queensland, Rockhampton, Australia.
6. Stone G, Bruget D, Carter J, Hassett R, McKeon G, Rayner D (2007) Land: Pasture production and condition. In *'State of the Environment Queensland 2007.'* Environmental Protection Agency, Queensland Government, Brisbane, Australia.
7. SoEC (State of the Environment Committee) (2011) *Australia State of the Environment 2011*. Independent report to the Australian Government Minister for Sustainability, Environment, Water, Population and Communities. DSEWPaC, Canberra, Australia.
8. Gleeson T, Martin P, Mifsud C (2012) *Northern Australian Beef Industry: Assessment of Risks and Opportunities*. ABARES report to client prepared for the Northern Australia Ministerial Forum, Canberra, Australia. ABARES project 43220.
9. Novelly P, Warburton-Northam D (2012) *A Report on the Viability of Pastoral Leases in the Northern Rangelands Region Based on Biophysical Assessment*. A report to the Department of Agriculture and Food, Western Australia (DAFWA).
10. Ash A, Hunt L, McDonald C, Scanlan J, Bell L, Cowley R, Watson I, McIvor J, MacLeod N (2015) Boosting the productivity and profitability of northern Australian beef enterprises: Exploring innovation options using simulation modelling and systems analysis. *Agricultural Systems* **139**, 50–65.

11. Crowley GM (2015a) *Trends in Natural Resource Management in Australia's Monsoonal North: The Beef Industry*. The Cairns Institute, James Cook University, Cairns, Australia.
12. Rolfe JW, Larard AE, English BH, Hegarty ES, McGrath TB, Gobius NR, De Faveri J, Srhoj JR, Digby MJ, Musgrove RJ (2016) Rangeland profitability in the northern Gulf region of Queensland: Understanding beef business complexity and the subsequent impact on land resource management and environmental outcomes. *The Rangeland Journal* **38**, 261–272.
13. Rural Bank (2016) *Australian farmland values 2015*. Rural Bank. Retrieved from https://www.ruralbank.com.au/news/view/australian-farmland-values-2015.
14. Thompson T, Martin P (2014) *Australian Beef Financial Performance of Beef Cattle Producing Farms, 2011–12 to 2013–14*. Research report 14.7. Australian Bureau of Agricultural and Resource Economics and Sciences (ABARES) research report prepared for Meat & Livestock Australia, Canberra, Australia, August. CC BY 3.0.
15. Kõlves K, Milner A, McKay K, De Leo D (2012) *Suicide in Rural and Remote Areas of Australia*. Australian Institute for Suicide Research and Prevention, Brisbane, Australia.
16. Caitcheon GG, Olley JM, Pantus F, Hancock G, Leslie C (2012) The dominant erosion processes supplying fine sediment to three major rivers in tropical Australia, the Daly (NT), Mitchell (Qld) and Flinders (Qld) Rivers. *Geomorphology* **151–152**, 188–195.
17. Hunt LP, McIvor JG, Grice AC, Bray SG (2014) Principles and guidelines for managing cattle grazing in the grazing lands of northern Australia: stocking rates, pasture resting, prescribed fire, paddock size and water points – A review. *The Rangeland Journal* **36**, 105–119.
18. Aubault H, Webb NP, Strong CL, McTainsh GH, Leys JF, Scanlan JC (2015) Grazing impacts on the susceptibility of rangelands to wind erosion: The effects of stocking rate, stocking strategy and land condition. *Aeolian Research* **17**, 89–99.
19. Crowley GM (2015b) *Trends in Natural Resource Management in Australia's Monsoonal North: The Conservation Economy*. James Cook University, Cairns, Queensland, Australia.
20. Edwards AC, Russell-Smith J, Meyer M (2015) Contemporary fire regime risks to key ecological assets and processes in north Australian savannas. *International Journal of Wildland Fire* **24**, 857–870.
21. Segoli M, Bray S, Allen D, Dalal R, Watson I, Ash A, O'Reagain P (2015) Managing cattle grazing intensity: Effects on soil organic matter and soil nitrogen. *Soil Research* **53**, 677–682.
22. Star M, Rolfe J, Long P, Whish G, Donaghy P (2015) Improved grazing management practices in the catchments of the Great Barrier Reef, Australia: Does climate variability influence their adoption by landholders? *The Rangeland Journal* **37**, 507–515.
23. Loughran RJ, Elliott GL, McFarlane DJ, Campbell BL (2004) A survey of soil erosion in Australia using Caesium-137. *Australian Geographical Studies* **42**, 221–233.
24. Lal R, Tims SG, Fifield LK, Wasson RJ, Howe D (2013) Applicability of 239Pu as a tracer for soil erosion in the wet-dry tropics of northern Australia. *Nuclear Instruments and Methods in Physics Research Section B: Beam Interactions with Materials and Atoms* **294**, 577–583.
25. Shellberg JG, Brooks AP, Rose CW (2013) Sediment production and yield from an alluvial gully in northern Queensland, Australia. *Earth Surface Processes and Landforms* **38**, 1765–1778.
26. DEHP (Department of Environment and Heritage Protection) (2012) *State of the Environment Queensland 2011*. Prepared by the Department of Environment and Heritage Protection, Queensland Government, Brisbane QLD 4000.
27. Brodie J, Pearson R (2016) Ecosystem health of the Great Barrier Reef: Time for effective management action based on evidence. *Estuarine, Coastal and Shelf Science* **183**, 438–451.
28. GBRWST (Great Barrier Reef Water Science Taskforce) (2016) *Clean water for a healthy reef. Full Report*. Queensland Government.
29. Abel N, Rolfe J (2009) Chapter 12: Public and private conservation of aquatic systems in northern Australia: Threats and opportunities. In *'Northern Australia Land and Water Science Rev*iew. Final report to the Northern Australia Land and Water Taskforce, CSIRO, Canberra' (Ed P Stone).
30. Zaar U (2009) *Gulf Water Study: Roper River Region. Report 16/2009D*. NT Government, Darwin NT.
31. DNREAS (Department of Natural Resources, Environment, The Arts and Sport), (2010) *Gulf Water Study: Final Report*. Natural Resources Division, NT Government, Australia.
32. Franklin DC (1999) Evidence of disarray amongst granivorous bird assemblages in the savannas of northern Australia, a region of sparse human settlement. *Biological Conservation* **90**, 53–68.
33. Woinarski JCZ, Ash AJ (2002) Responses of vertebrates to pastoralism, military land use and landscape position in an Australian tropical savanna. *Austral Ecology* **27**, 311–323.

34. Franklin DC, Whitehead P, Pardon G, Matthews J, McMahon P, McIntyre D (2005) Geographic patterns and correlates of the decline of granivorous birds in northern Australia. *Wildlife Research* **32**, 399–408.
35. Legge S, Kennedy M, Lloyd R, Murphy S, Fisher A (2011) Rapid recovery of mammal fauna in the central Kimberley, northern Australia, following the removal of introduced herbivores. *Austral Ecology* **36**, 791–799.
36. DoE (Department of the Environment) (2015) Australian Greenhouse Gas Emissions Information System: National Greenhouse Gas Inventory–Kyoto Protocol Accounting Framework. Retrieved from http://ageis.climatechange.gov.au; accessed on 28 August 2016.
37. Costanza R, d' Arge R, de Groot R, Farber S, Grasso M, Hannon B, Limburg K et al. (1997) The value of the world's ecosystem services and natural capital. *Nature* **387**, 253–260.
38. Millennium Ecosystem Assessment (2005) *'Ecosystems and human well-being: Synthesis.'* Island press, Washington, DC.
39. Granek EF, Polasky S, Kappel CV, Reed DJ, Stoms, DM, Koch, EW, Kennedy et al. (2010) Ecosystem services as a common language for coastal ecosystem-based management (Los Servicios del Ecosistema como un Lenguaje Común para el Manejo de Costas Basado en Ecosistemas). *Conservation Biology* **24**, 207–216.
40. Costanza R, Kubiszewski I, Ervin D, Bluffstone R, Boyd J, Brown D, Chang H et al. (2011) Valuing ecological systems and services. *F1000 Biology Reports* **3**, 14.
41. Ehrlich PR, Mooney HA (1983) Extinction, substitution, and ecosystem services. *Bioscience* **33**, 248–254.
42. Folke C (1991) Socioeconomic dependence on the life-supporting environment. In *Linking the Natural Environment and the Economy: Essays from the Eco-Eco Group*, (Eds C Folke, T Kåberger). Kluwer Academic Publishers, Dordrecht, the Netherlands.
43. de Groot R, Brander L, van der Ploeg S, Costanza R, Bernard F, Braat L, Christie M et al. (2012) Global estimates of the value of ecosystems and their services in monetary units. *Ecosystem Services* **1**, 50–61.
44. Ehrlich PR, Ehrlich AH (1992) The value of biodiversity. *Ambio* **21**, 219–226.
45. de van der Ploeg S, de Groot R (2010) The TEEB Valuation Database – a searchable database of 1310 estimates of monetary values of ecosystem services. Foundation for Sustainable Development, Wageningen, the Netherlands.
46. Worm B, Barbier EB, Beaumont N, Duffy JE, Folke C, Halpern BS, Jackson JBC et al. (2006) Impacts of biodiversity loss on ocean ecosystem services. *Science* **314**, 787–790.
47. Carins J, Pratt J (1995) The Relationship between Ecosystem Health and Delivery of Ecosystem Services. In *'Evaluating and Monitoring the Health of Large-Scale Ecosystems*, (Eds DJ Rapport, CL Gaudet, P Calow). Springer-Verlag, Berlin, Germany.
48. Farber SC, Costanza R, Wilson MA (2002) Economic and ecological concepts for valuing ecosystem services. *Ecological Economics* **41**, 375–392.
49. Costanza R (2000) Social goals and the valuation of ecosystem services. *Ecosystems* **3**, 4–10.
50. Kahneman D (2011) *Thinking, Fast and Slow*. Farrar, Straus and Giroux: New York.
51. Costanza R, Pérez-Maqueo O, Martinez ML, Sutton P, Anderson SJ, Mulder K (2008) The value of coastal wetlands for hurricane protection. *Ambio* **37**, 241–248.
52. Kubiszewski I, Costanza R, Anderson S, Sutton P (2017) The future value of ecosystem services: Global scenarios and national implications. *Ecosystem Services*, in press.
53. Gallopín G, Hammond A, Raskin P, Swart R (1997) Branch points: Global scenarios and human choice. PoleStar Series Report no. 7. Stockholm Environment Institute, Stockholm, Sweden. http://greattransition.org/archives/other/Branch%20Points.pdf.
54. Raskin P, Banuri T, Gallopin G, Gutman P, Hammon A, Kates R, Swart R (2002) *Great Transition: The Promise and Lure of the Times Ahead*. Stockholm Environment Institute, Boston, MA.
55. Australian Government (2015) *Our North, Our Future: White Paper on Developing Northern Australia*. Commonwealth of Australia, Canberra, Australia.
56. Costanza R, Kubiszewski I, Cork S, Patrick K (2015) Scenarios for Australia in 2050: A synthesis and proposed survey. *Journal of Futures Studies* **19**, 49–76.
57. Petheram C, Gallant J, Wilson P, Stone P, Eades G, Rogers L, Read A et al. (2014) *Northern Rivers and Dams: A Preliminary Assessment of Surface Water Storage Potential for Northern Australia*. A report to the Australian Government from the CSIRO. Commonwealth Scientific and Industrial Research Organisation, Canberra, Australia.

58. Tothill JC, Gillies C (1992) *The Pasture Lands of Northern Australia: Their Condition, Productivity and Sustainability*. Published by the Tropical Grassland Society of Australia, on behalf of Meat Research Corporation, Brisbane, Australia.
59. ACRIS (Australian Collaborative Rangelands Information System) (2011) *ACRIS livestock density update 2009–2011*. Prepared by Garry Bastin. ACRIS management unit. Department of the Environment, Australian Government.
60. ABARES (Australian Bureau of Agricultural and Resource Economics and Sciences) (2013) Customised data extracted from AAGIS dataset. ABARES. URL: http://www.agriculture.gov.au/abares/surveys frequently accessed from May until December 2016.
61. ABARES (2014) Regional farm debt: Northern Queensland gulf, south west Queensland and north west New South Wales. ABARES, Canberra, December. CC BY 3.0.
62. Griener R, Gregg D (2011) Farmers' intrinsic motivations, barriers to the adoption of conservation practices and effectiveness of policy instruments empirical evidence from northern Australia. *Land Use Policy* **28**, 257–265.
63. Hancock GR, Loughran RJ, Evans KG, Balog RM (2008) Estimation of soil erosion using field and modelling approaches in an undisturbed arnhem land catchment, Northern Territory, Australia. *Geographical Research* **46**, 333–349.
64. Bainbridge Z, Lewis S, Smithers S, Wilkinson S, Douglas G, Hillier S, Brodie J (2016) Clay mineral source tracing and characterisation of Burdekin River (NE Australia) and flood plume fine sediment. *Journal of Soils and Sediments* **16**, 687–706.
65. EPA (Environment Protection Authority) (2007) *State of Environment Report*. Department of Environment and Conservation. Western Australia. Retrieved from http://www.epa.wa.gov.au/state-environment-report-2007.
66. Hancock GR, Lowry JBC, Moliere DR, Evans KG (2008b) An evaluation of an enhanced soil erosion and landscape evolution model: A case study assessment of the former Nabarlek uranium mine, Northern Territory, Australia. *Earth Surface Processes and Landforms* **33**, 2045–2063.
67. Bray SG, Allen DE, Harms BP, Reid DJ, Fraser GW, Dalal RC, Walsh D, Phelps DG, Gunther R (2016) Is land condition a useful indicator of soil organic carbon stock in Australia's northern grazing land? *The Rangeland Journal* **38**, 229–243.
68. Shaw KA, Rolfe JW, English BH, Kernot JC (2007) A contemporary assessment of land condition in the Northern Gulf region of Queensland. *Tropical Grasslands* **41**, 245–252.
69. Crowley G, Garnett S (1998) Vegetation change in the grasslands and grassy woodlands of east-central Cape York Peninsula, Australia. *Pacific Conservation Biology* **4**, 132–148.
70. Crowley G, Garnett S (2001) Growth, seed production and effect of defoliation on an early flowering perennial grass, *Alloteropsis semialata* (Poaceae) on Cape York Peninsula, Australia. *Australian Journal of Botany* **49**, 735–743.
71. Department of Lands, Planning and the Environment, (2016) Department of Lands, Planning and the Environment. NT Government. Retrieved from https://nt.gov.au/environment/environment-data-maps, accessed on 22 December 2016.
72. Department of Parks and Wildlife (2016) Threatened species and communities. Government of Western Australia, 2016. Retrieved from http://www.dpaw.wa.gov.au/plants-and-animals/threatened-species-and-communities.
73. Firth R, Brook B, Woinarski J, Fordham J (2010) Decline and likely extinction of a northern Australian native rodent, the Brush-tailed Rabbit Rat *Conilurus penicilatus*. *Biological Conservation* **114**, 1193–1201.
74. Kutt A, Woinarski JCZ (2007) The effects of grazing and fire on vegetation and the vertebrate assemblage in a tropical savanna woodland in north-eastern Australia. *Journal of Tropical Ecology* **23**, 95–106.
75. Ludwig JA, Coughenour MB, Liedloff AC, Dyer R (2001) Modelling the resilience of Australian savanna systems to grazing impacts. *Environment International* **27**, 167–172.
76. Woinarski JCZ (1993) Australian tropical savannas, their avifauna, conservation status and threats. In *Birds and their Habitats: Status and Conservation in Queensland* (Eds C Catterall, P Driscoll, K Hulsman, A Taplin) pp. 45–63. Queensland Ornithological Society, Brisbane, Australia.
77. Charmley E, Williams SRO, Moate PJ, Hegarty RS, Herd RM, Oddy VH, Reyenga P, Staunton KM, Anderson A, Hannah MC (2016) A universal equation to predict methane production of forage-fed cattle in Australia. *Animal Production Science* **56**, 169–180.

78. Burrows WH, Carter JO, Scanlan JC, Anderson ER (1990) Management of savannas for livestock production in north-east Australia: Contrasts across the tree-grass continuum. *Journal of Biogeography* **17**, 503–512.
79. Burrows WH, Compton JF, Hoffmann MB, Dyason R, Dyason L, Garsden R (Eds.) (1998) Vegetation thickening and carbon sinks in the grazed woodlands of north-east Australia. Australian forest Growers Biennial conference Proceedings: Plantation and regrowth forestry: A diversity of opportunity. Lismore, NSW.
80. Burrows WH, Back PV, Hoffman MB (1999) Woodland management and woody weed control for Queensland's beef pastures. North Australia Program Occasional Publication No. 10, Meat and Livestock Australia.
81. Burrows WH, Hoffman MB, Compton JF, Back PV, Tait LJ (2000) Allometric relationships and community biomass estimates for some dominant eucalypts in Central Queensland woodlands. *Australian Journal of Botany* **48**, 707–714.
82. Dyer R, Stafford-Smith M (2003) Ecological and economic assessment of prescribed burning impacts in semi-arid pastoral lands of northern Australia. *International Journal of Wildland Fire* **12**, 403–413.
83. Gillieson D, Cook G (2008) The dynamic savanna project. Retrieved from http://environorth.org.au/downloads/Dynamic%20savanna.pdf accessed on 22 December 2016. Tropical Savanna CRC.
84. Murphy BP, Russell-Smith J, Prior LD (2010) Frequent fires reduce tree growth in northern Australian savannas: Implications for tree demography and carbon sequestration. *Global Change Biology* **16**, 331–343.
85. Scanlan JC (2002) Some aspects of tree-grass dynamics in Queensland's grazing lands. *The Rangeland Journal* **24**, 56–82.
86. Lonsdale WM (1994) Inviting trouble: Introduced pasture species in northern Australia. *Australian Journal of Ecology* **19**, 345–354.
87. Cook GD, Dias L (2006) It was no accident: Plant introductions by Australian Government agencies during the 20th century. *Australian Journal of Botany* **54**, 601–625.
88. Northern Territory Government (2010) *Weed Management Plan for Andropogon gayanus (gamba grass)*. Northern territory Government, Darwin, NT.
89. Petty AM, Setterfield SA, Ferdinands KB, Barrow P (2012) Inferring habitat suitability and spread patterns from large-scale distributions of an exotic invasive pasture grass in north Australia. *Journal of Applied Ecology* **49**, 742–752.
90. Rossiter NA, Setterfield SA, Douglas MM, Hutley LB (2003) Testing the grass-fire cycle: alien grass invasion in the tropical savannas of northern Australia. *Diversity and Distributions* **9**, 169–176.
91. Setterfield SA, Rossiter-Rachor NA, Hutley LB, Douglas MM, Williams RJ (2010) Turning up the heat: the impacts of *Andropogon gayanus* (gamba grass) invasion on fire behaviour in northern Australian savannas. *Diversity and Distributions* **16**, 854–861.
92. Brooks KJ, Setterfield SA, Douglas MM (2010 Exotic grass invasions: Applying a conceptual framework to the dynamics of degradation and restoration in Australia's tropical savannas. *Restoration Ecology* **18**, 188–197.
93. Adams VA, Setterfield SA (2013) Estimating the financial risks of *Androgon gayanus* to greenhouse gas abatement projects in northern Australia. *Environmental Research Letters* **8**, 025018.
94. Stein JL, Stein JA, Nix HA (2001) Wild rivers in Australia. *International Journal of Wilderness* **7**, 20–24.
95. Pusey BJ, Kennard MJ (2009) Chapter 3: Aquatic ecosystems in northern Australia. In *Northern Australia Land and Water Science Review (full report)* (Ed P Stone). Final report to the Northern Australia Land and Water Taskforce. CSIRO, Canberra, Australia.
96. Karfs RA, Trueman M (2005) *Tracking Changes in the VRD Pastoral District, Northern Territory, Australia—2005*. Report to the Australian Collaborative Rangelands Information System (ACRIS) Management Committee. NT Department of Natural Resources, Environment & the Arts, Darwin, NT (https://www.environment.gov.au/system/files/resources/780f289e-ccde-4ab2-b9c5-cdc4831745b9/files/nt.pdf).
97. van Doorn A, Woinarski JCZ, Werner PA (2015) Livestock grazing affects habitat quality and persistence of the threatened Purple-crowned Fairy-wren (*Malurus coronatus*) in the Victoria River District, Northern Territory, Australia. *Emu* **115**, 302–308.
98. Kaur K, Jalota RK, Midmore DJ (2007) Impact of tree clearing on soil attributes for a pastoral property in central Queensland, Australia. *Soil Science* **172**, 516–533.

99. Scanlan JC, MacLeod ND, O'Reagain PJ (2013) Scaling results up from a plot and paddock scale to a property–A case study from a long-term grazing experiment in northern Australia. *The Rangeland Journal* **35**, 193–200.
100. Eady SJ, Havard G, Bray SG, Holmes W, Navarro J (2016) Down scaling to regional assessment of greenhouse gas emissions to enable consistency in accounting for emissions reduction projects and national inventory accounts for northern beef production in Australia. *The Rangeland Journal* **38**, 219–228.
101. The Economics of Land Degradation (ELD) initiative (2015) *The Value of Land: Prosperous Lands and Positive Rewards through Sustainable Land Management*. ELD Secretariat, Bonn, Germany. Retrieved from www.eld-initiative.org.
102. Walsh D, Cowley R (2016) Optimising beef business performance in northern Australia: What can 30 years of commercial innovation teach us? *The Rangeland Journal* **38**, 291–305.
103. Foran BD, Stafford Smith DM (1991) Risk, biology and drought management strategies for cattle stations in central Australia. *Journal of Environmental Management* **33**, 17–33.
104. Landsberg RG, Ash AJ, Shepherd RK, McKeon GM (1998) Learning from history to survive in the future: Management evolution on Trafalgar Station, north-east Queensland. *Rangeland Journal* **20**, 104–118.
105. Purvis JR (1986) Nurture the land: My philosophies of pastoral management in central Australia. *Australian Rangeland Journal* **8**, 110–117.
106. McCosker T, McLean D, Holmes P (2010) *Northern Beef Situation Analysis 2009*. Meat and Livestock Australia, Sydney, Australia.
107. Woinarski J, Pavey C, Kerrigan R, Cowie I, Ward S (2007) *Lost from Our Landscape: Threatened Species of the Northern Territory*. NT Government Printer, Darwin, Australia.
108. Garnaut R (2008) *The Garnaut Climate Change Review: Final Report*. Cambridge University Press, Port Melbourne, Australia, p. 65.
109. Woinarski J, Russell-Smith J, Andersen AN, Brennan K (2009) Fire management and biodiversity of the west Arnhem Land Plateau. In *Culture, Ecology and Economy of Savanna Fire Management in Northern Australia: Rekindling the Wurrk Tradition* (Eds J Russell-Smith, PJ Whitehead, P Cooke) pp. 201–227. CSIRO Publications, Melbourne.
110. Woinarski JCZ, Legge S, Fitzsimons JA, Traill BJ, Burbidge AA, Fisher A, Firth RSC et al. (2011) The disappearing mammal fauna of Australia: Context, cause, and response. *Conservation Letters* **4**, 192–201.
111. Russell-Smith J, Gardener M, Brock C, Brennan K, Yates CP, Grace B (2012) Fire persistence traits can be used to predict vegetation responses to changing fire regimes at expansive landscape scales—An Australian example. *Journal of Biogeography* **39**, 1659–1668.
112. Clark E, Abel N, Measham T, Morison J, Rippin L (2009) Chapter 11: Water based tourism and recreation in northern Australia. In *Northern Australia Land and Water Science Review* (*Full Report*), (Ed P Stone). Final report to the Northern Australia Land and Water Taskforce, CSIRO, Canberra, Australia.
113. Warfe DM, Pettit NE, Davies PM, Pusey BJ, Hamilton SK, Kennard MJ, Townsend SA et al. (2011) The 'wet-dry' in the wet-dry tropics drives river ecosystem structure and processes in northern Australia. *Freshwater Biology* **56**, 2169–2195.
114. Brodie JE, Mitchell AW (2005) Nutrients in Australian tropical rivers: Changes with agricultural development and implications for receiving environments. *Marine and Freshwater Research* **56**, 279–302.
115. Oxford Economics (2009) *Valuing the Effects of Great Barrier Reef Bleaching*. Great Barrier Reef Foundation, Brisbane, Australia.
116. Stoeckl N, Hicks CC, Mills M, Fabricius K, Esparon M, Kroon F, Kaur K, Costanza R (2011) The economic value of ecosystem services in the Great Barrier Reef: Our state of knowledge. *Annals of the New York Academy of Sciences* **1219**, 113–133.
117. Butler JRA, Wong GY, Metcalfe DJ, Honzák M, Pert PL, Rao N, van Grieken ME et al. (2013) An analysis of trade-offs between multiple ecosystem services and stakeholders linked to land use and water quality management in the Great Barrier Reef, Australia. *Agriculture, Ecosystems & Environment* **180**, 176–191.
118. Finn M, Jackson S (2011) Protecting indigenous values in water management: A challenge to conventional environmental flow assessments. *Ecosystems* **14**, 1232–1248.
119. Stoeckl N, Jackson S, Pantus F, Finn M, Kennard MJ, Pusey BJ (2013) An integrated assessment of financial, hydrological, ecological and social impacts of 'development' on Indigenous and non-Indigenous people in northern Australia. *Biological Conservation* **159**, 214–221.

120. Burgess C, Johnston F, Bowman D, Whitehead PJ (2005) Healthy land: Healthy people? Exploring the health benefits of Indigenous Land Management. *Australian and New Zealand Journal of Public Health* **29**, 117–122.
121. Garnett ST, Sithole B, Whitehead PJ, Burgess CP, Johnston FH, Lea T (2008) Healthy country, healthy people: Policy implications of links between indigenous human health and environmental condition in tropical Australia. *The Australian Journal of Public Administration* **68**, 53–66.
122. Social Ventures Australia (2016) *Consolidated Report on Indigenous Protected Areas following Social Return on Investment Analyses.* Report to Department of Prime Minister and Cabinet, Australian Government, Canberra, Australia.
123. Sangha KK, Russell-Smith J, Morrison SC, Costanza R, Edwards AC (2017) Challenges for valuing ecosystem services from an Indigenous estate in northern Australia. *Ecosystem Services* **25**, 167–178.
124. Walsh D, Russell-Smith J, Cowley R (2014) Fire and carbon management in a diversified rangelands economy: Research, policy and implementation challenges for northern Australia. *The Rangeland Journal* **36**, 313–322.
125. Crowley G, Garnett ST, Shepherd S (2009) Impact of storm-burning on *Melaleuca viridiflora* invasion on grasslands and grassy woodlands on Cape York Peninsula, northern Australia. *Austral Ecology* **34**, 196–209.
126. Cowley R, Hearnden M, Joyce K, Tovar-Valencia M, Cowley T, Pettit C, Dyer R (2014) How hot? How often? Getting the fire frequency and timing right for optimal management of woody cover and pasture composition in northern Australian grazed tropical savannas. Kidman Springs fire experiment 1993–2013. *The Rangelands Journal* **36**, 323–343.
127. Russell-Smith J, Yates CP, Evans J, Desailly M (2014) Developing a savanna burning emissions abatement methodology for tussock grasslands in high rainfall regions of northern Australia. *Tropical Grasslands—Forrajes Tropicales* **2**, 154–166.
128. NAFI (North Australia Fire Information). (2017). SavBAT 3—Savanna Burning Abatement Tool, v3. Retrieved from www.firenorth.org.au. Accessed 12 January 2018.
129. Russell-Smith J, Yates CP, Evans J, Meyer CP, Edwards AC (2015) Application of a 'lower rainfall' savanna burning emissions abatement methodology. In *Carbon Accounting and Management in Fire-prone Australian Savannas* (Eds BP Murphy, AC Edwards, CP Meyer, J Russell-Smith), pp. 219–234. CSIRO Publications, Melbourne, Australia.
130. Russell-Smith J, Yates CP, Edwards AC, Whitehead PJ, Murphy BP, Lawes MJ (2015) Deriving multiple benefits from carbon market-based savanna fire management: An Australian example. *PLoS One* **10**, e0143426.
131. Cook GD, Meyer CP, Muepu M, Liedloff AC (2016) Dead organic matter and the dynamics of carbon and greenhouse gas emissions in frequently burnt savannas. *International Journal of Wildland Fire* **25**, 1252–1263.
132. Murphy BP, Russell-Smith J, Watt FA, Cook GD (2009) Fire management and woody biomass carbon stocks in mesic savannas. In *Culture, Ecology and Economy of Fire Management in North Australian Savannas: Rekindling the Wurrk Tradition* (Eds J Russell-Smith, P Whitehead, P Cooke), pp. 361–394. CSIRO Publications, Collingwood, Australia.
133. Wiedemann S, Telfer M, Cohn P, Russell-Smith J (2016) *Final Report: Opportunity Study—Engagement of Pastoral Properties with the Emissions Reduction Fund.* Project B.CCH.2099. Meat & Livestock Australia, Sydney, Australia.
134. Fitzroy Basin Association (2014) *Grazing Best Management Practice Evaluation.* Roberts Evaluation. Published by the Agforce, Queensland Government, FBA and Grazing BMP.
135. Douglass LL, Possingham HP, Carwardine J, Klein CJ, Roxburgh SH, Russell-Smith J, Wilson KA (2011) The effect of carbon credits on savanna land management and priorities for biodiversity onservation. *PLoS One* **6**, e23843.
136. Sangha KK, Sithole B, Hunter-Xenie H, Daniels C, Yibarbuk D, James G, Michael C, Gould J, Edwards A, Russell-Smith J (2017) Empowering remote Indigenous communities in northern Australia. *International Journal of Mass Emergencies and Disasters* **35**, 137–153.
137. Whitehead P, Oliver B (2014) Development by Design: Opportunities in northern Australia and the potential role of Indigenous people, with particular emphasis on the Northern Territory. A scoping study for The Nature Conservancy. *Working paper 01/2014.* North Australian Indigenous Land and Sea Management Alliance, Darwin, NT.

CHAPTER 6

Resilient Communities and Reliable Prosperity

Glenn James, Bentley James, Joe Morrison, and Douglas Paton

CONTENTS

Indigenous Australians are the first peoples of this land. As in all those parts of the world occupied by Indigenous peoples, every Australian Indigenous community has its own cultural features.

In the beginning, our *country* (including the creators) provided all that we required, and in exchange, we cared for the *country* (and the creators). Together we made successful systems of governance that connected people across vast networks of kin, and grew sophisticated knowledge, religious practices and language systems that evolved to changing circumstances and were passed down through countless generations. These factors allowed us to evolve and are the tools with which we endure and have the potential to prosper, through identity and connection to our *country*.

Unfortunately, this deep encyclopaedic knowledge has been discounted by *mununga* (non-Indigenous) society, who choose to ignore the lessons of adaptation that Indigenous people bring to the discussion about how to live in a world where survival and sustainability have never before been so challenged.

When the British colonisers took our lands in 1788 – without a formal settlement or acknowledgment – we began our campaign to wash the stain of ignorance from our nation's history, so we could coexist with the recent arrivals and improve the prosperity of our people through new relationships built on trust.

We continue to suffer the heavy hand of government and powerful industry interests who too often bring an unrelenting, unsophisticated, self-interested, and an ultimately destructive mind-set aimed at removing Indigenous 'uniqueness' and promoting mainstream modernity based solely on capitalism and wealth creation. Our particular Indigenous strengths, values, and interests are obscured and undervalued in this world of unequal distribution of wealth and political power. But Indigenous people endure and continue to struggle to be recognised for who we are and for the sorts of futures we desire.

Growing up in the 1970s and 1980s with my family and friends – both black and white – in and around Katherine, I saw the denial of a distinct identity, of connection to *country*, of the most basic of human rights and equal opportunity. But more recently I have worked at a different level, firstly as the founding chief executive officer of the North Australian Indigenous Land and Sea Management Alliance, and now as the CEO of the Northern Land Council. These roles have offered me a macro view of the socio-political landscape affecting Indigenous people, the unabated attempts to undermine the unique relationships of Indigenous people with their lands and waters, and continual pressure on Indigenous people in northern Australia as they continue to confront colonial attempts to substitute the white dream of northern development. This agenda is expressed in many forms, but perhaps the most insidious is the 'Our North, Our Future: White Paper on Developing North Australia', written by those who don't and wouldn't live here and whose interests lie only in globalisation. It promotes the paradox of shabby individualism dependent on sameness, shared narrow ideology and orthodoxy, rather than mutual support drawing on difference and innovation.

However, like many others drawing on deep, irrevocable culture-based strengths, I can see opportunities for trust-based relationships around mutual interests and for investing in our prosperity on our own terms. Aboriginal people are critical to the identity of Australia in a rapidly changing global community. Indigenous language, knowledge, religion, practice, and artistic expressions must be embraced as a key component of what it is to be Australian if Australians of all ethnic origins are to live sustainably. The nation must learn from Indigenous people how to live in Australia, to live with fire, to have kin relations with plants and animals, to embrace homelands as special places of learning and cultural reinvigoration as the essential underpinning for resilience and adaptation at the community level.

It is disingenuous to assume to 'know' – and therefore to own and speak for – Indigenous people. I believe that the best path for Indigenous futures and prosperity is empowerment of Indigenous people as individuals: but individuals who choose to work together bringing their unique cultures, languages, and skills to bear on all manner of engagement in modern Australia and in the global community. Yes: we are all 'Australians' now, but I am firstly Dagoman and Mualgal.

—Joe Morrison (CEO, Northern Land Council)

6.1 INTRODUCTION

The chapters in this publication build the case for Indigenous culture and society as central to the constructive future development of North Australia. Many Indigenous leaders and commentators have emphasised the need for greater recognition and respect for Indigenous rights, interests, and capabilities for the future prosperity of the North.[1–3]

Indigenous societies hold many unique qualities that underpin their resilience. These qualities are often obscured or have been misrepresented as barriers to Indigenous participation in development. This chapter contends that unique qualities and characteristics (e.g. networks of attachment to kin, customary estates, religious knowledge, cultural obligations) are key factors of resilience and assets for creative and effective participation in economic development, realisable through more equitable, respectful, and trusting relationship building around mutual interests.

This chapter takes a closer look at Indigenous values as resilience factors in relationship building with government and wider society. By resilience we mean the individual and collective ability to respond to adverse circumstances and adapt to change in ways that improve or at least maintain individual and collective well-being. Resilience emerges from how interdependent relationships among individuals, their communities, and the larger regional, national, and global systems are used over time.[4] It underpins the agency and identity that is critical to well-being.[5] Unfortunately, these qualities are often undervalued.

Government and other (e.g. non-government organisation [NGO]) perspectives on resilience fail to grasp that it develops from the accumulated, and diverse, life histories and experiences of people and communities.[4] They inappropriately assume that the imposition of top-down solutions focussing on social, physical, and other capital (e.g. physical infrastructure, social services, training, employment opportunities, legislated rights) will automatically translate into enhanced resilience. While such interventions can contribute to an empowering adaptive response to challenging circumstances, their success in this regard is muted by their failure to accommodate the social, resource and cultural diversity inherent in society. These problems are magnified when attempting to deploy these top-down strategies with Indigenous peoples; they fail to adequately capture subtle characteristics of what makes Indigenous people who they are, and, indeed, often see the social and cultural capital in Indigenous societies as erosive of local resilience. This is evident in policy and service disjuncture; for example, like the support of cultural tourism on one hand and the devaluing of bi-lingual education on the other, or the recognition of homelands centres (outstations) as operational nodes for fire management (including carbon farming) and their decapitalisation as 'unviable'.[6]

This chapter argues that resilience is not limited to coping, particularly with engagement in broader society, but about prospering. Although the perception of Indigenous community life from the 'outside' (including popular media) is bleak, often akin to that in refugee camps, it is illusory to disregard the richness of life and tradition on *country*. The chapter considers perspectives on resilience and prosperity in light of the recurring 'outside' narrative, then presents a simple conceptual framework as a lens through which to recognise the inherent qualities of resilience in this richness and complexity. It does so by adopting a 'strength-based approach'.[7,8] The latter approach argues that it is impossible to conceptualise resilience and adaptive processes without accommodating the role of the social, cultural and spiritual capital in the population of interest. It also argues that it is important to explore the challenging experiences that have underpinned how social, cultural, and spiritual resources are mobilized to facilitate resilience and adaptation.[4,9]

To illustrate this further, the chapter explores moments of Indigenous-led intervention and acts of resistance against erosion of societal values and well-being. From Aboriginal art to natural hazard management, characteristics of Indigenous resilience (uniquely embedded in social organisation, belief, and indubitable relationships to customary lands) have been expressed as key measures of well-being, in ground-up resistance to cultural erosion and intervention, and importantly, as assets to effective participation in the broader economy.

Structural inequities between state and Indigenous (minority) cultures may not be curable merely through better understanding and good-will, but productive relationships around mutually desirable outcomes (e.g. emergency management, carbon farming) substantially benefit from conceptual tools that recognise, value, invest in, and empower both the mobilization of existing Indigenous social capital and the development of new adaptive capacities. Additionally, this can show how continuities in local values extend across time and geography and point to opportunities to apply them at scale. Some such conceptual prompts are identified and explored to highlight the articulation of these local values, leading back to a discussion on recognition, respect, and investment in Indigenous resilience as a foundation for improved community governance.

6.2 PERSPECTIVES ON RESILIENCE AND PROSPERITY

> 'Reliable prosperity is delivered when economic systems are built to meet the fundamental needs of people and the ecosystems that sustain them over the very long term'.[2]

This is not a 'trickle-down' model for economic prosperity, thoroughly discredited elsewhere (see Chapters 1, 4, and 8). The rhetoric of trickle down is well known across Indigenous Australia. The suggestion here is rather that the characteristics of Indigenous life and well-being are part of the economic system 'build'. As James Scott observes, 'to the degree that the state's formal schema makes no allowance for local processes or actually suppresses them, it fails both its intended beneficiaries and ultimately its designers as well'.[10] Concomitantly, Bill Harney observes:

> 'We can justify everything you know, what we can see in front of us: bush medicine, rock formations and all that. ... We all got it from a song and a story from the creation and it's still there today and we're passing that information to the young ones from word-of-mouth,' Bill Harney.[11]

Indigenous landowners know that investment in the intergenerational transmission of values (through kinship networks, patterns of authority, obligation and responsibility) furnish reliable socioeconomic and ecological returns over time. Many Indigenous landowners are eager for economic development, where compatible with their cultural obligations. Even more so, where it can strengthen cultural precedents and knowledge necessary to care for local land, social, and religious assets in the face of change.[12] Indigenous people are developing novel and evolving forms of partnerships and coexistences, when not absolutely denied by the encompassing settler society.[13]

Local forms of resilience can enable adaptable and imaginative engagement with mainstream economies (e.g. savanna burning [Chapter 5], Indigenous Ranger groups, biosecurity services, emergency management [Box 6.1]). Where outcomes of mutual interest result, despite different worldviews and values driving the activity, a kind of prosperity agreeable to all can develop (Figure 6.1).[14]

Figure 6.1 *Where paths meet – two ways, one future.* (Courtesy of Duwun Tony Lee, Birrimilanga Nadine Lee, Ian Lee, Glenn James.)

Tensions lurk between differing motivations to achieve certain common outcomes. In environmental management, dependence on government support, interpreted as a kind of welfare rather than a legitimate ecosystem service (Chapters 5 and 8) threatens the continuity of economic engagement and fragile collaboration with the fickle state. It's telling that wider recognition of local Indigenous values in land management is empowering Indigenous land management groups to develop stronger relationships with the market. Collaborations in the market economy, where social, cultural, and conservation outcomes overlap to enhance payment for environmental services (PES) 'products', are providing other opportunities to strengthen prosperity, and relieving certain relationships of dependence on the state.

Many of the things that people say make them stronger or more resilient are things they carry with them or seek to draw from their heritage. These things (for example, local language learning and kinship networks) are often viewed by governments and others as cultural quirks of less importance to 'closing the gap' than mainstream development values (like standard Australian English [SAE], individualism, and entrepreneurship). It is this erroneous way of thinking that strengths-based approaches seek to redress. But they are not mutually exclusive. In the circumstances in which people really live – as distinct from the imaginary world of people without boots lifting themselves by their bootstraps – the latter may be impossible without the support of the former.

Can governments and others recognise important and subtle values in Indigenous resilience and respect them for their critical role in local and wider societal reliable prosperity? Indigenous metaphysics, knowledge of ancestral geography and social capital systems (e.g. kin networks) may be at best difficult to understand and even harder therefore to recognise as assets in the overpowering ethnocentric management of (Western) state and society.[12] However, the conceptual tools are all around if, as Joe Morrison urges, non-Indigenous agents respect cultural difference and the ability of Indigenous leaders to interpret across the cultural divide what's important to them as the basis for productive relationship building and for what constitutes success. This is reflected in the role of empowerment in strengths-based approaches and understanding the necessity of conceptualizing resilience as a multi-level process.[4]

Local expressions of the quality and richness of Indigenous societies contain important clues for relationship building, as to their persistence, the features of strength and resilience and the values that people draw on for prosperous cultural futures. Though there are many, we have selected a basic set of conceptual categories of value (features of resilience) that resonate in the stories and depictions of complex life. Concepts of connection, identity, knowledge, power, and seasonality acquire meaning in local settings but provide a useful lens to identify resilience features across time and geography.

BOX 6.1 GIRRINGUN ABORIGINAL CORPORATION CASE STUDY

RICKY ARCHER

NORTH AUSTRALIAN LAND & SEA MANAGEMENT ALLIANCE LTD., DARWIN

Girringun Aboriginal Corporation[1] represents the interests of nine Traditional Owner groups based in Cardwell, North Queensland. Girringun's area of operation is approximately 1.26 million hectares of land and sea *country* encompassing Indigenous Protected Areas, World Heritage sites, and freehold and leasehold lands. Girringun also manage a successful Indigenous Ranger team which manages the natural and cultural resources of both terrestrial and marine areas.

Girringun relies on a strong governance framework which is built on shared distinctive cultural practices of associated clan groups. This framework enables the corporation to operate as a mediator or arbitrator to support Traditional Owners to reach agreed outcomes. This integrated governance approach has enabled Girringun to develop Provision of Ecosystem Services opportunities and established Australia's first management plan for the traditional hunting of a protected species in the form of a Traditional Use of Marine Resource Agreement (TUMRA).

(*Continued*)

BOX 6.1 (Continued) GIRRINGUN ABORIGINAL CORPORATION CASE STUDY

- Represents interests of nine TO groups
- Establishes Cultural Authority for decision making process
- Fosters consolidated negotiations between TO's and third parties
- Supports TO's in sustainable economic development
- Operate a ranger group in a culturally appropriate way
- Brokers management agreements for TO's within Great Barrier Reef and Wet Tropics World Heritage areas
- Executes cultural and environmental initiatives within a culturally diverse IPA

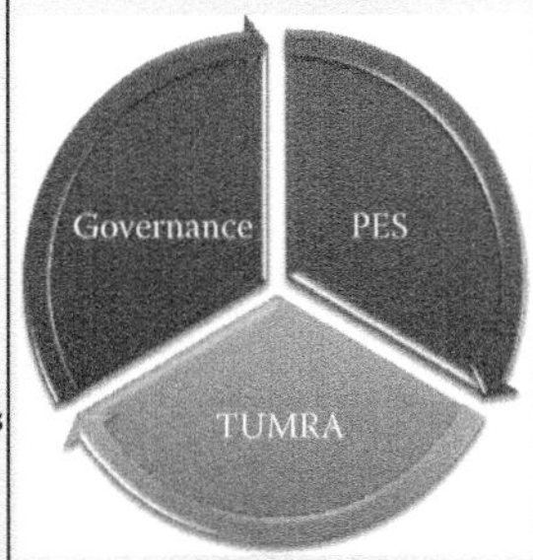

- Invests in Indigenous stakeholders to build capacity
- Develops fee-for-service capability
- Manages invasive threats for statutory bodies, i.e. Wet Tropics Management Authority
- Manages biosecurity risks for local and state governments
- Develops diversified economic interests, i.e. road maintenance/heavy machinery operation in state reserves
- Develops and brokers opportunities which meet both cultural and economic aspirations

- First ever agreement in Australia of its kind
- Formal agreement between saltwater TO's, GBRMPA*, and QPWS**
- Community members agreed to moratorium for traditional hunting of turtle and dugong in specified areas
- Successful institution which manages balance of traditional customary practice and threatened species management
- Enables research initiatives based on both science and traditional ecological knowledge
- TUMRA management is also supported through Girringun Aboriginal Ranger program
- Has created the platform for other Indigenous groups to develop similar agreements

* Great Barrier Reef Marine Park Authority
** Queensland Parks & Wildlife Service

REFERENCE

1. Girringun Aboriginal Corporation. Cardwell, Qld, 4849. http://www.girringun.com.au, 16 May 2018.

6.3 RESILIENCE: STRENGTH FROM *COUNTRY*, THE RICHNESS OF LIFE

Indigenous voices persistently celebrate the richness of their cultural heritage and its value in well-being. These reflections may appear idealised and overshadowed alongside horrific statistics about health, incarceration, poverty, poor education, and so on, but within them are expressions of strength, capability, aspirations, and the 'long view' of history. Common themes of Indigenous value, persistence, and adaptability include connection to land, kin and the wider society; sophisticated knowledge systems including new knowledge; identity from place, language, and heritage; power, authority, and personal confidence; and symbiosis with local resources tuned to seasonal cycles. These are also markers for how to strengthen opportunities for prosperity in relationship building.

Similar outcomes emerged from analysis of Indigenous and ethnic minority resilience in Taiwan.[8] This work highlighted how Indigenous peoples brought a suite of interrelated adaptive processes (e.g. frugality, persistence, social reciprocity, coexistence with the land) to bear on the challenges posed by having to confront environmental hazards. Although not exhaustive expressions of value, they are signposted in the expressions of cultural life below so as to be better recognised as pillars of resistance to the erosion of Indigenous lives and, importantly, promoted as assets to participation in wider society and economy.

6.3.1 Seasonality – Connection to *Country*

In recent times seasonal calendars have become a popular way of signifying the profound connection between people, *country*, and the yearly cycle of change on *country* (Figure 6.2). Distinctive local responses to northern monsoonal oscillations colour deeper continuities in societal organisation around seasons that endure across time and geography.

Figure 6.2 Examples of North Australian Indigenous seasonal calendars, each depicting six customary seasons: Wunambal-Gaambera in North Kimberley, WA. (Courtesy of the Wunambal Gaambera Aboriginal Corporation), Yan-nhangu, Crocodile Islands, NT. (Courtesy of Dr. Bentley James) and Yawuru *country*, around Broome, WA. (Courtesy of Nyamba Buru Yawuru Ltd).

The change of seasons dominates life in the North.[15] From the Kimberley to the Cape, each season (more subtle than the crude wet/dry dichotomy) is punctuated and understood with the onset of particular environmental and other indicators relating to particular *country*. Seasonal patterns of movement are remembered in the songs and journeys of the creator/spirits. These myths and legends link people to the network of sites constituting a pre-existing ancestral geography.[16] These are seasonal drivers for visits to *country*; the inspiration for ceremonies of renewal and customary re-engagement with sacred sites and places, fuelling art, dance, and songs of celebration that defines people. For the Yan-nhangu people of the Crocodile Islands the seasonal winds direct the yearly pattern of inter-island travel and access to ripening natural resources on distant islands, the fatness of fish, the timing of ceremonies, and the burning of grasses of the outer islands.

> When the wind *l̲irra-ŋän̲'ka* blows, the *Gurrmirriŋu* lights the grass fires of Gurriba Island. When we see this smoke, it reminds us that we must also light our fires. Old people would cry when they saw the smoke and sing ('Kinsman always living there … lights the fire on the islands … reminds us of the old ways, of the islands long ago, long past …'). The Gurrmirriŋu lights up the daisy grass (*dhambaŋaniŋ*) on Gurriba each year, then sings the magpie lark. The magpie lark (*d̲il̲ili*) sings in the morning at the beginning of the cold time (Dharratharraway), to tell the story of the wind called *rika* (the westerly from Dhämungurra at Murruŋga). When the magpie lark sings the song of rika, the turtles come home.[17]

Across the North, seasonal local abundances are signalled by calendar plants. In Yawuru *country*, *Marrul* season begins when the *gardgu* (long-fruited bloodwood) buds burst into flower, and ends when the rain stops and the winds start to blow from the south-east. In *Marrul*, Yawuru people go into the *birra* (bush) while lizards are fat and there are plenty of bush fruits'.[18] There are no places across the North were the people of the land do not have knowledge of the changing of the seasons and significance of the signs. The named winds, cyclic changes, and reappearing seasons remain an elemental, undeniable, and significant part of Indigenous life in the North of Australia.

6.3.2 Connectedness – Knowledge and Authority

Wunambal-Gaambera people from the north-west Kimberley say, 'We talk about living things in *ngauwudu* (*country*), to show how rich it is in animals and plants that were there before *Kartiya* (Europeans) came, and are still here today'.[19] Local authority and detailed knowledge of place underpin the remarkable linguistic, cultural, and biodiversity values celebrated across the Indigenous estate. The exchange of this knowledge, authority, and of natural resources connected to *country* amplifies the health benefit to these natural systems and people that rely on it.[20]

Senior landowner Jacob Nayinggul (deceased) and linguist Murray Garde spoke in 2003 about the guiding cultural principles relating to land tenure and land management in the Kakadu region and western Arnhem Land – specifically the Kunmokurrkurr system of patrilineal clan membership and land affiliation.

KINSHIP AND *COUNTRY* – A DISCUSSION BETWEEN JACOB NAYINGGUL AND MURRAY GARDE, 2003

Jacob: Bu how bu, nangale, na-ngale kabi-djawan, kayime konda nangale nuye kunred. Bu ngudda ke kunred yiman Kunbarlanja o yiman kunred kure kunmokurrkurr kayo, yiman Mirarr clan. Bu ngadjawan ngayime ngudda Mirarr, konda kunred Mirarr because kunred and that kunmokurrkurr kabenere. Kunred and kunmokurrkurr, kunred and kunmokurrkurr. Kubolk-Mirarr, kubolk-Mandjulngunj Kunbarlanja. Bu Kunbarlanja ngarre, if kan Kunbarlanja ngarrbendjawan bininj ngadjawan and bininj nakohbanj minj nakohbanj but yawurrinj, minj kakodjbulu but kabengkan. Ngarridjawan konda nangale nuye kunred, 'Mandjulngunj' kayime. Straight away namekke kanwon o ngarriyime nangale Mandjulngunj? "Nani kankan ngarrku kumekke kayime nanih Mandjulngunj nuye kunred. Wanjh ngarrburrbun bu ngarrkarrme yiwarrudj already, bu nangale after kabolkmang bu nungka karrowen. Namekbe Mandjulngunj bininj o kaworhnan dabbolk. Balanda ngurriyime 'Senior traditional owner'. Nawu benbornang ngalbeywurd o nabeywurd.

Whoever asks another person [about clans] will say something like, 'Who owns this place here?' Maybe your own land or a place like Kunbarlanja or any place which has a patri-clan name, such as the Mirarr clan. If I'm discussing this with someone I might say, 'You are a Mirarr clan member, and your land is therefore Mirarr land', because land and people go together. Every place has a clan name, every place a clan. Mirarr land, or the Mandjulngunj clan land at Kunbarlanja for example. If we go to Kunbarlanja, if I take you to Kunbarlanja and we ask people there, senior people, well not necessarily senior people, but young people too. They may not yet have grey hair, but they still know. If we were to say, 'Who owns the land here?' The immediate answer will be 'Mandjulngunj'. Straight away they will give us an answer and show us who are the Mandjulngunj people they are talking about. We have knowledge about the system that already exists for who inherits that land when a particular senior landowner passes away. A Mandjulngunj clan member (for the Kunbarlanja example), a senior adult person who in English is called 'a senior traditional owner'. His children [or a woman's brother's children], a son or a daughter will take over from that person.

Steadfast attachment to one's *country* and responsibilities for its laws and ceremonies extend through focused local networks of kin via marriage, ritual, and economic relationships to broader regional systems of cooperation. The sense of place and spiritual connection embedded in these networks of exchange is celebrated in ceremonies, sung in songs, danced, and painted on the body, and in so doing, yield an ethic of enrichment evident in persistent care for kin and *country*. This unique value-adding metaphysics is a crucial driver of local and regional prosperity, and a key to Indigenous survival.

The following diagram depicts an idealised rendering of the local geographical arrangements of kinship, marriage, and ritual of the Yan-nhangu people focused on the Crocodile Islands (Figure 6.3). These connubial links engender strings of connectedness with wider regional kin networks that extend over many generations.[21]

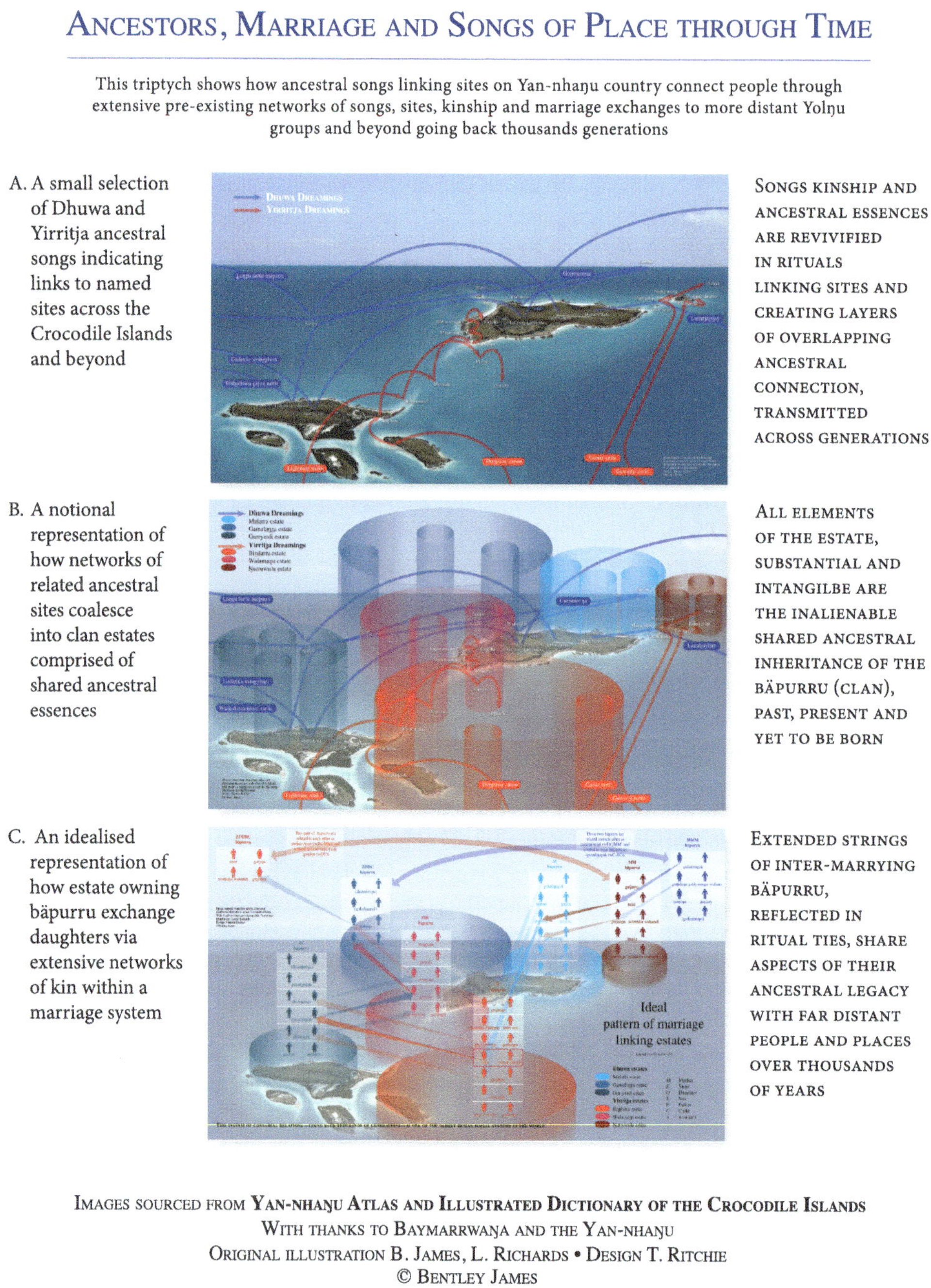

Figure 6.3 Ancestral geography, kin, *country*, and generations. (From Yan-Nhaŋu Atlas and Illutrated dictionary of the Crocodile Islands.)

Ancestral connections imbue commitment and resilience to life projects on *country* with a spiritual authority. The homelands movement has proved to be an extremely adaptive Indigenous project.[16] The durability of local population structures constituted through attachment to *country* demonstrate the cohesiveness, functionality and sustainability of life on homelands.[16] Old people say that residence close to sacred sites infuses life with ancestral power.[22] 'This one the same place, we were born here, our granddad born this place. We born this place, we get old this place. Our father, our mother they get old and died in this *country*. We still the same, we can't go any other place'.[23] Little surprise that these localised population patterns have been demonstrated by genes and thermoluminescence to have persisted for 50,000 years: 'Aboriginal people have always known that we have been on our land since the start of our time' (Kaurna Elder Lewis O'Brien).[24] Handing on the unique metaphysics of ancestrally inherited laws and knowledge of place is recognised as crucial.

6.3.3 Knowledge – Yiŋiya Teaching from *Country*, Learning from *Country*

In the 'Teaching from Country, Learning from Country' project Mark Yiŋiya Guyula (Member of the Legislative Assembly for the Northern Territory seat of Arnhem) describes the local environment around his homeland with added metaphysical dimension, as having a unique agency in knowledge.[25] Yiŋiya is a Liya-Dhälinymirr man and previous lecturer in Yolŋu studies at Charles Darwin University. When teaching from his *country* he says education is

> 'different to the education you get in the classrooms because the classrooms don't talk to you. We're learning out there under a tree, we're learning out there in the bush walking around, the trees are always communicating with you. The hills, the land, the air are always communicating, teaching you, and understands every need that Yolŋu children have to go through ... I can turn around, the *bulunu* (east wind) wind blows gently and gives me the feeling of what the stories are and the stories are automatically being told by the land itself, through me'.

The land, in other words, has a particular agency in the work of bringing up Yolŋu children in the knowledge of their environment.[25] Of Yiŋiya's homeland, he talks about winds, ancestral knowledge, and his determination to centre his 'Teaching on Country' on Yolŋu philosophies of knowledge, place, pedagogy, and technology. 'When Yiŋiya talks about the Bulunu wind, or the reef or the sound of the waves or the clouds ... these particular images are chosen because they link Yiŋiya in with his own particular ancestral songs – the particular totems and places and connections which make him who he is'.[25] These are powerful extant principles of knowledge through connection, of an undeniable ancestral edict to care for *country*, offering real motivation, driving local opportunity for sustainable, reliable, and prosperous life projects, many of which quietly endure 'under the radar' across the North.

6.3.4 From Time Immemorial

The people of Mapoon, a community in north-western Cape York, recall the connections amongst regional networks cooperating in care for *country*. They are the river people of Mapoon and, 'from time immemorial, they have known each other, married each other, celebrated together and hunted together on their land, rivers, and sea'.[26]

How every hunting place or patch of yams had someone to protect it. Harry Toeboy talked of his tribal home. 'The stick is the letter you write to another man. I would write Jerry, or to other places, saying "All right, the grass is to be burnt at my home." Jerry added, "Harry is talking about his home, that Pine River, and that been kept by his old man and he is the boss of the old place there now". It's a place where they can get a lot of kangaroo, wallaby, and many other things. When the time comes to burn, they get messengers and send them everywhere to different tribes. They send them up to Weipa, up to Morton (50 miles inland), they send them to the Batavia, they send them up to the Pennyfather, and send them up to port Musgrove. ...' Mrs Jimmy added. 'Generations survived ... they live and they know the water and the animals. They know each other's surroundings ... and they know their bush medicines and they pass it on; my mother passed it on to me, I passed it on to my daughter, my daughter passes to her daughter, and it will go for generations'. Mapoon Story.[26]

In some places, care for *country* has evolved into innovative customary and contemporary ecological services and natural resource-management programs. In addition to environmental health gains, caring for *country* positively influences the social determinants of health.[20] Skilled and detailed use of fire was responsible for the long-term productivity and biodiversity of this continent.[27] For example, Dean Yibarbuk, consultant and Nawarddeken traditional owner, has steered the alignment of local and broader community interest in fire into a million-dollar industry partnership producing social benefits on local terms. He says, 'We are much happier. We feel better. We like being back on *country* where our ancestors are'.[28] Developments in 'carbon farming' (Chapter 5) are seeing these carbon farming partnerships reproduced across the north of Australia. Without significant synergies and the social and cultural drivers of Indigenous land management (e.g. with cultural obligations to care for ancestral, familial *country*), development of Indigenous fire management-based carbon economy in North Australia would have little chance of success (Box 5.5). In the nexus between looking after *country* and *country* caring for its people, resilience is growing and prosperity for Traditional Owners and land managers improving.[29]

The concepts of connection, identity, knowledge, power, and seasonality are useful lenses through which local expressions of resilience can be seen as both unique to 'place' and related across the diversity of the North (Boxes 6.6 and 6.7).

'Under our law, the creators authorise and make us responsible for looking after the *country*. In our law, this authority for *country* is called Uunguu. We are happy to share our *country* with visitors, the government, or companies. This must be done, however, in a way that does not break our law'.[30]

The remarkable durability of attachment to *country* and the cohesive locality based values invested in local populations constitute significant assets, and are drivers for long-term future investment by Indigenous people in the northern economy, in ways that do not break the 'law'.

6.4 INTERVENTIONS FROM *COUNTRY*: RESISTANCE AND RESILIENCE

'I bin thinkin' this bin Gurindji country. We bin here longa time before them Vestey mob'. Vincent Lingiari and the Gurindji campaign. This famous campaign and many others, such as the 1946 Pilbara strike and the 1963 Yirrkala bark petition, for example, express similar, poignant though apparently separate resistance, in this latter case contributing to the design and passing of the *Aboriginal Land Rights (Northern Territory) Act 1976*[31] (Chapter 3) – the high-water mark of Indigenous land rights in Australia. Other examples of resistance, whether in the public eye, flying under the radar, or misunderstood, are part of the substance of every Indigenous community.

6.4.1 Yirrkala Bark Petition 1963

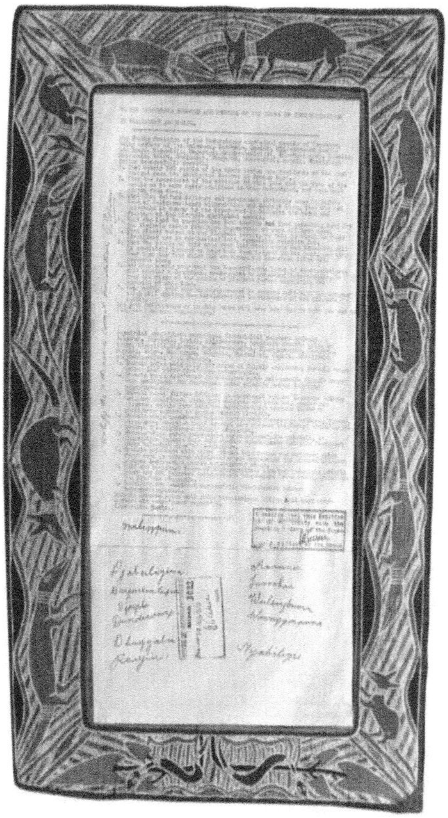

Bukudjulni gonga'yurri napurrunha Yirrkalalili yulnunha malanha Balamumu, Narrkala, Gapiny, Miliwurrwurr, nanapurru dhuwala mala, ga Djapu, Mangalili, Madarrpa, Magarrwanalmirri, Djambarrpuynu, Marrkulu, Gumaitj, Galpu, Dhaluangu, Wangurri, Warramirri, Naymil, Riritjingu malamanapanmirri djal dhunapa.

1. Dhuwala yolnu mala galki, 500 nhina ga dhiyala wanganura. Dhuwalawanga Arnhem Land yurru djaw'yunna naburrungala.
2. Dhuwala wanga djaw'yunna ga nhaltjana yurru yolnunundja dhiyala wanga nura nhaltjanna dhu dharrpanna yolnu walandja yakana lakarama madayangumuna.
3. Dhuwala nunhi Welfare Officers ga Government bungawa lakarama yolnuwa malanuwa nhaltjarra nhuma gana wanganaminha yaka nula napurrungu lakarama, walala yaka lakarama, walala yaka lakarama Governmentgala nunhala Canberra nhaltjanna napurru ga Guyana yolnuyu Yirrkala.
4. Dhuwala wanga napurrnyu balanu larrunarawu napurrungu nathawu, guyawu, miyapunuwu, maypalwu nunhi napurru gana nhinana bitjarra yinathilimirri, napurru dhawalguyanana dhiyala wanganura.
5. Dhuwala wanga yurru dharrpalnha yurru yolnuwalandja malawala, ga dharrpalnha dhuwala bala yolnuwuyndja nhinanharawu Melville Bay thurruwanga balandayu djaw'yun nyumukunin.
6. Dhuwala yolnundja mala yurru nhamana balandawunu nha mulkurru nhama yurru momaga daranun yalalanumirrinha nhaltjanna dhu napurru bitjarra nhakuna Larrakeahyu momara walalanguwuy wanga.
7. Nuli dhu bungawayu House of Representatives djaw'yun yulnuwala nathili yurru nha dhu lakarama interpreteryu bungawawala yolnu matha, yurru nha dhu djaw'yun wangandja.
8. Nunhiyina dhu marrlayun marrama'ndja nhinanharawu yolnuwu marrnamathinyarawu.

Dhuwala napurru yolnu mala yurruliyamirri yama bitjan bili marr yurru napurru nha gonga'yunna wagarr'yu.

To the honourable speaker and members of the House of Representatives in Parliament assembled:

The Humble Petition of the Undersigned aboriginal people of Yirrkala being members of the Balamumum, Narrkala, Gapiny, Miliwurrwurr people and Djapu, Mangalili, Madarrpa, Magarrwanalinirri, Djambarrpuynu, Gumaitj, Marrakulu, Galpu, Dhaluangu, Wangurri, Warramirri, Naymil, Rirritjingu tribes respectfully showeth.

1. That nearly 500 people of the aforementioned tribes are residents of the land excised from the Aboriginal Reserve in Arnhem Land.
2. That the procedures of the excision of this land and the rate of the people on it were never explained to them beforehand and were kept secret from them.
3. That when Welfare Officers and Government officials came to inform them of decisions taken without them and against them, they did not undertake to convey to the Government in Canberra the views and feelings of the Yirrkala aboriginal people.
4. That the land in question has been hunting and food gathering land for the Yirrkala tribes from time immemorial; we were all born here.
5. That places sacred to the Yirrkala people, as well as vital to their livelihood are in the excised land, especially Melville Bay.
6. That the people of this area fear that their needs and interests will be completely ignored as they have been ignored in the past, and they fear that the fate which has overtaken the Larrakeah tribe will overtake them.
7. And they humbly pray that the Honourable the House of Representatives will appoint a Committee, accompanied by competent interpreters, to hear the views of the people of Yirrkala before permitting the excision of this land.
8. They humbly pray that no arrangements be entered into with any company, which will destroy the livelihood and independence of the Yirrkala people.

And your petitioners as in duty bound will ever pray God to help you and us.

(Petition of the Yolngu people of Yirrkala 14 August the Australian House of Representatives 14 and 28 August 1963 Parliament House, Canberra)

6.4.2 Laurie Baymarrwaŋa's Legacy

Laurie Baymarrwaŋa (1917–2014),[32] born on the outer Crocodile Island of Murrungga, did not speak English, and was recognised as Senior Australian of the year 2012 for her unceasing work to teach kin the ancestrally inherited language of *country* and site-based local knowledge – lessons clearly only she could give:

'Nhaŋu dhäŋuny yuwalkthana Yolŋu mit̲tji marŋgiyini mana dhana mayili mana dhäŋuny mana limal-amagu ganatjirri maramba barrathalayuma gurrku mana waŋgalaŋga'.

'We will pass on the stories (wisdom) of our sea *country* for the new generation to make it strong'.

Laurie Baymarrwaŋa (1999)[17]

Baymarrwaŋa's vision created a homeland, a bilingual school, and employment programs to care for her sites, family, and *country*.[33] She knew that kin living on homelands situated on *country* are embedded in a pre-existing ancestral geography that naturally enhances the reproduction of local knowledge, language, identity and local power. Alive at hundreds of homelands across the North, people continue to teach local knowledge through local language, where the connection between ancestral languages and *country* is strongest, and ancestral identities and kinship matter in relationships. Baymarrwaŋa's homeland, bilingual school, and Ranger and heritage programs are designed specifically to pass on the ancestrally inherited local knowledge inhering in the Yan-nhangu language of the Crocodile Islands. Her motivation to enhance the linguistic, cultural, and biological diversity drove her to develop Indigenous ecological knowledge (IEK) maps, language nests, an online IEK teaching program, a trilingual dictionary, and an atlas, to promote her language, which, while not officially banned, is still effectively discriminated against on a number of levels.[34]

With the wisdom of one in her nineties she could see the necessity to foster local resources and modern technologies to build livelihood activities and a culture-based economy for her kin on the islands.[35] Her indefatigable energy to drive a family of projects based on continuities and innovations in local knowledge, local language, local identity, local governance was near endless, demonstrating extraordinary resilience prospering with insubstantial resources.[35] Baymarrwaŋa's vision gives us a glimpse of the potential power of local developments, realisable through respectful and trusting relationships around mutual interests. Joe Morrison celebrates her thus: 'We must support nation-building leaders like her who work to save Australia's unique languages and culture, champion the deep knowledge of *country*, give courage and conviction to those who live on their homelands and struggle against immense hardship'.[33]

6.4.3 Yuendumu Doors

The school at Yuendumu (an aberration of the Warlpiri place name Yurntumu – a honey ant Dreaming area), about 300 km north west of Alice Springs, has a strong history of community (predominantly Warlpiri and Anmatjerre) participation in its education practices; a strong Aboriginal school council; annual school *country* visits to customary homelands; historically strong bilingual/bicultural program; very hard working, highly productive literacy centre; and more. These attributes have all been hard fought resistance to the overwhelming influence of western style education, its modes of delivery and ethnocentric policy settings. Concern that local lore, language and culture be given at least equal status to English has fuelled this resistance but its strength and influence in the school ebbs and flows in the face of (fickle) outside control.

> 'They [our old people] instructed us in Warlpiri law and told us not to forget what we had been taught; told us to hold on to our law and follow it the right way' (Tess Ross[36]).

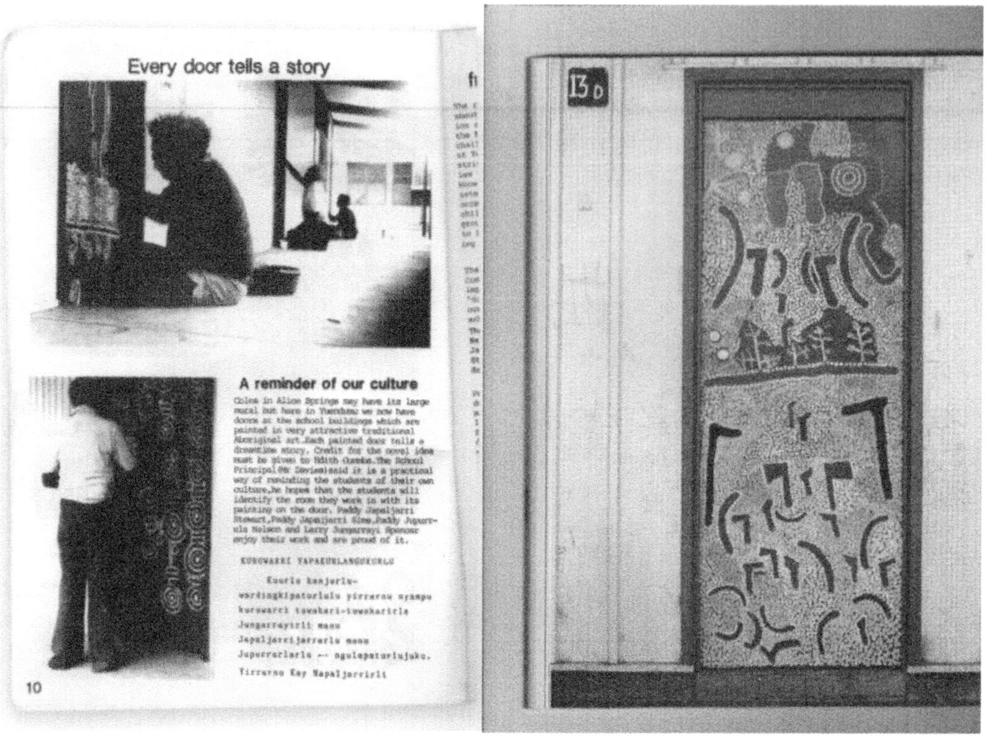

Warlpiri men painting the Yuendumu school doors, and *Purlka-purlka kurlu*.[38]

Responsibility and concern for learning comes right from the top. Senior Warlpiri and Anmatjerre men, worried that education at school did not involve enough lore, took an opportunity to colour the school with *Jukurrpa* (all-encompassing system of belief, meaning and lore) by painting *kuruwarri* (symbolic designs), imbued with stories of creation, that amongst other things reinforce identity of all the kids by showing connection with their ancestors and their essence in the land. *Kuruwarri* were painted on the classroom doors to be part of the kid's everyday experience of learning at school. Imbued in the public process of painting the doors were the key kin and moiety relationships of *kirda* and *kurdungurlu*, the place-based context of meaning, related (higher) forms of language, nuanced stories of ancestral creation with messaging about discipline and behaviour and other valuable moments of unique and resilient Warlpiri-ness.[37]

The school doors hold many messages for Yuendumu people, some about the adaptability of forms of customary pedagogy and those men and women of high degree responsible for it, and some about resisting normalisation processes. There's strong messaging there also for the education department (that has undermined bilingual education for example), that the things that make kids Warlpiri and Anmatjerre, and that make them confident, strong, and adaptable like their old people, are not being passed on to them well enough through the mainstream school system. Language (a critical ingredient) is about identity and survival.[38] Indigenous voices repeat with piercing clarity what Yothu Yindi put so distinctly 'Power to the people, power to the land, power for cultural revival, power, power, power for survival…'.[39]

6.4.4 Barunga Statement

We, the Indigenous owners and occupiers of Australia, call on the Australian Government and people to recognise our rights

- To self-determination and self-management, including the freedom to pursue our own economic, social, religious, and cultural development.
- To permanent control and enjoyment of our ancestral lands.
- To compensation for the loss of use of our lands, there having been no extinction of original title.
- To protection of and control of access to our sacred sites, sacred objects, artefacts, designs, knowledge, and works of art.
- To the return of the remains of our ancestors for burial in accordance with our traditions.
- To respect for and promotion of our Aboriginal identity, including the cultural, linguistic, religious, and historical aspects, and including the right to be educated in our own languages and in our own culture and history.
- In accordance with the universal declaration of human rights, the international covenant on economic, social and cultural rights, the international covenant on civil and political rights, and the international convention on the elimination of all forms of racial discrimination, rights to life, liberty, security of person, food, clothing, housing, medical care, education, and employment opportunities, necessary social services, and other basic rights.

We call on the Commonwealth to pass laws providing

- A national elected Aboriginal and Islander organisation to oversee Aboriginal and Islander affairs.
- A national system of land rights.
- A police and justice system which recognises our customary laws and frees us from discrimination and any activity which may threaten our identity or security, interfere with our freedom of expression or association, or otherwise prevent our full enjoyment and exercise of universally recognised human rights and fundamental freedoms.

We call on the Australian Government to support Aborigines in the development of an international declaration of principles for indigenous rights, leading to an international covenant.

And we call on the Commonwealth Parliament to negotiate with us a treaty recognising our prior ownership, continued occupation and sovereignty, and affirming our human rights and freedom.

Council of Aboriginal Reconciliation (1998)

6.4.5 Yirrkala 2008: Petition Presented to the Prime Minister of Australia

Nhanaburru, wangkanmala bapurru dhimirrunguru, arnhemland, nganaburrungu ngurrngu delak mala, nganthun yukurra nhuna 26th Prime Minister Australia-wu. Nhukala ganydjarr'yu nhunhi nhe ngurrungu walalangu malangura nhuma walala rrambangi, Australian Parliament-ngura, ga ngurrungu Dharuk-mirri nhangu Garraywu Queen Elizabeth-gu, yurru nhandarryun-marama djinawa-lili Australian-dhu luku-wu rom-dhu yurru dharangan ga galmuma nganapurrungu dhangang ga bukmak nha-mala nhanapurrungu:

(Continued)

- *Nhanapurrungu walnga-mirri dhukarry ngudhudal-yana.*
- *Nhanapurrungu, wanga, wanga-ngaraka ga nguy gapu, ngunhi dhimirrunguru, arnhemland.*
- *Dharrima gungnharra, warkthunara, lukunydja rrupiya-yu wanga-wuy-ga gapu-wuy ga dhangangnha-yana ga lukunydjana yana.*
- *Dharray walnga-wuy ga djaka yurru nhanapurrung-gala-nguwu djamarrkuli-wu yalalangu-wu.*

Dhuwalanydja rom dhuwalana bilina.

Dhuwalanydja rom wawungu wanga-wuy ngandarryunmarama Australian-gala bapurrulili.

Nganapurru marrliliyama nhukula ngurru-warryun-narayngu, marr yurru Commonwealth Parliament ngurru warrwun ga dharangan dhuwala rom ga marryuwak gumana dhayutakumana lukunydja rom.

We, the united clans of East Arnhem land, through our most senior Delak, do humbly petition you, the twenty-sixth prime minister of Australia, in your capacity as the first amongst equals in the Australian Parliament, and as the chief adviser to Her Majesty Queen Elizabeth the Second, to secure within the Australian Constitution the recognition and protection of our full and complete right to:

- Our way of life in all its diversity
- Our property, being the lands and waters of East Arnhem land
- Economic independence, through the proper use of the riches of our land and waters in all their abundance and wealth
- Control of our lives and responsibility for our children's future

These rights are self-evident.

These rights are fundamental to our place within the Australian nation.

We ask for your leadership to have the Commonwealth Parliament start the process of recognition of these rights through serious constitutional reform.

6.4.6 Milingimbi Makarraṯa – 'Bringing the Past into the Future'

The cultural houses of colonial society have traditionally objectified Indigenous cultures through the collection, study, and display of artefacts. New generations are demeaned by what Said described in another cultural context as 'orientalism',[40] and although some museums etc. are anxious to return cultural items or learn more about them, the appropriate means to do this are not always obvious or easy from afar – particularly considering poorly understood impact of spiritual agency in objects and local dynamics around knowledge and custodianship. In Milingimbi, Arnhem Land NT, customary conflict resolution strategies are being deployed to help explain and resolve relationships with institutions holding items of significance in their collections. An invitation to museum and art gallery directors to participate in a Makarraṯa conflict resolution ceremony with the title 'Bringing the past into the future' was used to encourage new relationship building.[41]

Makarrata' ga maŋutji-lakaram baman'ŋuwuy ŋurrunaŋalpuy ganydjarr bukmakku bapurru'wmalaŋuw ŋurrunaŋalpuy ganydjarr ŋunhi ŋilimurruny ga waŋgany manapana waŋany mit'tji limurruŋy ga yarratalil waŋgany manapana ŋilimurruŋgal ŋalapalmirriwal ga ŋalapalmirr ŋunhi gupaganyumirrnha mala ŋunhi walal ŋäthilmirryinan birr'baman.

Dhuwalatjana buku-luŋmaranhamirrwurr ŋilimurr ga lap'marram marr-dharaŋanamirr dhukarrnydja dhuwalajana mar-dharaŋanamirrgurr romgurrnytja. Nhanamirra bala'raliyunmirra ŋilimurrdhu rrambaŋi buku-law'marraman ga dhawal'yuna yu_tanha ŋayaŋunhirrpanminyaraw ŋunhi nhaltjan ŋanapurrdhu guŋga'yunamirr ga _lap'marram yu_tawnha ga ralk'thunamirriw romgu.

Dhurrwara ŋilimurr ga _lap'marraman ŋunha bala birr baman'ŋuwuy ga dhuwana bala dhiyaŋubala ŋilimurr ga gupa-garin ga dhawal'yun yu_taw'nha guyaŋanharay ralyuny malaŋuy.

Rrambaŋi ŋilimurr yaka warray rom ŋamaŋamayunmirra ga ral-manapanmirra dhuwalatjan romgurr ŋilimurruŋ yolŋu'yulŋu dharranhayŋu nhiniŋu ral-manapanmirrwurr ga dharaŋananmirriwurr romgurrnydja.

Limurrdhu dhukarr malŋmarramyutan marr-dharaŋananmirryirra marr ga limurrdhu walŋa limurruŋuwuy limurr lap'marrama ga bala raliyunmirra guŋgayunmirra walŋa yuta'thu ŋanbaw'yun mak'makthunawurra romgurrnydja. Ga marr-dharaŋanamirrgurr romgurrnytja limurruwuynha limurr ga bawalamirriny.

Makarrata signifies the ancient power of the Yolngu clan nation; the ancient power that really connects us not only to our forefathers and their forefathers before them, but us to our ancestral law.

Through this gathering, through this dialogue and this event, through listening together we can come to peel off a new layer of hope around how we share.

We are opening a door from the past to the present and are now entering into a realm of new possibilities.

Here from now on let us all come together in the spirit of Makarrata to light up the ancient flame which is profoundly laid under the core essence of our Yolngu cycle of life.

Together we will not only celebrate this unique and ancient event in Yolngu culture but through engaged and authentic dialogue, we can redefine new meanings for sharing our lives and new, respectful ways of being, in understanding ourselves and each other.[41]

6.5 REINFORCING CULTURE THROUGH LANGUAGE AND LEARNING

Language is the repository of and vehicle for the transmission and preservation of customary knowledge connected to *country*. In the triangulation between language, land, and people, language is the medium of communication and an important signifier of identity and authority. As Murray Garde observes (Box 6.2), grammatical categories reflect unconscious cultural patterns of thought and action, with insights into how humans structure perceptions of the world. As language contains and expresses the worldview of its speakers, so too the language of choice is the means of empowerment; a key component of building resilience. Locality-based Indigenous languages, often associated with homelands, provide positive health outcomes and psychological resilience, and protect traditional law and local cultural and biological diversity.[20]

BOX 6.2 THE IMPORTANCE OF LANGUAGE

MURRAY GARDE

Land management in northern Australia is a cross-cultural exercise. It might seem trite however to state that a language encodes the cultural worldview of a people. We all know this, so it makes sense to explore what languages can tell us about how people variously conceive of landscapes, ecological processes and natural species. Linguistic inquiry is central to an understanding of the cognitive and social practices of a society in relation to land management. Grammatical categories are seen as reflecting unconscious cultural patterns of thought and action, and provide insights into how humans structure perceptions of the world. Understanding how Aboriginal people in northern Australia have managed land requires an understanding of how human ecology is encoded in language. Does your land management program have a place for the maintenance of environmental insights expressed through the language of the traditional owners?

Here are some issues for consideration, based on experiences managing *Bininj* (local word for person – now meaning Aboriginal person) lands in Kakadu National Park and Indigenous Protected Areas in western and central-north Arnhem Land:

- How are different ecological zones referred to in the local languages? What are the major categories and are these categories further divided into more fine-grained terms for various aspects of the landscape? Be aware that the cognitive categories are most likely to be very different to those in English.
- Do terms for types of *country* encode human interaction with it? These can be terms that express affordances or hindrances. Here are examples from *Bininj Kunwok*:
  Bambarr: 'A dead-end gorge or valley where fire can trap kangaroos ready for spearing'.
  Kangalkwarre: [literally: it-cliff-bad] 'impassable cliff/scree'
  Kawarddedubbe: [literally: it-rock-blocked] 'impassible rock, rocky country that cannot be negotiated'

Language about social categories and relationships to land, and the inherent responsibilities certain social groups have to manage land, is full of richly metaphoric terminology. In *Bininj Kunwok*, a traditional owner of a clan estate is known as 'the right person' *na-redweleng* (if a male), *ngal-redweleng* (if a female) and *birri-redweleng* (plural). Such a person would call an estate their 'father's *country*' and they collectively exercise the ultimate authority over such an estate. However, people also have certain rights in relation to their mother's *country* and both a person who calls an estate their mother's *country* and the *country* itself are referred to by the noun *karrardwarreken*, 'that of my mother's'. An alternative way of referring to one's mother's *country* is the verb [pronoun.prefix]+*bolkkuykme*. This latter verbal form of reference literally means '+to spray [from the mouth] on the *country*', the etymology of which refers to the ritual of a new visitor to a site being introduced to the ancestral spirits of that place by being sprayed with water by a person whose mother is a member of the land owning clan. An alternative expression is: *nga-bolknunjyo:* literally 'I-place-saliva-lie' = my mother's land' which recollects the saliva of a baby on its mother's breasts.

Another area of enquiry is what linguists call sign metonymy. These are words or expressions normally used for a particular referent, which are also used to denote something associated with it. These associations might involve the co-occurring seasonal

(*Continued*)

BOX 6.2 (Continued) THE IMPORTANCE OF LANGUAGE

availability of species, such as a fish that is plentiful when a particular tree is flowering or fruiting. Here are three examples from *Bininj Kunwok*:

Bokorn: 1. *Syzygium eucalyptoides* ssp. *bleeseri* (a fruit tree) and 2. Spangled grunter fish, *Leiopotherapon unicolor*. The fruiting of the *Syzygium* tree is associated with the appearance of schools of spangled grunters which are plentiful at this time of year as they migrate to breed.

Ngarnarrh: 1. black cockatoo, *Calyptorhynchus banksii*, and 2. *man-ngarnarrh*, *Owenia vernicosa* (a tree with fruit that is non-edible for people). The connection between the bird and the tree is that black cockatoos eat the fruit of *Owenia vernicosa*.

Alyurr: 1. Lightning, lightning spirit, being; 2. Leichhardt's grasshopper, *Petasida ephippigera*; and 3. *Pityrodia* shrubs. The connection is that Leichhardt's grasshoppers appear only in the build-up season when lightning is prevalent and they are coloured red, which is believed to attract lightning. Thirdly, the *Pityrodia* shrubs, also called *alyurr*, are the sole food of Leichhardt's grasshoppers.

Australia has the largest and most rapid loss of languages anywhere in the world over the last century.[42,43] Despite official policies on language diversity,[34] culturally inclusive methodologies,[44] and bilingual education,[45] there is inconsistent and insubstantial support.[46] Newly emerging place-based education initiatives linked to Indigenous land management (ILM) (including Ranger groups) and school-based *learning on country* programs are engaging students and local knowledge authorities in bi-cultural education – improving attendance, literacy, and numeracy, local cultural capital and assisting local languages[47] with outstanding success.[48] The success of bilingual education programs, attached to recognised beneficial ILM projects, and *new* education initiatives, is in the scaffolding of new information on the basis of existing linguistic and cultural knowledge. New information is being contextualised within local knowledges, authority structures, identity and connectedness, augmenting resilience and prosperity (Boxes 6.2 through 6.5). Sadly, the high level of Indigenous multilingualism present in the North and other remote parts of Australia goes unrecognised and neglected by the commentators, experts and policy writers of the staunchly monolingual settler society.

Languages, like worldviews, are systems of coordinates that divide, discriminate, and order reality with profound implications for cognition, social relations, navigation, memory, and relations to *country*.[49] In Yuendumu, the old men painted the school doors to instruct children with nuanced stories of ancestral creation, discipline, behaviour, and other valuable moments of unique and resilient Warlpiri-ness.[37] In his *Teaching on Country*, Yiŋiya understands Yolŋu language to be the medium through which an added metaphysical dimension of agency in knowledge on *country* is communicated by the land itself.[25] Baymarrwaŋa's life project was to pass on ancestrally inherited knowledge and law living in the Yan-nhangu language of the Crocodile Islands.[17]

At hundreds of homelands across the North, people continue to teach local knowledge through local language, where the connection between ancestral languages and *country* is strongest.[50] Ancestral endowment of language is not just associative but a fundamental ontological truth.[22] Ranger programs, bilingual education in ILM (Box 6.5), and work on homelands are mechanisms extant and emerging to support the transmission of language and its ineffable metaphysical linkages to *country*.[46] This is true also where local languages are fragmentary and under pressure, such as where towns and cities dominate the cultural landscape. Linguists have long recognised that 'Aboriginal English' practices, grammars, and lexicons constitute a living continuity with the language of *country*.[51] The will to maintain identity through connection with kin and customary lands is focusing on use and reengagement with local language – evident at many levels across the North.

BOX 6.3 RECOGNISING INDIGENOUS KNOWLEDGE IN A WESTERN EDUCATION PARADIGM

ERICA MCCREEDY

Indigenous Education

Despite the implementation of numerous Government policies for Indigenous engagement in universities, Indigenous Australians remain under-represented. Remote and very remote Indigenous communities have access to fewer education services and those that exist inadequately recognise the strengths and abilities of Indigenous children.[1] Over reliance on direct classroom based instruction devalues other approaches such as Indigenous traditional modes of teaching.

Land and Sea Management as a Tool for Further Learning

The North Australian Indigenous Land and Sea Management Alliance Ltd (NAILSMA) uses land and sea management based activities to achieve employment-ready training and higher education participation outcomes. Growing interest by Indigenous participants in leading and delivering research and management programs on-*Country* combine traditional and scientific knowledges. Through participatory action research, appropriate locally driven educational activities that promote Indigenous knowledge systems become a vehicle for the effective transfer of Western scientific and technical skills for Indigenous students.

The significance of local and traditional knowledge to higher education lies in its capacity to engender confidence, enhance psychological resilience and promote positive personal identity in learners. This is particularly significant where mainstream language and culture are effectively foreign to the local learning environment.

Case Study: Trialling the Certificate III Indigenous Land Management

With agreement from community members, the Certificate III Indigenous Knowledge units would be trialled to identify six units within this course where Indigenous knowledge could be formally recognised. Several senior Traditional Owners living in Maningrida (Northern Territory) and surrounding homelands, Indigenous school staff and Rangers shaped the pilot project - as was the case in Yirrkala (NT). Each Indigenous Knowledge unit states:

> 'Assessment of this unit must be conducted by recognised and appropriate Community Elders and/or Custodians with appropriate assessor qualifications or co-assessed by a qualified assessor in cooperation and discussion with appropriate Community Elders and/or Custodians.'

This directive provides a strong and clear statement that appropriate community members *must* remain the trainers and assessors of these units, identifying the content and methods of delivery well into the future.

ILM Training Framework

The development of an ILM Framework involved participants breaking down each of the unit elements, translating them into their preferred language and then developing a culturally appropriate learning framework focused on delivering nationally accredited

(*Continued*)

BOX 6.3 RECOGNISING INDIGENOUS KNOWLEDGE IN A WESTERN EDUCATION PARADIGM

training in first language. Further, it supports Indigenous leaders to be trainers and training co-facilitators to provide employment and engagement of senior local people as consultants and co-assessors of local knowledge.

Indigenous land management training framework				
Step 1 Involvement • Community Elders and traditional owners are involved in the delivery and assessment of the units • Qualified trainer and assessor to work with community Elders • Community participants involved in the training • Note: Participation may depend on where the training is held and whose *country* it is	Step 2 Community directed • Community Elders and traditional owners identify the workshop topic • Workshops are held in preferred language and then translated to English as required	Step 3 Indigenous land Management (ILM) unit • Choose the ILM traditional knowledge unit that the workshop is based around • In groups, break down the ILM unit, translate into preferred language and discuss the meaning in a local context	Step 4 Practical activity • Community Elders and traditional owners will identify the practical activity for students to participate in and demonstrate their learnings and knowledge	Step 5 Assessment • Community Elders identify if people have the knowledge • Note: Delivery and assessment against this standard must comply with community protocols and guidelines and be supported by elders and custodians of *country*.

As a result of this pilot project, the Certificate III in Indigenous Land Management was put on scope in the NT in June 2016.

REFERENCE

1. Lamb, S, Jackson, J, Walstab, A and Huo, S (2015). *Educational Opportunity in Australia 2015: Who Succeeds and Who Misses Out*. Centre for International Research on Education Systems, for the Mitchell Institute, Melbourne, Australia: Mitchell Institute.

6.6 HOW CAN REMOTE COMMUNITY GOVERNANCE BE REDISCOVERED TO BUILD RESILIENCE?

6.6.1 Rediscovering Resilience in Complex Community Life

Remote Indigenous towns, settlements, or communities are normally comprised of Traditional Landowner groups making up a small minority amongst numbers of others, attracted or forced off their homelands elsewhere in the region – whether by missions, government rations, or buffalo camps and tobacco.[52] The same kind of ancestral stories and manifestations animate townscapes as they do wider *country*, creating extraordinary social adaptations and vulnerable behavioural protocols. These communities come under various types of governing legislation (e.g. Aboriginal Land Rights [1976] Act, Local Government Acts, Aboriginal Associations Act). Their councils and shires are generally poor, providing substandard services; a plethora of government, NGO, and other agencies fill gaps and undertake works in largely uncoordinated and ad hoc manners, bringing their own authority and mandated agendas. Each service provider, funder and representative organisation adds a layer of governance with all the requisite committees, KPIs, and reporting requirements to the already complex intercultural environment. Over-governance is an issue both for Indigenous people whose homes these places have been now for many generations, and for agencies wanting to deliver services more effectively and build resilience and capability. Recognising this complexity is the first step to recognising Indigenous resilience beneath it. The tendency of government and local organisations to only engage the visible and oft superficial community structures rather than 'real' underlying community governance arrangements, inevitably invests resources and time in processes and structures likely to fail, erode local authority structures, and exacerbate conflict and apparent dysfunction, with the community characteristically being 'blamed' for the failure.[53]

This reflects the predominant policy perspective that resilience results from what is done to or for people. This fails to accommodate the fact that, when dealing with complex social and environmental challenges, the key lies in engaging with people in an empowering framework. For this framework to be developed and enacted appropriately, it is essential to understand that it is a two-way process; it reflects the co-creation of solutions and their sustainable implementation through the interaction between empowered communities (and their members) and empowering settings (provided by broader societal agencies). Empowerment is essential to creating trust.[4] The former, the factors that underpin empowered communities, is a function of the social, cultural, spiritual, and environmental capital of Indigenous peoples. This must be accommodated in planning and intervention, and to do so it is essential to adopt a bottom-up approach that can accommodate local diversity (e.g. regarding needs, goals, and sociocultural characteristics, and so forth). In this context, the propensity for attributing failure to local communities is thus more accurately understood as a function of mainstream societal agencies failing to understand the empowerment concept or choosing to focus on only one aspect thereof. An example of the benefits of an empowerment-based approach is evident in the work of the Aboriginal Research Practitioners Network (ARPNet) and NAILSMA (see below).

The success of the ARPNet (Box 6.4) in navigating and respecting complex community life in their research contracts is no surprise. Working in the local languages enables all the nuances of meaning (mentioned earlier) to be captured, immediately valuing local authority, identity, and relationship to the underlying customary landscape and local history. Such active demonstrations of trust, not always recognised by outside partners, provide local community-led relationship initiatives based on and able to build on inherent resilience features – such as the capacity to adapt cultural learning to new education spaces and technological capacities (Boxes 6.2 through 6.6).

BOX 6.4 LIVING WITH *WIDDIJITH* – PROTOCOLS FOR BUILDING COMMUNITY RESILIENCE

BEVYLINE SITHOLE, HMALAN HUNTER-XENIE, DEAN YIBARBUK, CHERRY DANIELS, GRACE DANIELS, AND OTTO BULMANIYA CAMPION

ABORIGINAL RESEARCH PRACTITIONER'S NETWORK (ARPNET)

Building community resilience poses some conceptual as well as practical problems and raises important philosophical and ethical questions with regard to the role of government and outside organisations. Community-based researchers from ARPNet explored community resilience in two remote communities in the 'Top End' of the Northern Territory using ARPNet's participatory tool box by/for community based *Bininj* researchers – 'dilly bag' and surveys.[1]

Findings from the study indicate that *Bininj* people consider their resilience within the broader scope of their living traditions, stretching back thousands of years, and surviving natural disasters and harsh environments. Secondly, *Bininj* felt that their ability to build resilience in the present is often challenged, or inhibited by government ineptitude, failure to engage genuinely, and damaging policies. Overwhelmingly, communities, families and groups in remote areas talk about the need for government and outside organisations to leave space for them to build resilience, their way, in their spaces (sometimes in hidden arenas) and on their own terms. These communities see a need for government involvement (and want them to be involved) but, they say government can, and needs to, engage in a more enabling and empowering way than current policies allow (Figure B6.4.1).

Figure B6.4.1 ARPNet team members, from left to right, back row: Bev, Christine, Calissa, Grace, Geraldine, Patrina, Marjorie, Hmalan, Antony, Howard; front row: Debra, Cherry, Serina, Cherry-Anne, David, Dean and Otto (missing 7 team members). (Courtesy of Nathan Maddock.)

(Continued)

BOX 6.4 (Continued) LIVING WITH *WIDDIJITH* – PROTOCOLS FOR BUILDING COMMUNITY RESILIENCE

There is wide acceptance among policy makers and researchers that current government services are ill-equipped to deal effectively with bushfires and natural hazards in remote areas.[2–4] International and Australian research shows that there is dissatisfaction with top-down, technocratic approaches to disaster planning and response. Questions include: Who will plan and deliver emergency services in remote communities; how can local needs be addressed within the existing framework of disaster relief; and what are local expectations? Are these policies producing the right outcomes, and closer to home, which groups, families and people are working together on these policies and how?

Building Community Resilience Needs to Be Based on Mutually Agreed Principles between Remote Communities et al. and Government

The first of these principles is mutual trust. Indigenous people feel that government and outside organisations have a 'problem with trust'. People say government does not trust their culture or lore, government is dismissive of important things like ceremony, they don't trust Aboriginal ways of doing things. The elders sit, watch and most 'cry for their people and country' because no one is coming to talk to them about things they know about and have lived with.[5]

The second big principle is mutual respect. To work together, Indigenous people say there should be a recognised interdependence, a respect of one's knowledge system and practices which is missing at the moment.

The third big principle is mutual recognition. Currently, Indigenous people feel they have to ignore or hide their lore and practices because existing frameworks neither recognise nor give room for them. 'Ceremony and lore' are a massive part of managing *Widdijith* (cyclones); people feel the lack of recognition of this crucial element in their management/ and response to emergencies – as an impressive example of *'a lack of recognition'*. Notwithstanding the tremendous work done by rangers, it is important to recognise that rangers are not in a position (and are not expected) to fulfil all the requirements for proper caring for *country* – this is rightfully the role and responsibility of traditional landowners.

Conclusion

Resilience is not possible without strengthening the core of society – the family, the belief system and relationships with and responsibilities to *country*; creating adequate and appropriate spaces that will allow families to regroup, reconnect and become stronger, ensure social capital and provide the foundation for resilience. Our bold productive suggestion is for government to consider alternative, more constructive ways of engaging, creating space/opportunities/support for communities to find themselves again, and rebuild as only the people, families and groups inside Indigenous communities can. Recognising *and engaging with the* value and relevance of *Bininj* belief systems, of kinship and ancestral laws of living on and caring for *country*, is crucial for bolstering these key pillars.

The big message to government: it is not only about addressing Aboriginal wellbeing, about fixing the fundamentals of how Indigenous policy and action is formulated and applied.

(Continued)

BOX 6.4 (Continued) LIVING WITH *WIDDIJITH* – PROTOCOLS FOR BUILDING COMMUNITY RESILIENCE

It is about the uneasy tension between world-views where one overwhelms rather than accommodates the other. Current experience indicates that affected communities would like to see a more open, inclusive and responsive types of engagement consistent with the tenor of national and international strategic policies on disaster risk reduction.

ACKNOWLEDGEMENTS

ARPNet co-researchers included S Namarnyilk, E Narorroga, O Dann, K Dirdi, G Nayilibibj, E Phillips, K Daniels, A Daniels, G Daniels, G Turner, CA Daniels, T Daniels, P Thomas, D Thomas, T Rami, C Brown. We are so very grateful for the welcome, interest, permission and participation of the Traditional Owners, Djungai and communities of Ngukurr and Gunbalanya.

REFERENCES

1. ARPNet, Sithole B (2012). *The ARPNet Dilly Bag: A Practical Field Guide to Participatory and Other Research Tools for Use by Aboriginal Research Practitioners in Australia.* ARPNet, Research Institute for Environment & Livelihoods, Charles Darwin University, Darwin.
2. COAG (2004). *Natural Disasters in Australia: Reforming Mitigation, Relief, and Recovery Arrangements.* Commonwealth Department of Transport and Regional Services, Canberra.
3. Gould J, Sithole B, Campbell A, James G, Sutton S (2014). Building community resilience to natural hazards in Northern Australia. *Proceedings of the Research Forum, Bushfire and Natural Hazards CRC & AFAC Wellington Conference*, 2 September 2014.
4. Sangha KK, Sithole B, Hunter-Xenie H, et al. (2017). Empowering remote Indigenous communities in natural disaster prone northern Australia. *International Journal of Mass Emergencies & Disasters*, in press.
5. Seiffert M (2013). The trauma of the government taking control: Roper River Mission (Ngukkur), Northern Territory. *Journal of the Anthropological Society of South Australia* **37**, 90–106.

BOX 6.5 NORTH AUSTRALIAN BUSHFIRE & NATURAL HAZARD (BNH) TRAINING: A CASE STUDY IN BUILDING REMOTE COMMUNITY RESILIENCE

STEPHEN SUTTON

SCHOOL OF PSYCHOLOGICAL AND CLINICAL SCIENCES, CHARLES DARWIN UNIVERSITY, DARWIN

DARWIN CENTRE FOR BUSHFIRE RESEARCH, CHARLES DARWIN UNIVERSITY, DARWIN

This case study describes a journey to building remote community resilience as a component of the broader project *Building Community Resilience in Remote North Australia.* The origin of this BNH project comes, in a sense, from the *Sendai Framework for Disaster Risk Reduction 2015–2030*, and the *Australian National Strategy for Disaster Resilience.*

(*Continued*)

BOX 6.5 (Continued) NORTH AUSTRALIAN BUSHFIRE & NATURAL HAZARD (BNH) TRAINING: A CASE STUDY IN BUILDING REMOTE COMMUNITY RESILIENCE

This training project evolved from discussions with community members about the inadequacy of current fire management training. The interview team conducted dialogue in language around 'what sort of training do you want and why', at several locations in western Arnhem Land.

Respondents expressed appreciation for training provided by Bushfires NT and others, since discontinuity in educating young people about traditional fire management had resulted in increased wildfires, until recently unknown, threatening traditional estates, greenhouse gas abatement enterprises, and causing fire danger, making safety a priority.

Participants said that 'fire training should be here' on-*country*, incorporating the best of western science and local wisdom. In short, training that incorporated, in their words, a 'two-toolbox' approach.

In response, a range of course units has been developed. Some are adaptations of existing 'necessary' training (e.g. *Working Around Aircraft*). However, newly developed units examine the Aboriginal 'world view' and the way fire and its management is so deeply embedded in Aboriginal culture emphasising that the Aboriginal 'world view' can't be extricated from it in any coherent way.

Interviews undertaken by ARPNet (Box 6.5) in local Indigenous language, especially respectful of local values, provided nuanced insight into a 'strategic plan' for different habitats. These plans, articulated in the language of the 'Dreaming', are embodied in songs carried by initiates – the informants themselves – making for effective multi-modal bi-cultural transmission.

The need to tie together the core elements of BNH technical skills and empowerment of local Aboriginal people was achieved in units tailored for and delivered by them (see also Box 6.6 through 6.8).

What was new and constructive was effective inclusion of clearly established systems for making decisions about land and events according to ancestrally inherited connections, knowledge and authority central to Bininj practices and culture (Box 6.8).

The following diagram is a first approximation of the leadership arrangements for making decisions about land and BNH in several Bininj communities in western and southern Arnhem Land (Figure B6.5.1).

It does not purport to represent all Aboriginal communities, but it does show how different *Bininj* leadership is to 'standard' agency models. It shows a complex series of interactions that individuals are required to *fulfill* in specific roles attending the decision-making process. The *darlnyin* is 'the big boss', but must refer to the *djungkayi*, 'the manager', the *mingkirrinji*, the 'traditional owner', and the wider family in taking a decision. If the decision relates to fire then the 'fire men' – senior men with totemic affiliation and deep practical knowledge of fire – are critical players. The *Njirri* is like an auditor and ensures that proper protocols are observed. Unlike western auditors however, the *Njirri* may also effect punishment for breaches of protocol.

In addition to these 'statutory' considerations, decisions need to be made that reflect the universal dichotomy of people and land into *Dhuwa* and Yirritja. All people and all parts of the landscape are designated *Dhuwa* or *Yirritja* and no decision is made without encompassing a balance between the two.

For non-Aboriginal observers these arrangements are further complicated because they are limited to specific clan estates. The *djungkayi*, and *mingkirrinji* will speak for a clearly

(Continued)

BOX 6.5 (Continued) NORTH AUSTRALIAN BUSHFIRE & NATURAL HAZARD (BNH) TRAINING: A CASE STUDY IN BUILDING REMOTE COMMUNITY RESILIENCE

defined area of land (*country*), but their authority will not be acknowledged beyond those bounds. On another clan estate 'leaders' become 'followers', taking a subordinate role.

This 'fluid' state of leadership is part of an Indigenous world view sometimes incomprehensible to western eyes. Perhaps these protocols would be more useful if incorporated into training programs for the statutory BNH agencies, rather than, or in parallel with, training in remote communities. This would make sense in terms of building community resilience, and understanding through better integration of different levels of government and familiarity with of the scope and nature of *intercultural* 'shared responsibility'.

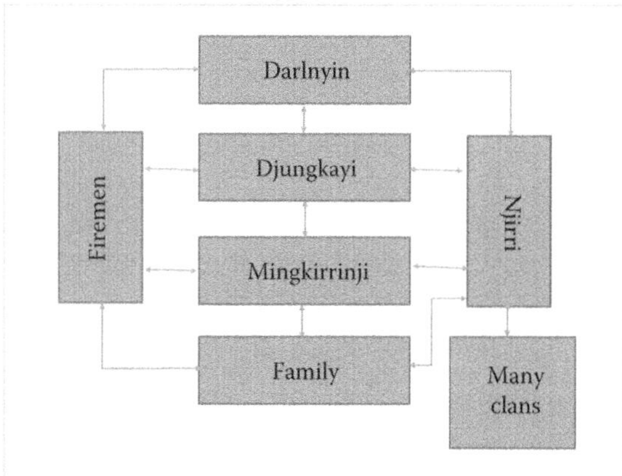

Figure B6.5.1 Notional Bininj BNH leadership and decisions-making rubric.

ACKNOWLEDGEMENTS

The North Australian BNH Training Project was jointly funded by the Bushfire and Natural Hazards CRC and the Higher Education Participation Partnerships Program - Whole of Community Engagement project. A very long list of generous people have contributed their time and knowledge.

6.6.2 Capturing Difference and Continuity in Building Resilience

These lessons can be transposed to the regional level. Consider the potential for regional scale emergency management approaches based on the framework, process, and trajectory of the Yolngu-driven Burrmalala cyclone project at Galiwin'ku (Box 6.6). The initial post-cyclone community resilience project was small scale and appeared narrow in focus, but community leaders rapidly saw the value of their own research in helping them to address the legacy of compounded disempowerment and cultural erosion. The emerging proposition to (re)create a community interface run by Yolngu leadership to manage government and other agency engagement with their community is a challenging one and one that not only expresses but seeks to enhance the qualities that make Yolngu who they are and who they want to be in their future town setting (Box 6.6). This story resonates with complementary research by ARPNet at Gunbalunya and Ngukurr (Box 6.4) and by Girringun in North Queensland (Boxes 6.1 and 6.7). With respect to natural disaster management and community resilience, for example, there are many communities across the North that already do or would get value from locally driven research into disaster resilience and resilience more generally, as with Galiwin'ku. There is a significant opportunity for EM and other agencies to learn from the process and outcomes and adopt regional or even transregional approaches to engagement and relationship building with Indigenous communities (Box 6.8). The benefits sought are more efficient cost-effective service delivery, improved local resilience, and capability to prepare, respond, and rebuild. The first step may be knowing how to recognise and relate to both obvious and obscured resilience qualities.

Similar ground-up processes over a number of years (from about 1995) led to the development of the North Australian Indigenous Land and Sea Management Alliance (now NAILSMA Ltd) in 2004.[54] Motivated by a lack of Indigenous representation across jurisdictions, NAILSMA has brought a strong 'culture-based economy' focus to Indigenous land management, economic development, and policy fora at regional and national scales. Some of the successful initiatives (such as Savanna Burning programs, Chapter 5) are discussed in other chapters, and all are the result of strong relationship building and collaboration amongst Indigenous leaders, their representative groups and businesses, and variously NGOs, governments, and industry.

The relative success and importance of the NAILSMA model (Figure 6.4) is through bottom-up collective voices, strong partnerships, respect for diversity, and facilitating understanding and relationship building from Traditional Owner level to policymakers, industry, and others. This broad approach supports economic innovation and engagement in more conventional economic activity, as Peter Yu describes:

> We see the growth of the emerging cultural economy as a basis to enable Indigenous people to live on their traditional *country* and infuse indigenous values in northern Australia's economic and political fabric.
>
> However, it is important to be realistic and appreciate that for many years to come the old extractive and environmentally exploitative industries will dominate northern Australia's overall economy. Indigenous people cannot afford to see the old economy and the new economy as mutually exclusive.[1]

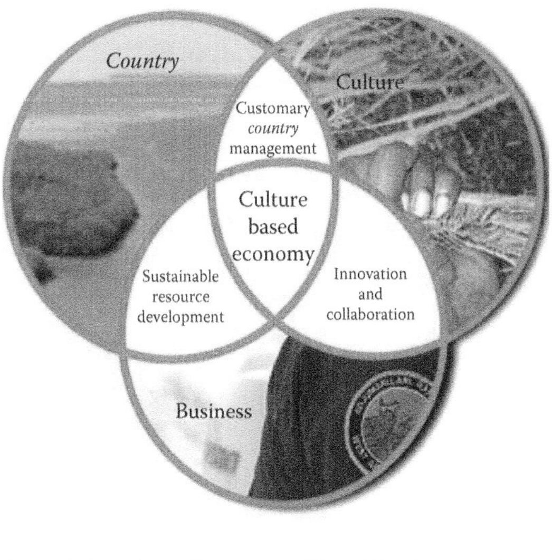

The culture-based economy framework provides a way of conceptualising the relationships amongst the three main sector interests, *country*, culture and business, flagging also the need for effective relationships with a variety of interested investors, researchers, funders, regulators and other partners.

Figure 6.4 Culture-based economy. (From Morrison, J. Managing for the future: A culture-based economy for northern Australia, A paper presented to the Charles Darwin *Symposium Presentation*, Charles Darwin University, Darwin, Australia, May 11, 2006.)

BOX 6.6 GALIWIN'KU BURRMALALA – STRONG WINDS OF CHANGE

ELAINE LAWURRPA, ALAN MARATJA, AND DANNY BURTON
YALU MARNGGITHINYARAW INDIGENOUS CORPORATION GALIWIN'KU

GLENN JAMES
NORTH AUSTRALIAN LAND & SEA MANAGEMENT ALLIANCE LTD., DARWIN

In 2015 tropical cyclones (TC) Lam (19 February) and Nathan (22 March) struck the north coastal communities of Galiwin'ku, Milingimbi, Ramingining, and Gapuwiyak, with devastating effect, thankfully with no loss of life. The category 4 TC Lam did the most damage with wind gusts estimated at 230 km/hr. At Galiwin'ku (population ~2,500) for example, infrastructure damage was significant with several hundred houses damaged and over 50 needing reconstruction or replacement. Power services to many houses were damaged, as was communications infrastructure, roads, water supply, and other municipal services. Hundreds of people were displaced and housed in temporary accommodation until rebuilding could occur. Understandably, the social fabric of the community was also put under enormous stress, adding to individual post-traumatic stress.

In this case, as with others nationally, emergency management (EM) agencies are particularly interested in building resilience through their response and recovery efforts; and increasingly interested in local factors that impact and may be drawn on to maximise community resilience, preparation, and response capability, and adaptability in natural hazard management. Ongoing cost-benefit assessment is also key to EM strategic investment. To this end, community resilience research has been supported by state and territory EM agencies and particularly through the Bushfires and Natural Hazards Cooperative Research Centre (BNHCRC).

Several months after Lam and Nathan, when infrastructure repairs and rebuilds were done, a group of senior *Yolngu* (a regional term for an Aboriginal person) saw an opportunity to help fellow community members talk through the post-event trauma, uncertainty, and disempowerment evident at Galiwin'ku. Two senior Yolngu researchers with years of experience in participatory research techniques and three junior researchers worked under the auspices of *Yalu Marnggithinyaraw* Indigenous Corporation (Yalu) – the local Yolngu research organisation – to develop a locally appropriate project, following local principles including local ethics approval.

lundu-nhina: Establish a good environment for the right people to sit together
rom-lakaranhamirr: Reminding each other of the lawful ways of doing things
ral-mirriyanhamirr: Sharing the tasks to be done
ral-gama: Bringing back what we must share
rulangdhuna: Putting what we have produced in place

Yolngu research project planning framework, developed by Yalu. (Burrumalala: 2016)

This was supported at arm's length by the North Australian Indigenous Land and Sea Management Alliance Ltd (NAILSMA) through the community resilience cluster of projects under the 'northern hub' of the BNHCRC, and Australian Red Cross (ARC).

(*Continued*)

BOX 6.6 (Continued) GALIWIN'KU BURRMALALA – STRONG WINDS OF CHANGE

The research project was aptly named the *Burrmalala* (strong wind) project. It began in part to help residents express their emotional and other distress, and secondly to collect local information and views about vulnerabilities and strengths in community governance, including how agencies performed in preparing for and responding to the cyclones. This was carried out by familiar Yolngu, speaking the local lingua-franca (*Yolngu-matha*), following local social and cultural protocols, with understanding, respect, and sensitivity to the circumstances of others in the process. The research team gathered nuanced information and messaging through observing complex shared kin relationships and protocols, and through the lens of cultural knowledge embedded in local and historical experience intrinsic to them. They could access and interpret 'meaning' that would likely be obscured for non-Yolngu researchers.

Comments and concerns around the resilience of built capital such as old and structurally weak housing and overcrowding problems were common, and although infrastructure provision was seen as the responsibility of 'outside' or non-Yolngu agents, effective planning, building, and managing were seen as requiring proper engagement with Yolngu to be most effective. The marriage of top-down with bottom-up solutions was a key message from research respondents and on many accounts EM agencies and local organisations like the Arnhem Land Progress Association (ALPA) store worked well under very difficult circumstances to include Yolngu in their on-ground work and achieved good results.

The Northern Territory Emergency Services in effect helped create a momentary relationship with Yolngu leaders and others where mutual interests (basic quality of life needs) aligned. Respondents emphasised that other processes in their history, associated with the 'mission days', the Northern Territory Emergency Response ('The Intervention'), and the introduction of the East Arnhem Shire, for example, created relationships around community governance tending to lack effective engagement or continuity with Yolngu, weakening social and cultural capital (Burrmalala Report [in press]):

> *Balanda* [predominantly Anglo Australians] *who have lived at Galiwin'ku and learnt a few things, develop relationships and understandings, leave and take with them the knowledge; about agencies, systems, processes, policies and programs; about Matha* (*language*)*, Gurrutu* (*relationships*)*, and understandings about Yolngu values including developed and trusted connections to Yolngu at Galiwin'ku.*
>
> *Two-way knowledge is supposed to be transferred at places like the local authority, DELAC, Yolngu Wanganamirri Mittji, Makarr dhuni, but it's more like a one-way cultural transfer – the Yolngu story and Yolngu agenda is missing* [*in the decision making*]...
>
> *Yolngu need a system that Yolngu control. Too many Balanda come to Galiwin'ku and think they already understand Yolngu and Galiwin'ku and want to tell Yolngu what's wrong with them and how to fix it but they don't know anything...*
>
> *Yolngu must sort out a system. Yolngu should choose Balanda staff.*
>
> *Balanda who work at Galiwin'ku should demonstrate respect for Yolngu rights, interests and land.*

The cyclones were tragic and frightening but were not seen as a dominant influence on community resilience and capability. The research inspired more detailed discussion around (re)creating a 'community interface' ... a place where Yolngu leaders can meet outside agencies to assist them in delivering their services better, direct them to follow

(*Continued*)

BOX 6.6 (Continued) GALIWIN'KU BURRMALALA – STRONG WINDS OF CHANGE

appropriate protocols, collaborate with them, ensure Yolngu pick up capability and responsibilities through their interaction, ensure economic benefits accrue, and ensure Yolngu values are invested in.

> *The community interface should be the place information is exchanged and everyone (Yolngu clans as well as the agencies) get to have their say. The* [current] *community interface is mixed-up, all over the place, controlled by the Balanda* [non-Yolngu].
>
> *Yolngu don't have control, don't set the agenda. All agencies and Yolngu clans need Yolngu and Balanda staff to help them with their own governance so they can all meet at the community interface to share knowledge and information.*
>
> *The interface needs some resources plus good Balanda and Yolngu to fix it up and make it work – based on Yolngu ways – Language, Law, Histories.* (op sit)

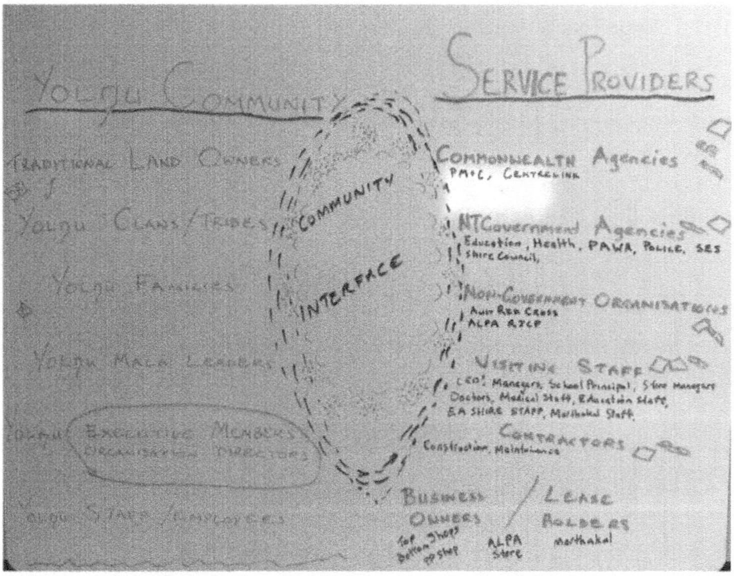

Participatory Action Research diagram, depicting the conceptual and physical space envisaged for a Community Interface - with 'u' shapes representing people meeting inside that space.

The previous explanation is mirrored in other research on resilience done by ARPNet in Gunbalunya and Ngukurr (see Box 6.4):

This Burrmalala Project offered a rare insight into what makes Yolngu feel 'strong' or 'weak': a lens through which to recognise that local residents have long established and successful ways of responding to problems and challenges, significantly untapped capabilities, essential traditional and local knowledge, and unique qualities undervalued in cross cultural relationships. Although in its infancy and not yet locally mandated, Yolngu leaders see this structure as a way of taking the lead on better community governance; of asserting their rights, skills, values and wisdom up front. There is no illusion that this be a simple or quick task, after many years of cultural erosion, but also with new sources of resilience and imagination to draw upon.

BOX 6.7 INDIGENOUS EMERGENCY RESPONSE PROSPECTUS: CYCLONE YASI, GIRRINGUN ABORIGINAL CORPORATION

RICKY ARCHER

NORTH AUSTRALIAN LAND & SEA MANAGEMENT ALLIANCE LTD., DARWIN

Cyclone Yasi (a Category 5 system) crossed the North Queensland coast in February 2011 and hit Cardwell which suffered infrastructure and environmental destruction, causing an estimated $3.5 billion in damages. A substantial recovery effort was vital for the community and the region – enter Girringun Aboriginal Corporation.[1]

Girringun coordinated emergency response efforts with the help of several Indigenous Ranger groups from north Queensland: Ewamian Rangers, from Georgetown; Carpentaria Land Council Aboriginal Corporation Rangers, from Burketown & Normanton; Yuku Baja Muliku Rangers, from Cooktown. These ranger groups were neither financially supported as emergency respondents nor was it their core business. They attended regardless. Girringun felt their resources and skills were under-utilised and devalued in the formal state emergency response framework and identified a need to provide a service to the community in partnership with other Indigenous Ranger groups.

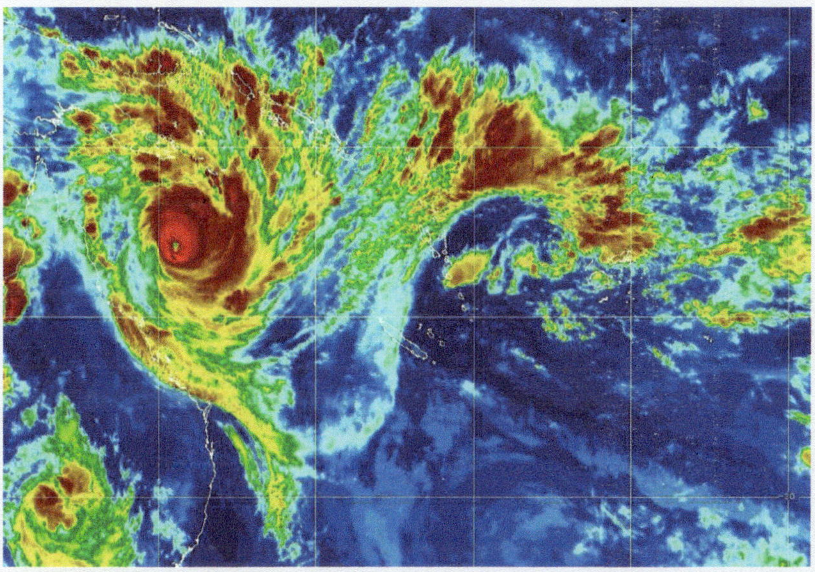

Cyclone Yasi (www.goes.noaa.gov)

Throughout the recovery efforts, Girringun saw opportunities to harness the capabilities of Indigenous ranger groups in emergency response activities. Phil Rist (Executive Officer at Girringun) is now raising awareness and collaboration to develop an Indigenous Emergency Response Plan. This Plan would use Indigenous Ranger resources and strengthen capabilities of Indigenous rangers to respond to disasters in their region or state. This would also enable Indigenous groups/organisations to negotiate fee-for-service arrangements with State and Commonwealth Governments. There are opportunities for other ranger groups across the remote North to adopt similar approaches and alleviate common barriers to emergency response effectiveness.

(*Continued*)

BOX 6.7 (Continued) INDIGENOUS EMERGENCY RESPONSE PROSPECTUS: CYCLONE YASI, GIRRINGUN ABORIGINAL CORPORATION

OPPORTUNITIES

- Existing Ranger skills and capacity
- Fee-for-service opportunities for Indigenous organisations
- Build relationships amongst Indigenous sector and broader community
- Strengthen roles of Ranger groups
- Provide new funding opportunities for Ranger groups

Indigenous emergency response framework

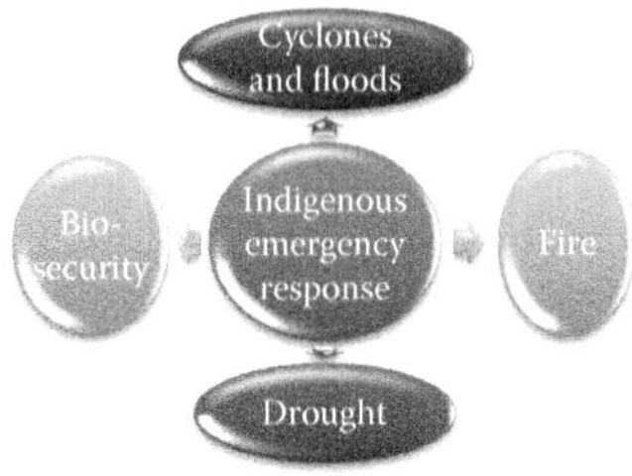

REQUIREMENTS

- Support for collaboration and networking—i.e. reinstate National Indigenous Land & Sea Forums
- Program funding reflect need for Indigenous emergency response capability
- Support from state and Cwlth bodies/institutions
- Strategic development in the Indigenous sector

REFERENCE

1. Girringun Aboriginal Corporation, Cardwell, Qld, 4849. http://www.girringun.com.au.

BOX 6.8 KEEPING TRACK – A TOOLKIT FOR MONITORING AND DECISION MAKING BASED ON LOCAL VALUES[1]

GLENN JAMES

NORTH AUSTRALIAN LAND & SEA MANAGEMENT ALLIANCE LTD., DARWIN

The following is a brief description of a local management toolkit, designed with Indigenous custodians of the land and sea who are looking back to core local values to steer and strengthen their *country* management enterprise.

After working through this toolkit, you will have a monitoring, evaluation, and decision-making system that:

1. Keeps track of important local values
2. Uses local information collected about those values
3. Lets you know and tell others about how well you are looking after the things you and they value
4. Reports back to inform local enterprise decision makers

Background

Indigenous land managers and non-Indigenous people are often involved in environmental management work for mutually desired outcomes (for example, conservation and employment) but from different foundational perspectives (e.g. for strengthening traditional ownership values compared to abating Greenhouse gas). Yet it is often non-Indigenous environmental and

(*Continued*)

BOX 6.8 (Continued) KEEPING TRACK – A TOOLKIT FOR MONITORING AND DECISION MAKING BASED ON LOCAL VALUES

enterprise outcomes which are given priority, that are the focus of 'external' project investment and support, and are used as the key measures of success. Local values may be recognised but are often seen as co-benefits and rarely prioritised as fundamental measures of success.

This toolkit seeks to help focus local management back onto local core values and provide better balance in keeping track of them for decision-making. The framework works on the principles of free, prior and informed engagement in decision-making, using locally collected data about self-defined interests to inform decisions.

The approach of this toolkit is to guide the group / facilitator though a series of 7 workshop sessions to tailor a project monitoring and evaluation system (including all activities and tools) which focuses on measuring indicators of success based on traditional owner/ custodian values. When finished, the local framework should empower Indigenous enterprise owners to emphasise local values in their decision making, giving confidence to balance donor, market and other 'outside' interests and values in the way they manage enterprise. The toolkit can be adapted and applied to a broad range of natural resource management or other development activities focussing explicitly on traditional owner and local group values as measures of success. This is an asset-based approach to the process, which becomes the core of the whole framework as it is built through the following sessions.

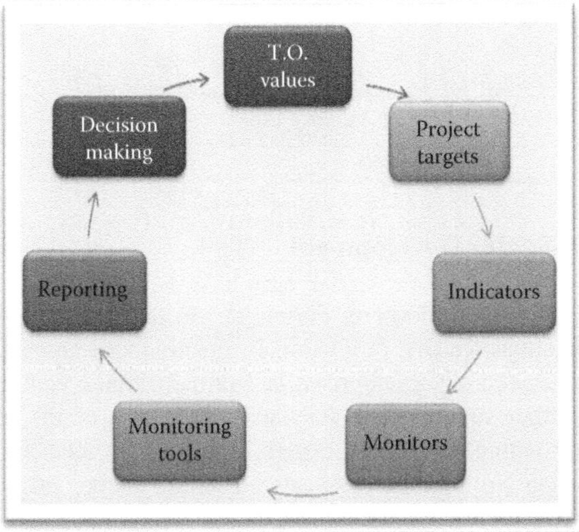

Monitoring and decision-making framework – Workshop series

Each of these steps is populated locally through participatory workshops. Each session will look back to these sets of values to ask "has our work promoted our stated values?" and will look forward to ask "how can we adapt our activities to enhance the benefits that we want that to achieve?".

The values were grouped into five broad areas: Connection, Identity, Knowledge, Power, and Seasonality. These value sets are not definitive but resonate with Indigenous land managers and so can be interpreted locally. They are considered important types of value to be maintained and enhanced. People give distinct local meaning to these value sets (particularly in cross-cultural, e.g. business settings), can use them in goal setting and can measure and monitor their condition.

(Continued)

BOX 6.8 (Continued) KEEPING TRACK – A TOOLKIT FOR MONITORING AND DECISION MAKING BASED ON LOCAL VALUES

Connection: Often expressed in terms of one's place in networks of kin, by implication also to specific inherited customary land estates. This establishes obligations and responsibilities, protocols governing relationships in local and even wider networks of people and *country*. Social organisation and rules for behaviour for example, depend on such connectivity. Competence with and respect for this is inherently valuable and by extension so too (appropriate) connection to non-Indigenous society is of measurable value.

Identity: A value with many and varied qualities, but often expressed in relation to one's own language, ancestry, 'dreamings', affiliation to *country* and community etc. which in turn underpins confidence and authority to act, locally and in wider society.

Knowledge/skill: Local and traditional knowledge systems enable effective management of land and sea *country*, connecting the physical, social and spiritual world. This knowledge foundation generates confidence to take on 'Western style' knowledge and skill, enhancing adaptive capacity to, for example, participate in the wider contemporary economy.

Power/empowerment: The extent to which Aboriginal people feel able to draw on spiritual affiliation; apply customary, local and acquired knowledge to decision-making; and to manage their interests locally and further afield is a common expression of empowerment and authority.

seasonality: Knowledge of and synchronicity with the land and its seasonal cycles is a strong expression of health and adaptability of Indigenous societies. The dominant influence of seasonal patterns and processes on life is reflected in social and spiritual systems that have grown up, explain it and work with it.

REFERENCE

1. NAILSMA (2017). *Keeping Track: A Toolkit for Monitoring and Decision-making Based on Local Values.* North Australia Land & Sea Management Alliance Ltd, Darwin.

6.6.3 Indigenous Experts Panel and Sector Development

In June 2012, the North Australian Indigenous Experts Forum on Sustainable Economic Development at Mary River attracted Indigenous leaders, professionals, practitioners, and experts to discuss Indigenous participation in key aspects of development in North Australia: economic, political, research and development, and other strategic engagement in the future of the north. Advice from the IEF and from the corresponding Indigenous experts panel was to feed directly to the North Australian Ministerial Forum on Northern Development. The key theme and vision statement was *towards resilient communities through reliable prosperity*.[2]

A significant challenge for the forum was dispelling the received wisdom that unique cultural norms and diversity of first peoples are weaknesses or barriers to effective participation in mainstream economies and socio-political life. The forum emphasised that Indigenous social and cultural capital are unrecognised assets to Indigenous economic success and to the prosperity of the North as a whole – not irrelevant or obstructions to it. Throughout the forum report are references to ancestral connection between people and *country*, unmatched knowledge and commitment to care for *country*, ownership of land and resources, initiatives based on cultural norms, community leadership, and the like.[2] The forum reasserted that these endemic, locally grown, and sustainable qualities demand recognition above and beyond legal rights (current and future) to land and sea. The IEF reaffirmed at regional scale what traditional land owners strive for in their local settings: that social, cultural, and other capital they hold, grow, and bring to the table is inseparable from more obvious assets deferred to in development speak – like land rights – and are key ingredients for prosperity.

Issuing from common ground and interconnectivity the IEF (draft) Framework for Community Resilience through Reliable Prosperity confidently asserts the collective voice to higher order concerns in the relationship with the state, identifying, for example, issues, goals, strategies, and action plans under five key areas:

1. Economics and commercial priorities
2. Culture and heritage
3. Conservation
4. Employment and infrastructure
5. Governance[55]

Exploring notions of resilience provides a means to grasp strengths, values, and assets of people so that better relationships may be developed around better mutual understanding. As noted by (the now Senator) Patrick Dodson in the second IEF forum report, 'The new paradigm for northern development, put simply, is to establish the right relationship between governments and Indigenous people to create better policies for all Australians'.[3] Such a task seeks to address current inequity and unbalanced knowledge and power relationships between Indigenous people and the settler state, and not, as Joe Morrison stresses, through 'assum[ing] to know'.

As with the examples given earlier in this chapter, the strongest initiatives for advancing Indigenous interests at scale are most often hard fought by Indigenous leaders against the tide of policy, and economic opportunism, at the risk of top-down infrastructure being ineffective, disempowering, and alienating (unless properly married to and designed with local Indigenous leadership at community level).[52] The important thing is the active relationships around the infrastructure that give it meaning. On the ground, people often say the failure of so many top-down initiatives are the fault of "outsiders to sit down, spend time, listen, and work things out together[56]... coming in like a cyclone to implement plans, without negotiation".[56]

Understanding how to negotiate and build active relationships across cultural traditions requires recognition of the strengths and interests. Conceptual categories like connection, identity, knowledge, power, and seasonality can be useful markers of continuity of value across even distant communities, providing a foundation for active relationship building in the any particular one.

Denial or lack of recognition and support for these endemic local social capital assets is a key factor in creating vulnerability,[57] disempowerment, tensions over local authority and decision making, and stultifying capability building.[25] We must focus on local strengths, the ones that make people who they are,[58] that drive aspirations, imagination and action, bolster adaptive capacity, and help explain why, in Jack Green's words, 'we're still here' (Chapter 7).

The relationship between resilience and prosperity is played out between adaptive knowledge and value systems in complex interacting economies, for example, as described by Altman in the Hybrid Economy model.[59] The old 'white paper' thinking of resilience and prosperity as outcomes determined in a single (mainstream) economy limits local innovation, opportunity and self-reliance. To (re)paraphrase Scott's 'Seeing like a state', 'The degree to which the state's formal schema makes no allowance for local processes or suppresses them, it fails both its intended beneficiaries and ultimately its designers as well'.[10]

6.6.4 So Much to-do about Resilience

'Resilient communities can adapt, change, and learn from problems or challenges. For example, resilience may be actively manifest by community members or leaders influencing and/or preparing for economic, social and environmental changes'.[60] They do so by being able to mobilize and apply their social, cultural, spiritual, and environmental values and local authorities, to the resolution or accommodation of the social and environmental challenges they have encountered. That is, the notion of resilience has purchase because strength and adaptability persistent in Aboriginal cultures continues to drive the struggle for recognition and prosperity, despite the erosive impact of recent history – consistently pointing to the need for reciprocity in relationship building and practice.

International responses to derogation of Indigenous rights (by colonising states and corporations) include the UN Declaration on the Rights of Indigenous Peoples[61] and the Convention on Biological Diversity (particularly articles 8j and 10c).[62] Basic human rights underwrite the key enablers for Indigenous prosperity, resilience, and social equity. There is a sharp connection between cultural, religious, economic, customary land, and governance characteristics with rights, to be recognised and protected. Repeatedly, the discourse of rights has been co-opted to undermine a fair go for Indigenous people and distracted from common sense in collective action.

> 'Every government agency or contractor that comes here to Galiwin'ku brings their own authority with them. We don't know what they're all doing or why they're here. We've lost control of what's happening in our community.'
>
> **—Elaine Lawurrpa (2016)[57]**

In Australia, Indigenous leaders have fought hard for rights and opportunities. Despite relative success with land rights and so on, forums like the IEF are more necessary than ever, to argue for recognition and primacy of Indigenous heritage and cultural qualities. They are key assets to local and broader national development and prosperity. The IEF privileges 'bottom-up' resilience features (e.g. spiritual connection to *country*) and is clear about 'top-down' requirements for promoting resilience (e.g. supported engagement and material substance in economic prosperity). Bottom-up cultural requisites exist for empowering community-based resilience to develop robust and prosperous enterprises. These local indigenous strengths must be considered in the vision for future northern development – business as usual will fail all Australians. Clearly, coexisting cultural traditions can maximise collective benefit through recognition of and active engagement around intersections of mutual interest, whether the cultural or other drivers of those interests can be fully understood.

REFERENCES

1. Yu, P (2013) Keynote address: Scoping the new paradigm for northern Australian development. *World Indigenous Network conference*, Tuesday 28th May 2013 North Australian Land and Sea Mangers Alliance Ltd (NAILSMA), Darwin, NT.
2. Morrison J (2012) Towards resilient communities through reliable prosperity. North Australian Indigenous Experts Forum (IEF) on Sustainable Development – First forum report Mary River Park, NT, June 19–21. NAILSMA Knowledge Series 013/2012, North Australian Land and Sea Mangers Alliance Ltd (NAILSMA), Darwin, NT.
3. Dodson M Chairman's Address NAILSMA (2013): An Indigenous prospectus for participation in the sustainable development of north Australia. North Australian Indigenous Experts Forum (IEF) on Sustainable Economic Development – Second forum Report Kakadu National Park, NT, 30 April–2 May 2013. NAILSMA Knowledge Series 019/2013, North Australian Land and Sea Mangers Alliance. Ltd. Darwin, NT.
4. Paton D, Johnston D (2017) *Disaster Resilience: An integrated approach* (2nd Ed). Charles C. Thomas, Springfield, IL.

5. Kirmayer LJ, Dandeneau S, Marshall E, Phillips MK, Jenssen Williamson K (2011) Rethinking resilience from indigenous perspectives. *Canadian Journal of Psychiatry* **56**(2), 84–91.
6. Martin DF, Martin BF (2016) Outstations as manifestations of Aboriginal self-determination: Wik strategic engagement and disengagement over the past four decades. In *Experiments in Self-determination: Histories of the Outstation Movement in Australia* (Eds. N Peterson, F Myers F), pp. 201–228. ANU press, Canberra. ACT.
7. Dass-Brailsford P (2010) *Crisis and Disaster Counselling*. Sage Publishing, Los Angeles, CA.
8. Paton D, Jang L-J, Liu L-W (2016) Long term community recovery: Lessons from earthquake and typhoon experiences in Taiwan. In *The Consequences of Asian Disasters: Demographic, Planning and Policy Implications* (Eds. H James, D. Paton). Charles C. Thomas Publisher, Springfield, IL.
9. Norris FH, Stevens SP, Pfefferbaum B, Wyche KF, Pfefferbaum RL (2008) Community resilience as a metaphor, theory, set of capacities, and strategy for disaster readiness. *American Journal of Community Psychology* **41**, 127–150.
10. Scott JC (1998) *Seeing Like a State: How Certain Schemes to Improve the Human Condition Have Failed* (p. 7). Yale University Press. London, UK.
11. Harney B (2016) *Yidumduma Bill Harney: Bush Professor* (Curator M West). Godinymayin Yijard Rivers Arts and Culture Centre. Katherine, NT.
12. Morphy F (2008) Invisible to the state: Kinship and the Yolngu moral order. Centre for Aboriginal Economic Policy Research (CAEPR) seminar series, 29 October. ANU, Canberra, Australia.
13. Peterson N (2005) What can the pre-colonial and frontier economies tell us about engagement with the real economy? Indigenous life projects and the conditions for development. In *Culture, Economy and Governance in Aboriginal Australia* (Eds. D Austin-Broos, G Macdonald), p. 7. Sydney University Press, Sydney, NSW.
14. McRae-Williams E, Gerritsen R (2010) Mutual incomprehension: The cross cultural domain of work in a remote Australian aboriginal community. *The International Indigenous Policy Journal* **1**(2), 1–27.
15. Jones R (1980) Australian coastal savannah hunters. In *Human Ecology in Savannah Environments* (Ed. DR Harris), p. 111. Academic press, London, UK.
16. Morphy F (2010) (Im)mobility: Regional population structures in Aboriginal Australia. *Australian Journal of Social Issues* **45**(3), 363–382.
17. Baymarrwaŋa L, James B (2014) *Yan-nhaŋu Atlas and Illustrated Dictionary of the Crocodile Islands*. The Tien-Wah press, Singapore & Sydney Australia.
18. Galiya EJ, Chairman, Yawuru Culture and Heritage Sub-committee (2013) *Walayjala-jalaburu Jayida Jarringgun Buru Nyamba Yawuru Ngan-gamirlimirli: Planning for the Future: Yawuru Cultural Management Plan*. The cultural management plan for Yawuru coastal country and the Yawuru Conservation Estate (2nd Ed.), p. 80. Broome, WA.
19. Wunambal-Gaambera Aboriginal Corporation–Land and Sea Management Unit (2000) *Ngauwudu Management Plan* (p.14). *Kimberley Land Council* Aboriginal Corporation. Kununurra, WA.
20. Burgess CP, Johnston FH, Berry HL, McDonnell J, Yibarbuk D, Gunabarra C, Mileran, Bailie RS (2009) Healthy country, healthy people: The relationship between Indigenous health status and caring for country. *Medical Journal Australia* **190**(10), 567–572.
21. Keen ID (1995) Metaphor and the metalanguage: "Groups" in northeast Arnhem Land'. *American Ethnologist* **22**(3), 502–527.
22. James B (2015) The language of spiritual power: From mana to märr on the Crocodile Islands. In *Strings of Connectedness: Essays in Honour of Ian Keen* (Ed. PG Toner), pp. 235–263. ANU Press, Canberra, Australia.
23. Northern Land Council (NLC) (2016) *Kenbi Land Claim: Title Deed Handover*. 21 June, NLC printers, Darwin, NT.
24. Tobler R, Rohrlach A, Soubrier J, Bover P, Llamas B, Tuke J, Bean N et al. (2017) Aboriginal mitogenomes reveal 50,000 years of regionalism in Australia. *Nature* **544**, 180–184.
25. Christie M, Guyula Y, Gurruwiwi D, Greatorex J, Garŋgulkpuy J, Guthadjaka K (2010) Teaching from Country, learning from country. *International Journal of Learning in Social Contexts*, **2**, 6–17.
26. Roberts JP, Russell BM, Parsons (1975) *Mapoon Story*. International Development Action. Amber Press. Sydney, NSW.
27. Rose D (1996) *Nourishing Terrains: Australian Aboriginal Views of Landscape and Wilderness*. Published by Australian Heritage Commission, Canberra, Australia.

28. Social Ventures Australia (2016) Social Return on Investment analysis of the Warddeken Indigenous Protected Area and associated Indigenous ranger programme (full report). Social Ventures Australia. Canberra, Australia.
29. Putnis A, Josif P, Woodward E (2007) *Healthy Country, Healthy People: Supporting Indigenous Engagement in the Sustainable Management of Northern Territory Land and Seas.* CSIRO Publications, Melbourne, Australia.
30. Ngauwudu Management Plan (2000) *Wunambal-Gaambera Aboriginal Corporation*, Land and Sea Management Unit, pp. 14. *Kimberley Land Council* Aboriginal Corporation, Kununurra, WA.
31. Hardy F (1968) *The Unlucky Australians.* One Day Hill, Melbourne. VIC Australia.
32. Baymarrwaŋa L (2012) Senior Australian of the Year 2012. https://www.australianoftheyear.org.au/honour-roll/?view=fullView&recipientID=780
33. Morrison J (2012) North Australian Land and Sea Mangers Alliance (NAILSMA) https://www.nailsma.org.au/laurie-baymarrwangga-senior-australian-year-2012.html
34. Truscott A, Malcom I (2010) Closing the policy-practice gap: Making indigenous language policy more than empty rhetoric. In *Re-Awakening Languages* (Eds Hobson J, Lowe K, Poetsch S and Walsh M), p. 6. Sydney University Press, Sydney, Australia.
35. James B (2011) https://drbentleyjames.wordpress.com/2011/10/04/working-for-culture-and-country-territory-quarterly-2011/.
36. Warlukurlangu Artists Aboriginal Association (WAAA) (1987) *Yuendumu Doors. Kuruwarri*, p. 127. Australian Institute of Aboriginal Studies. Canberra, Australia.
37. Jones P, Warlukurlangu Artists Aboriginal Association (WAAA) (2014) *Behind the Doors: An Art History from Yuendumu*, p. 33. South Australian Museum. Adelaide, Australia.
38. Devlin B (2011) The status and future of bilingual education for remote indigenous students in the Northern territory. *Australian Review of Applied Linguistics* **34** (3), 260–279.
39. Yothu Y (2002) Homeland movement song #3 *Homeland Movement Album*, Mushroom Records 26th November 2002. Festival Mushroom Records. Melbourne, Australia.
40. Said EW (1978) *Orientalism.* Penguin books. London, UK.
41. Lapulung K, Porteous S (2016) Makarrata: Bringing the past into the future. *Booklet Companion to Conflict Resolution Ceremony* (Eds L Allen, L Hamby), unpublished, pp. 7–8. August 11–14, Milingimbi, NT.
42. Nettle D, Romaine S (2000) *Vanishing Voices: The Extinction of the World's Languages.* Oxford University Press, New York.
43. Australian Institute of Aboriginal and Torres Strait Islander Studies (AIATSIS) and Federation of Aboriginal and Torres Strait Islander Languages (FATSIL) (2005) National Indigenous Languages Survey (NILS) Report, pp.255. Department of Communications Information Technology and the Arts. Australian Government, Canberra, Australia.
44. Ministerial Council on Education, Employment Training and Youth Affairs (MCEETYA) (1995) *A National Strategy for the Education of Aboriginal and Torres Straits Islander Peoples 1996–2002*, p. 1. Department of Education, Employment Training and Youth Affairs. Canberra, Australia.
45. The Australian Labor Party (2007) Australian Labor Party web site: http://www.alp.org.au. Canberra.
46. Nicholls C (2005) Death by a thousand cuts: Indigenous language bilingual education programmes in the Northern Territory of Australia, 1972–1998. *International Journal of Bilingual Education and Bilingualism* **8**(2–3), 160–177.
47. Fogarty W, Schwab RG J (2015) Land, learning and identity: Toward a deeper understanding of Indigenous learning on country. UNESCO *Observatory Multi-Disciplinary Journal in the Arts* **4**(2), 1–16.
48. Fogarty B, Schwab R (2013) Learning through country: An exploration of new opportunities in Indigenous education. *Ngiya: Talk the Law* **4**, 98–121.
49. Levinson SC, Wilkins DP (2006) Background to the study of the language of space. In *Grammars of Space Explorations in Cognitive Diversity* (Eds SC Levinson, DP Wilkins), pp. 1–22. Cambridge University Press. London, UK.
50. Biddle N, Swee, H (2011) Sustainability—Land, language and culture, *Lecture 10, Measures of Indigenous Wellbeing and Their Determinants Across the Life Course 2011.* Centre for Aboriginal Economic Policy Research (CAEPR) Lecture Series, CAEPR, ANU, Canberra, Australia.
51. Eades D (1993) Language and the law: White Australia versus Nancy. In *Language and Culture in Aboriginal Australia* (Eds M Walsh, C Yallop), p.185. Aboriginal Studies Press, Canberra, Australia.

52. Cook PM (2009) Buffalo and tin, baki and jesus: the creation of the modern wilderness. In *Culture, Ecology and Economy of Savanna Fire Management in Northern Australia: In the Tradition of Wurrk* (Eds J Russell-Smith, P Whitehead, PM Cook), pp. 69–85. CSIRO publishing Collingwood Vic, Clayton, Australia.
53. Sithole B, Hunter-Xenie H, Yibarbuk D, Daniels C, Daniels G, Campion OB, Namarnyilk S et al. (2016) *Living with Widdijith—Protocols for Building Community Resilience.* Aboriginal Research Practitioners' Network (ARPNet). Charles Darwin University, Darwin, NT.
54. NAILSMA and Morrison J (pers. comm.)
55. Morrison J (2006) Managing for the future: A culture based economy for northern Australia. *A paper presented to the Charles Darwin Symposium Series*, Charles Darwin University, Darwin, NT, May 11.
56. Christie M (ND) *The Box of Vegies: Method and Metaphysics in Yolngu Research.* Unpublished paper.
57. Lawurrpa E, Maratja A, Burton D (2016) Burrmalala (Strong Winds). *Research Project Report prepared by Yalu Marnggithinyaraw Indigenous Corporation and the people of Galiwin'ku North East Arnhem Land.* Galiwin'ku, NT.
58. Christie M, Guthadjaka K (2008) Teaching when nothing is lying around kathy guthadjaka talking with Michael Christie in Darwin (October 25, 2008). *International Journal of Learning in Social Contexts*, **2**, 25–31.
59. Altman JC (2005) The Indigenous hybrid economy: A realistic sustainable option for remote communities? *Paper presented to the Australian Fabian Society for Aboriginal Economic Policy Research*, Australian National University, Canberra, Australia. http://caepr.anu.edu.au/sites/default/files/Publications/topical/Altman_hybrid.pdf
60. North Australian Land and Sea Mangers Alliance (NAILSMA) (2012) *Towards Resilient Communities Through Reliable Prosperity.* North Australian Indigenous Experts Forum on Sustainable Development-First forum report Mary River Park. June 19–21, NAILSMA, NT.
61. Reconciliation Australia (2007) The United Nations Declaration on the Rights of Indigenous Peoples. Retrieved from: https://www.reconciliation.org.au/wp-content/uploads/2016/09/Reconciliation-Australia-United-Nations-Declaration-on-the-Rights-of-Indigenous-Peoples-UNDRIP.pdf
62. Convention on Biological Diversity (1992) United Nations. URL: http://www.biodiv.be/convention/

CHAPTER 7

'Like a Rusty Nail, You Can Never Hold Us Blackfellas Down'; Cultural Resilience in the Southwest Gulf of Carpentaria

Seán Kerins and Jacky Green

CONTENTS

> It doesn't matter how hard you hammer us Blackfellas down you can never hold us down. We always be comin' back up just like a rusty nail. We'll always be back. It doesn't matter what white people try to do to us to take our Law and culture – we'll be back. We might stay down for a while but we'll be back.
>
> **—Jacky Green**
> *Garawa, Borroloola, 2015*

7.1 BUILDING RESILIENCE

The Garawa, Gudanji, Marra, Yanyuwa, and Waanyi are the Indigenous peoples of the southwest Gulf of Carpentaria region of the Northern Territory (NT). Their countries extend from south of the Roper River in the west, towards the NT/Queensland border region in the east. They include the littoral waters of the Gulf of Carpentaria and encompass the offshore Sir Edward Pellew Islands. They extend inland about 200 km south towards the stony escarpment *country* that forms the edge of the Barkly Tableland.

The Garawa, Gudanji, Marra, Yanyuwa, and Waanyi people call the period from the 1870s to around 1910 the 'wild time'. For them it was a time of murder, mayhem, and chaos. It was the time they were dispossessed of their lands and waters and dispossessed of their ability to determine their own futures. It began when white settler colonisers attempted to clear them from their lands with guns, poison, and intimidation, to make way for the first wave of European 'development'.[1] Indigenous groups fought back to protect their *country*, especially their water resources, from the invasion of stockmen and their cattle.[2] This first development project is estimated to have cost the lives of 'at least 600 men, women, children, and babies, or about one-sixth of the population' throughout the region.[3] The Binbinga and Wilangarra peoples whose countries were situated around what would become the township of Borroloola ceased to exist as discrete peoples. Once settler colonisers took their lands and other natural resources, many Aboriginal people engaged in 'unequal labour relations in order to stay put, fulfil customary obligations, and survive as a distinct group'.[4] Others endured life in the Christian Brethren mission at Doomadgee, in Queensland, described as 'a history of infantilising surveillance underpinned by Australian law'.[5]

Although engaged in unequal labour relations or corralled into missions, the Garawa, Gudanji, Marra, Yanyuwa, and Waanyi never surrendered. While life on some pastoral stations was one of paternalism and often violent, many families maintained connections to their ancestral *country*, and maintained their social capital – their laws, cultures, knowledge, ceremonies and songs – to survive as distinct groups. When legal opportunities became available they regained ownership of some of their ancestral lands and then over a period of 40 years they slowly rebuilt their natural capital; their lands, waters and other natural resources.

The Indigenous peoples of the Gulf have long exhibited cultural resilience, which is the 'capacity of a distinct cultural system to absorb disturbance and reorganise while undergoing change so as to retain key elements of structure and identity that preserve its distinctness'.[6] Their cultural resilience has been vital to their survival as Indigenous peoples and vital for their long-term life project,[7] to live on and manage their ancestral lands, to protect their *country* and livelihoods, and, eventually, to determine their own future as Indigenous peoples.[8]

Indigenous livelihoods include the diverse activities in which people engage to sustain themselves. They 'incorporate tangible economic activities associated with the customary economy such as hunting, fishing and gathering, as well as those of the cash economy including work, welfare and commercial enterprise'. They also incorporate 'intangible aspects of social life that are reliant not only on physical resources, but also on symbolic resources associated with relatedness to knowledge of *country*'. Importantly Indigenous livelihoods rely on 'networks of relatedness of people to kin and *country* and entail a complex of obligations defined by a corpus of Indigenous law and custom'.[9]

Indigenous cultures in the Gulf are inseparable from the environment. People imbue the environment with 'meaning-making practices [that] define the terms and values that regulate social life concerning economy, ecology, personhood, body, knowledge, and property'.[10] Their unique ethno-ecological identity, a social identity shaped by their unique human-environment relationships, is central to understanding their life project. Their ethno-ecological identity is also critical for understanding the challenges and opportunities for a model of development for Indigenous people in the region constructed not 'from above and the outside', but rather 'from below and within'.[6]

7.2 PEOPLE, CULTURE, ECONOMY, AND *COUNTRY*

Yanyuwa and Marra people talk of themselves as saltwater people. Their countries are situated in the Gulf coastal bioregion characterized by undulating plains, meandering tidal rivers, coastal floodplains, and significant mangrove forests.[11] Their economies, knowledge, and cultural identities are enmeshed in the coastal and marine environment, including the littoral waters and the off-shore islands.[12] Garawa economy, knowledge, and cultural identity centres on the coastal environment, river systems, billabongs, and the vast plains of the tropical savanna.[13] Further inland, in the stone and hill *country* around Cape Crawford, the Gudanji

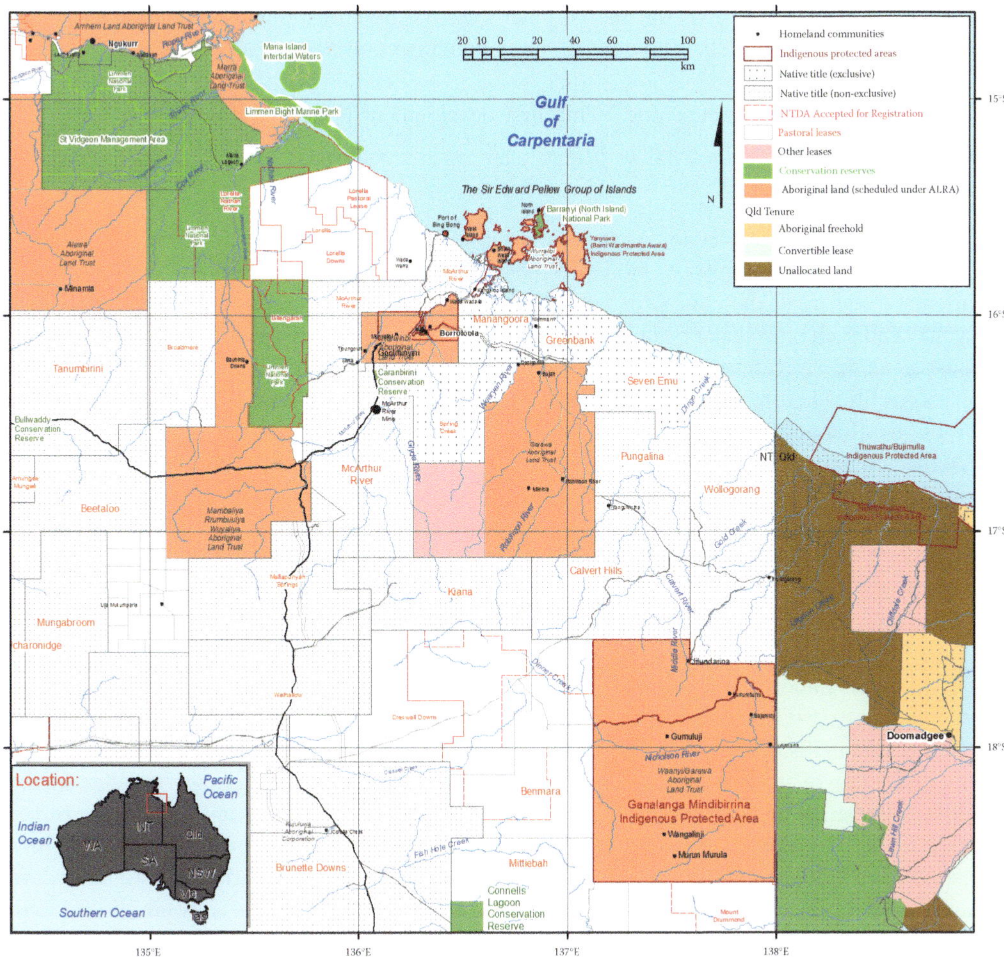

Figure 7.1 Land tenure in the south-west Gulf *country*. (Courtesy of Andrew Edwards.)

economy, knowledge, and cultural identity centres on the escarpment *country* and the waters of the mighty McArthur River system.[14] The Waanyi people's economy, knowledge and cultural identity centres on the Nicholson River and surrounding drier savanna plains.[15] Garawa, Gudanji and Waanyi countries are mostly situated in the Gulf Fall and Uplands bioregion that includes spectacular gorges, waterholes, and dissected sandstone plateaus with eucalypt woodlands growing over spinifex grasslands.[16]

Although speaking their own languages, people are closely related through marriage, ceremony, law, and their connections to *country*.[8] People can belong to one or more Indigenous group and maintain complex social networks across the region.

Country is not seen primarily, as represented in Figure 7.1, as a set of discrete named blocks, but rather as an open network of named places. Some places have 'big names'; that is, they are powerful and sacred places, while others are of local significance. The 'big names' places are linked to the places of local significance by the pathways taken by ancestral beings that travelled in the *Wankarla* (The Dreaming; 'an ever-present condition of existence').[17] The travels of the ancestral beings and the places where they rested, transformed themselves or performed significant feats are celebrated in ceremonies and in the *kujika* (sacred songs). Their pathways form a network of important cultural and sacred places across the entire region, including the littoral waters, before extending into, and beyond, neighbouring countries.

> These sacred sites were put down in the Dreamtime when the old people travelled through the land. Those Dreamings travelled like human beings, and their spirit is still there in the *country*. We talk to them as our own relations and we believe their spirits come back into our families in the new generations that are born.[18]

As well as being connected to the places of local significance, some people may also be connected to, and have customary law responsibilities for, places across the region. For example, people belonging to different countries on the same Dreaming path through their fathers are said to have 'one ceremony'. They may call each other 'fathers' or 'brothers' and 'sisters', even if they speak different languages.

Just as an up-river development, such as an in-river mine, may have a negative impact on the people's livelihoods downstream, so too does a development project that impacts on a 'big name' place in one area negatively affect the cultural rights of numerous peoples across the region who are connected to that place through the Dreaming path and its related law, songs, and ceremonies.

7.3 GAINING PROPERTY RIGHTS TO PROTECT *COUNTRY*

After 100 years of having their lives and *country* dominated by pastoralists, police, missionaries, welfare officers, and government bureaucrats, things began to change in the 1970s with the arrival of the Whitlam Labor Government and the enactment of the Aboriginal Land Rights (Northern Territory) Act 1976 (Cth) (ALRA).

In the mid-1970s, as soon as Aboriginal people in the Gulf heard that they would be able to regain ownership of some of their ancestral land, and live on it legally, they began to lodge some of the first land claims under the newly minted ALRA. These claims were vital to their life project. The return of land along with the implementation of the Community Development Employment Projects (CDEP) scheme, established in 1977 primarily as an income support and community development program for remote Aboriginal communities, enabled many Indigenous families to return to live on their ancestral lands. To assist the newly formed homeland communities with day-to-day administration, and the management of the CDEP scheme, Indigenous groups established their own small governance institutions in the form of Indigenous corporations, often called 'outstation resource agencies'. For a period of time, with the support of the Australian Government in relation to housing, health, education, and social security, life on their ancestral *countries* was again possible. Families lived on their land in small communities often near 'big name' places. Most Indigenous families set about running small herds of cattle, undertaking community development projects, keeping gardens, hunting, fishing, and gathering, and taking care of their sacred places.

While the Australian Government initially supported the homeland communities, the NT Government provided little support. The NT Government maintained an antagonistic relationship towards Indigenous people. They opposed all land claims and quickly vested substantial areas of un-alienated Crown land in the Gulf region (St Vidgeon, Nathan River, and Billengarrah) in the Northern Territory Land Corporation and the Conservation Land Corporation, thereby making the land unavailable for claim. The NT Government also proposed the township of Pellew over some of the Sir Edward Pellew Islands to also exclude the area from the land claim process because land claims under the ALRA excluded claims within town boundaries. They also marginalised Indigenous people in both regional planning and policy development, and then working with the Federal Government they fast-tracked the McArthur River Mine development to gain legislative approval before the arrival of the Native Title Act (Cwlth) 1992.

At the same time Mt Isa Mines, the original owners of the McArthur River mining lease, purchased three pastoral properties (McArthur River, Tallawah, and Bing Bong), beating Indigenous groups who were in the eleventh hour of negotiations to purchase the leases to secure their ancestral *country*. Both these actions ensured that Indigenous people would have no opportunity, however small, to be able to negotiate certain social, cultural, and environmental conditions associated with the proposed mining development at McArthur River or the port development at Bing Bong.

Indigenous people, their land claims and aspirations, have long been seen as an obstacle to development. For decades the NT Government failed to invest Commonwealth Grants Commission funds from goods and services tax (GST) revenue identified for remote area spending in the provision of citizenship services to remote communities. The lion's share of the funds is spent in Darwin, where most of the non-Indigenous population lives. In remote regions, where 80% of the Indigenous population lives mostly on Indigenous owned lands, there's little trickle-down.[19] Decades of under-spending have seen the entrenchment of poverty amongst the Indigenous population across the region. With little investment in Indigenous development initiatives incarceration rates in the NT have skyrocketed. For example, in 2005, 566 people per 100,000 were incarcerated; in 2015 this had increased by 56% to 885 people per 100,000. In 2013, Aboriginal and Torres Strait Islander prisoners made up 86% of the prison population.[20] The impact of these extreme incarceration rates is especially prevalent in remote Indigenous communities where it is estimated that up to 14 per cent of men and up to 2% of women between 20 and 39 years from individual remote Indigenous communities might be in prison at any given time.[21]

7.4 INDIGENOUS PEOPLES AND THEIR *COUNTRY* TODAY

Together, the Garawa, Gudanji, Marra, Yanyuwa and Waanyi own about 37,000 km^2 of their ancestral lands as inalienable Aboriginal freehold title through the ALRA. Along with this, they own a further 9,000 km^2 of land also held in common as exclusive possession native title, which gives them rights to possess and occupy the area to the exclusion of all others. Over the remaining 26,000 km^2 of land throughout the region, mainly pastoral leases, they hold non-exclusive native title rights, which means they have no right to control access to, and use of, the area, but hold rights to live on the area, hunt, fish, gather food, and teach law and custom on *country* (Figure 7.1). Together, their total land holdings and native title rights in the region encompass a massive 64,000 km^2 of land. An emerging challenge is to recognise the rights of native title holders to enjoy and benefit from their customary resources and lands or be compensated for damage caused to their *country* and property.

There are only about 1,200 people who live in southwest Gulf region. A thousand of these people are Indigenous. It's a young population with about 60% aged under 24 years old. About 700 Indigenous people live in overcrowded and poor-quality housing at the four 'town camps' that immediately surround the Borroloola township. Another 240, mostly Garawa people, live at Robinson River, with the remainder of the Indigenous population living at over 25 small homeland communities on Indigenous-owned land throughout the region.[22] With only a small portion of Waanyi *country* situated in the NT and the majority, including Boodjamulla (Lawn Hill) National Park and the Bidunggu Land Trust situated in Queensland, many Waanyi people reside in Doomadgee, an Indigenous township of about 1,300 people of whom over 90% are Indigenous.[23]

The majority of the small, non-Indigenous population of about 200 people live predominately in Borroloola, with many employed in government and non-government agencies that service the Indigenous population. Some own small enterprises such as tourist accommodation, garages, food stores, or contracting businesses, few of which employ local Indigenous people. Others are employed on a few of the pastoral leases or at McArthur River Mine, where workers mostly fly in and out from Darwin.

Despite Indigenous groups holding property rights over vast areas, maintaining a significant reliance on the environment for their livelihoods and well-being, and being by far the most significant population in the region they have little voice in how the region's land, water, and resources are used, how they are valued, or how they will be managed in the future.

The marginalisation of Indigenous peoples in regional planning not only stands in the way of their own development aspirations, but results in the erosion of Indigenous cultural decision-making institutions, the desecration of sacred sites, and the contamination of ecosystems and food resources, all of which threaten their life project.

7.5 DEVELOPMENT – FROM ABOVE AND OUTSIDE

Beginning in the late 1970s, the southwest Gulf region underwent what geographer John Holmes describes as a multifunctional transition. This transition saw a decline of the pastoral hegemony governing land use, and the emergence of a more complex interplay, or contestation of values, between production (pastoralism and mining), Indigenous, tourism, conservation and environmental values.[24] In the early 1990s, responding to the multifunctional transition, the NT Government started a significant strategic land use planning process across the Gulf region. It was driven by an expectation that the region could benefit from the potential for development associated with tourism, recreation, amateur and commercial fishing, mining, and the creation of various types of parks and conservation areas as well as associated service infrastructure.[25]

Despite the increasing size of Indigenous-owned lands in the region and the research signalling the need for supportive policy frameworks to reconcile these competing values,[26] Indigenous issues received only passing attention in the planning process. In the whole of *Gulf Region Land Use and Development Study 1991* less than one page of text is devoted to Indigenous aspirations. This action was described by Holmes as 'a monumental failure to recognise the pivotal role which participatory coordinative regional planning could play in ensuring the engagement of Aboriginal interests'.[27]

Instead of supportive policy frameworks to reconcile these competing values and ensure the inclusion of the region's Indigenous peoples, both the NT and Australian Governments imposed another kind of development, one from above and from outside. It was policy that favoured the governments' own development agendas, giving primacy to the market, mining and energy resources, tourism, and food export.

This top-down development approach saw the NT Government allocate 200 ha of river frontage near Batten Point – a strategic area on the bank of the McArthur River – for recreational fishers to develop a township (King Ash Bay) and tourist enterprises to access fisheries resources from the Yanyuwa *country* in the McArthur River delta and around the Sir Edward Pellew Islands. Today, it is estimated that more than 10,000 people visit King Ash Bay, with thousands of people fishing in Yanyuwa Sea *country*. Then in the early 1990s, this same top-down development approach saw McArthur River Mine begin its mining operation about 60 km southwest of Borroloola with the promise of employment for local Indigenous people, a promise that has never really eventuated.

Despite the claims made by mining companies and governments, mining projects provide few benefits to the Indigenous people, who instead bear the adverse impacts of these projects. There are numerous instances across the Gulf region where large-scale developments have left legacy issues for Indigenous people. These include Redbank copper mine, which was abandoned in the mid-1990s with an estimated 54,000 tonnes of partially treated acid-forming material, including vat residues and unsecured waste rock piles that allowed affected waters to bleed underground. The massive stockpile at Redbank Mine was left exposed to the monsoonal rains for 17 years, resulting in highly toxic water extending up to 7 km from the mine site. The copper sulphide leaching forms concentrations so high in some waterways that there is no longer any aquatic life.[28] Clean-up costs are estimated between $10 million and $100 million[29] (Figure 7.2).

Of all the development projects in the southwest Gulf region, McArthur River Mine poses the most serious and long-term legacy issue for Indigenous people. In 2006, with government support,

Figure 7.2 Hanrahan's Creek with copper sulphide leaching from Redbank Mine. (Courtesy of P. Taplin.)

Glencore, the current owner of the mine, diverted a 5 km section of the McArthur River to make way for the massive open-cut pit in the bed of the river. Before the expansion of the mine and the diversion of the river, the public were told that only 12% of waste rock would be potentially acid-forming material. But reports in 2014 provided by McArthur River Mine's independent monitor, the ERIAS Group, now provide evidence that the massive waste rock pile contains around 80% of the potentially acid-forming material.[30] The independent monitor warned that over time, monsoonal rains will penetrate the waste rock's clay cap, reaching the potentially acid-forming rock below the surface before leaching acid, saline, and metalliferous drainage into the groundwater.[30]

In 2013, the mine's independent monitor also warned that levels of lead found in the fish in the mine's diversion channel exceeded the maximum permitted by Food Standards Australia New Zealand. The permitted lead level was exceeded in 9 out of 10 fish caught, with the lead testing similar to that found at the mine site.[30] Then, in 2015, it was reported that 400 cattle had been shot near McArthur River Mine because some had tested positive for lead.[31]

In 2015, from above and outside, the Australian Government launched its plan for developing northern Australia. No Indigenous groups in the Gulf were consulted during the drafting process. Again, Indigenous peoples' development aspirations were excluded from regional planning processes. For them the *Our North, Our Future: White Paper on Developing Northern Australia* had little to do with their north or their future but rather was another policy tool which would not assist Indigenous peoples to achieve their own development aspirations, but rather a legitimising strategy for the government's 'open for business' developmental agenda[32] (Figure 7.3).

As well as the contamination of their waters and food resources and desecration of sacred sites Indigenous people worry that when the miners have gone, and the investors have made their profit, it will be their children and grandchildren who will bear the final cost of this development project.

Figure 7.3 Clouds of sulphur dioxide being emitted from the gigantic waste rock pile at McArthur River Mine 2014. (Courtesy of D. Morris.)

7.6 DEVELOPMENT – FROM BELOW AND WITHIN

Being left out of regional planning processes meant that the potential economic opportunities for Indigenous people such as fishing, cultural tourism, and environmental management in the region were slim. In the late 1990s, the Australian Government slowly began to withdraw service delivery (health, housing, and education) from homeland communities in some places. The wind-back of services along with the lack of policy to support Indigenous people living in homeland communities resulted in many families having to move to overcrowded and stressful living conditions in regional townships.[33]

In the mid-2000s, the policy environment deteriorated further. Under the Howard Coalition Government the Indigenous policy environment became highly political, combative, and erratic. Nationally, the Aboriginal and Torres Strait Islander Commission – the peak national Indigenous representative body – was dismantled with bipartisan support, and its policies and programs 'mainstreamed' into various government departments. Small remote Indigenous communities were coerced into 'mutual responsibility agreements' and then 'shared responsibility agreements' to secure basic citizenship services. Homelands were portrayed as nonviable, archaic 'cultural museums'.[34] In the Northern Territory the Racial Discrimination Act (CTH) 1975 was suspended and the Northern Territory Emergency Response (NTER), an intervention using the military was swiftly imposed to 'stablise, normalise, and exit'[35] Indigenous communities, before morphing into Stronger Futures under the Rudd-Gillard Labor Government in 2011.

With people 'yard[ed] up like cattle'[36] in regional townships, some lands were left semi-occupied with landowners only able to make short trips back to their homelands. Once tracks were washed away it became even more difficult to visit and care for *country*. Without people on *country* there was no fire management, which saw the beginning of a cycle of destructive wildfires across the region. For example, in the Nicholson River region in the early 2000s, these fires burnt in excess of 16,000 km^2, threatening homeland infrastructure and feed for cattle on the region's pastoral stations. Relations between pastoralists and their Indigenous neighbours during this time were tense. The long-term result of these uncontrolled hot late-season wildfires was clearly evident in areas of *country* that were beginning to lose significant areas of vegetation classes. The loss of vegetation leads to the loss of feeding and breeding habitats for many native – especially endemic and threatened – species in the region, such as the Carpentaria grass wren. Homeland infrastructure was also at risk along with important cultural and rock art sites. Such hot fires were also emitting additional greenhouse gases that marginally exacerbate global warming.

> 'Once the government cleared us off our lands by shooting us and putting chains around our necks and dragging us off. Then, long time later, they said "Here's your land back, we don't need it". That tall man he poured the sand through that old man's hands. That made us real happy and we began to move back home. Government gave us a bit of help to get back and set ourselves up. But you know what? They never really took those chains off from round our necks, 'cos now they slowly pullin' on them. They pullin' us off our lands again and yardin' us up like cattle in town. They pullin' us off our land by not giving us schooling, health, and housing services on our homelands. They not helpin' us. They sayin' to the parents if you don't send your kids to school we gonna stop your money and send you to prison. But there aren't no schools, so the parents have to move off their *country* to live like white man in town with no culture' (Figure 7.4).

Despite the erratic top-down policy, environment Indigenous groups remained resilient and continued to pursue their life project. Responding to the ecological threats from wildfires, feral animals and the decline of important cultural species, as well as wanting to create meaningful employment in a region with few mainstream opportunities, Garawa, Waanyi, and Yanyuwa began building within their cultural groups small-scale cultural and natural resource management initiatives such as ranger programs. These 'bottom-up' initiatives aimed to provide social, economic, and environmental benefits to themselves and to the wider Australian community.[37] This is an example of Indigenous land owners utilising their common property resources to encourage labour and investment in their land, biodiversity, and communities. It also reminds us that the economy is embedded in social relations[38] (Box 7.1).

Figure 7.4 Yardin' us up like cattle in town, Jacky Green 2014.

BOX 7.1 BUILDING ON NATIVE TITLE: THE RIGHT TO TRADE

SEÁN KERINS
CENTRE FOR ABORIGINAL ECONOMIC POLICY RESEARCH
AUSTRALIAN NATIONAL UNIVERSITY, CANBERRA

PETER WHITEHEAD
DARWIN CENTRE FOR BUSHFIRE RESEARCH
CHARLES DARWIN UNIVERSITY, DARWIN
NORTH AUSTRALIAN LAND & SEA MANAGEMENT ALLIANCE LTD. DARWIN

In June 2016 the Federal Court of Australia recognised Indigenous people's rights to trade in the Borroloola region of the NT.*

In the court, Indigenous people argued that at the time their sovereignty was usurped by the British, their ancestors possessed rights to access, and take for any purpose, the resources of their estate, and to control access to, and use of, the estate and its resources by others. Indigenous people based their argument on their customary law illustrated by their ancestors' long-term trading relationship with Macassan traders, who travelled from what is today Indonesia to northern Australia to trade with Indigenous people until stopped by the Australian Government.

Despite opposition from the NT Government, Justice Mansfield found in favour of the Indigenous people. He said, 'In my view, it is … appropriate to accept that the provision of goods by the Macassans to Aboriginal people was not random and that it was in return for access to their land and waters and the resources'.† This decision built on the 2008 High Court (Blue Mud Bay) decision that Indigenous people retained the rights to control access to their land estates down to the low tide mark.

The Minister of Indigenous Affairs, Nigel Scullion, responded to the federal court decision, saying, 'This is a great result for the Rrumburriya Borroloola people', and that, 'with native title affirmed, focus can now turn to maximising the opportunities and benefits that can be derived from … native title rights and building a prosperous future for the next generation of Rrumburriya Borroloola people'.‡

Here we focus on how Indigenous people in the Borroloola region might maximise those benefits from their lands, waters, and natural resources, just like their ancestors did.

Properly recognising rights could be advanced in at least two ways. First, by providing a financial return, such as environmental use fees for access to their estates and natural resources. And, second by implementing a community-based fisheries management system in the region.

Community-based fisheries management is where coastal communities and fishers, being the most dependent on coastal and marine resources, have a primary role in deciding how these resources are managed. It is built on the understanding that management decisions of all sorts are often best made at the most local level possible.§

Considering that in the south-west Gulf *Country* Indigenous people's livelihoods, cultural identities, well-being, and customary rights are enmeshed in the marine environment, and

* Rrumburriya Borroloola Claim Group v Northern Territory of Australia 2016 FCA 776.

† Ibid., 46.

‡ Scullion N (2016). Native title recognised in Borroloola, Media Release Department of Prime Minister and Cabinet, 11 August 2016, https://ministers.dpmc.gov.au/scullion/2016/native-title-recognised-borroloola.

§ Graham J, with Charles A and Bulla A (2006). Community-based fisheries management handbook. Gorsebrook, Research Institute, Canada.

(*Continued*)

BOX 7.1 (Continued) BUILDING ON NATIVE TITLE: THE RIGHT TO TRADE

that where a significant amount of fishing activity occurs, it makes sense that they play a primary role in managing the region and its resources, taking up the rights established in Australia's courts. It also makes sense that they are not denied opportunities to develop long-term sustainable economic opportunities that can help them build a prosperous future.

However, in response to the Blue Mud Bay High Court decision the NT Government sought to avoid requirements for permission to access waters over Aboriginal controlled land. Shortly after the High Court decision was announced the Northern Land Council implemented an interim agreement allowing access without a permit, and this remains the situation.*

During negotiations, Traditional Owners emphasised their obligations to maintain the health of their seas and protect culturally-important sites and expressed frustration at their inability to influence activities in their waters. These concerns can be expected to be important in future negotiations, which resumed in 2017.

In four places along the NT coastline and river systems with high recreational/sports fishing activity, the NT Government negotiated with landowners modest payments to allow access without individual permission. One of these high activity fishing areas is the waters around the Sir Edward Pellew Islands and in the McArthur River system. Here, it is estimated that about 10,000 people per year fish in Indigenous estates and take their resources. Many fishers travel great distances from Darwin, Alice Springs, and interstate.

To ensure the continued access of all these fishers the NT Government, in addition to the modest fee, agreed to establish an Aboriginal fisheries consultative committee (the Wurrahiliba Management Committee).

Aboriginal fisheries consultative committees have operated in the NT for the past 25 years and, as demonstrated by the strong views of Traditional Owners, are long known to be ineffective in promoting the rights of Aboriginal people in the management of their estates and resources and provide few long-term benefits. These arrangements were agreed prior to the federal court decision about Aboriginal people's right to trade in the Borroloola region and so do not reflect this important development.

At best, Aboriginal fisheries committees provide opportunity for users of Indigenous marine estates and resources to meet with Aboriginal people who hold customary rights over the region. However, they do not provide real power to Aboriginal people to manage access to their estates and their fisheries or derive a return from the use of their natural resources. Systems that recognise Indigenous people's right to control access to their resources and their right to trade in them are vital to Indigenous economic development in the region. They require that Indigenous right holders exercise far more power and responsibility than merely having a right to be consulted.

These powers and responsibilities must offer opportunities for Indigenous people to make an income. They may want to do this by direct participation in industries like fisheries or through, for example, implementing fees for accessing lands and taking resources, as other landowners do. Such fees could play an important role in funding a community-based fishery and its management and compliance costs.

Playing a primary role in managing access to their estates and use and protection of natural resources requires a framework for the local community to have real influence over

- Opening and closing areas
- Establishing catch limits and gear requirements
- Gathering catch data

* At the time of writing in May 2017.

(Continued)

BOX 7.1 (Continued) BUILDING ON NATIVE TITLE: THE RIGHT TO TRADE

- Monitoring water and habitat quality
- Developing regulations
- Resolving disputes
- Imposing sanctions
- Pursuing aquaculture opportunities
- Playing primary roles in compliance
- Developing cost recovery processes

Much more is needed than impotent Aboriginal fisheries consultative committees that, by design, serve mostly to assert the interests of external players. Community-based management systems are urgently needed in the southwest Gulf region to create opportunities for Indigenous peoples to build prosperous futures for the next generation of Rrumburriya Borroloola people, just as the Minister of Indigenous Affairs envisaged.

Yanyuwa people, for example, were concerned primarily about their marine environment. Hunters were asking why many of the turtles and *dugong* they were catching contained an unusual black-coloured fat within their bodies. There was considerable fear expressed by people living on the islands about what may happen to their sea *country* and food resources should McArthur River Mine be flooded in the monsoon season. The Yanyuwa were concerned also by the increasing number of recreational fishers in the area, by trespassers camping on their land and the interference with both sacred and burial sites, especially on the Sir Edward Pellew Islands. They were determined to make their own decisions about how their *country* should be managed and they set about to secure financial resources from Natural Heritage Trust (NHT) and Aboriginals Benefits Account (ABA) to enable them to manage their environments. Yanyuwa families began to develop a series of management plans for their land and sea *country*.[39,40] Then, with the assistance of Mabunji Aboriginal Corporation to manage their accounts and provide office space, they formed the *Li-Anthawirriyarra* Sea Rangers. The Rangers' work focuses on, amongst other things, managing important turtle and dugong breeding and feeding areas, the eradication of feral cats from the offshore islands, removal of ghost nets and marine debris, and the transfer of cultural knowledge between generations and fee-for-service work for government agencies.[40]

In 2011, the Yanyuwa people declared the Sir Edward Pellew Islands and the Narwimbi Aboriginal Land Trust as the Yanyuwa Indigenous Protected Area (IPA). (An IPA is an area of Indigenous-owned land or sea where Indigenous owners have entered into an agreement with the Australian Government to promote biodiversity and cultural resource conservation.) The Yanyuwa IPA is managed within International Union for Conservation of Nature (IUCN) standards by a team of Yanyuwa Rangers that employs 21 men and women. Yanyuwa people are also engaging with market opportunities by developing an innovative cultural tourism enterprise based on their cultural knowledge and the breeding cycle of marine turtles.

Garawa and Waanyi people began working in both the Nicholson and Robinson River regions in 2005. They undertook a series of planning meetings and established two ranger groups, the Garawa and Waanyi/Garawa Rangers. Since their inception the Northern Land Council (NLC) in Darwin has administered the Garawa and Waanyi/Garawa Rangers. Almost immediately they developed a relationship with Bushfires NT, especially the research staff, and, together, they began implementing more than 50 fuel-load monitoring plots to measure greenhouse gas emissions from fires on their *country*. Their aim was to take over fire management in the region in order to both protect *country* and create meaningful employment, and then once they had reduced the severity and size of fires in the region they would be able to trade their abatement/carbon reduction in markets and then use the funds to progress their life project (Box 7.2).

BOX 7.2 LAND SECTOR OPPORTUNITIES IN THE NT GULF REGION

KAMALJIT K. SANGHA
DARWIN CENTRE FOR BUSHFIRE RESEARCH
CHARLES DARWIN UNIVERSITY, DARWIN

DOMINIQUE LYNCH
DARWIN CENTRE FOR BUSHFIRE RESEARCH
CHARLES DARWIN UNIVERSITY, DARWIN

The 44,000 km^2 NT Gulf region adjoining QLD in the south east of the Gulf of Carpentaria (Figure B7.2.1) affords a useful regional land sector case study given its mostly low pastoral and current economic potential, few properties, and small, substantially Indigenous populations (~1,200) of which ~1,000 people are resident in the adjoining township of Borroloola and most are unemployed.

Annual rainfall across the region ranges from 1,000 to <600 mm in the south. Despite low pastoral productivity, the region is richly endowed with high cultural and biodiversity values – the latter

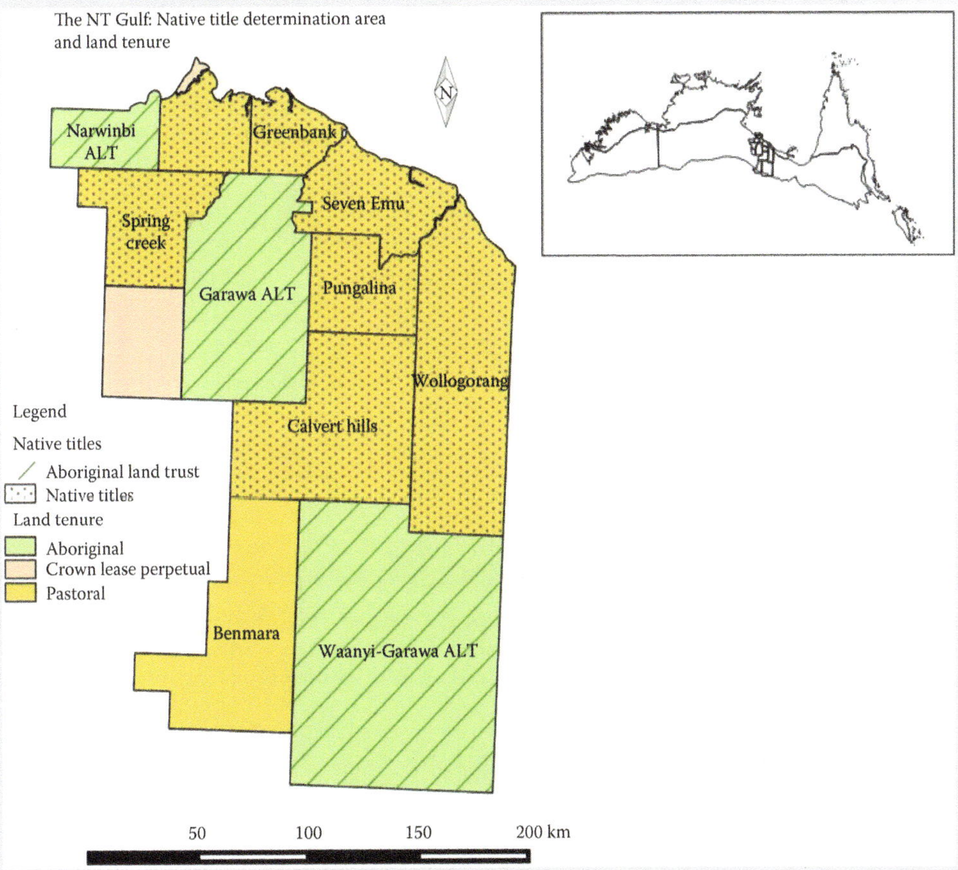

Figure B7.2.1 Land tenure in the NT Gulf region, south of Borroloola.

(Continued)

BOX 7.2 (Continued) LAND SECTOR OPPORTUNITIES IN THE NT GULF REGION

reflecting mostly rugged terrain intersected by a number of major perennial watercourses, and an extensive maritime boundary. Regional environmental service opportunities for Indigenous people associated with marine stewardship and fishing enterprises are showcased in Box 7.1.

Pastoral Opportunities

Of the region's properties, three operate as non-Indigenous conventional pastoral enterprises supporting an estimated 75,000 head of cattle between them, four Indigenous properties support small herds (600–1,500 head), three operate as Aboriginal Land Trusts (of which two are declared Indigenous Protected Areas), and a further pastoral lease operates effectively as a privately funded conservation reserve. The number of cattle on the non-pastoral properties is unknown but likely to be very low. Regional Indigenous interests in land are reflected in the observation that Native Title currently exists over most of the region (Figure B7.2.1).

Based on assessments of stock carrying capacity derived from land systems,[1] and expert opinion, it is evident that carrying capacity in the NT Gulf region is very low, with 79% of the region capable of supporting just 0–3 head/km^2 (Figure B7.2.2). While acknowledging that

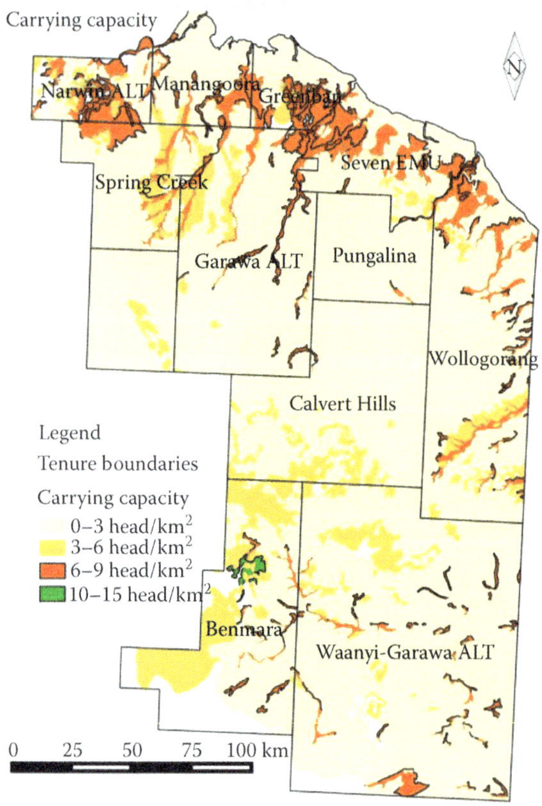

Figure B7.2.2 Carrying capacity of pastures based upon land systems with 79% of the total area affording 0–3 head/km^2, ~10% affording 3–6 or 6–9 head/km^2 each, and rest 1% affording 10–15 head/km^2.

(Continued)

BOX 7.2 (Continued) LAND SECTOR OPPORTUNITIES IN THE NT GULF REGION

carrying capacity is influenced also by factors other than pasture potential (e.g. management practices, infrastructure, availability of resources, herd performance), it is likely that two of the three non-Indigenous pastoral properties are substantially overstocked, whereas others with smaller herds are likely to be significantly understocked. Overstocking has evident significant flow-on impacts on a range of ecosystem services, including land degradation and water resources.

Using the same approach as adopted in chapter 5 for assessing economic returns for typical pastoral businesses,[2] we estimated earnings before interest and tax (EBIT) for regional properties based on respective cattle numbers. Even without taking debt into account, all pastoral enterprises were shown to be (negatively) economically marginal.

The environmental costs associated with regional cattle production are very substantial. Based on a regional estimate of 18% of pastoral land in the region being degraded,[3] estimated costs associated with lost pasture production and broader impacts of land degradation run into tens of millions of dollars when applying global land degradation procedures.[4] Additionally, environmental costs associated with annual greenhouse gas (GHG) emissions from cattle can be costed at $1.6 million, when applying average region-specific GHG emissions of 1.78 t CO_2-e/head/year,[5] at a current Australian Government regulated market price of $10.23 t CO_2-e. Obviously, costs of GHG emissions are sensitive to fluctuating market prices.

By any objective measure, NT Gulf pastoral enterprises currently are at best economically marginal, and those with large cattle numbers would incur significant environmental impacts. Despite this, the three largest regional pastoral enterprises have each sold recently for many millions of dollars.

Diversifying the Land Sector

The earlier assessment illustrates the need to diversify land sector opportunities in the NT Gulf region, and especially for Indigenous people given their evident connections to, and associated cultural custodial responsibilities for looking after, *country*. Such opportunities include managing the outstanding cultural and biodiversity values of the region (Figure B7.2.3) and building on current Australian Government investments in local Indigenous ranger and IPA programs. Currently 12 local Indigenous Rangers are employed full time, and about 40 casually, on the Garawa (Robinson River) and Waanyi-Garawa (Nicholson River) Land Trusts – the latter of which was recently declared the Ganalanga-Mindibirrina IPA.

As well as stewardship responsibilities, significant employment and land sector enterprise opportunities exist through developing a variety of contractual services, including cultural site and infrastructure maintenance; road works management; degraded land and mine site rehabilitation; and fire and associated GHG emissions management activities. For example, despite limited resources, the Garawa and Waanyi-Garawa Rangers have developed substantial regional fire management expertise over recent years. Such expertise could be applied to help better manage regional fire regimes for a range of community, pastoral, biodiversity, and environmental outcomes. Regional fire regimes contribute more than 120,000 t CO_2-e of GHG emissions annually. Based on experience elsewhere, fire management could reduce regional emissions by around a third, realizing around $500,000 annually (based on current carbon prices) and an additional $2–4 million when formal carbon sequestration methodologies become available.

(Continued)

BOX 7.2 (Continued) LAND SECTOR OPPORTUNITIES IN THE NT GULF REGION

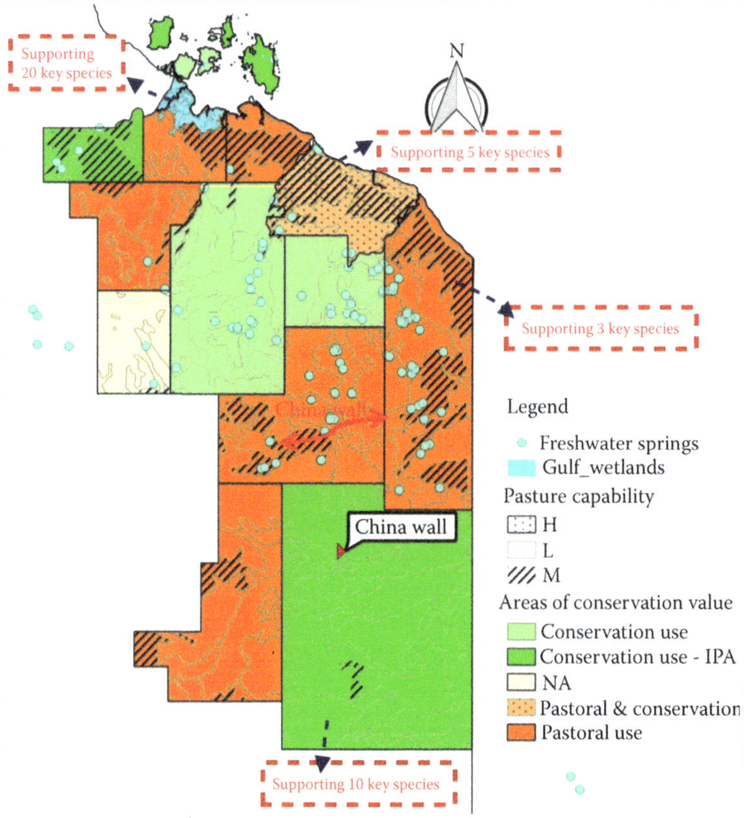

Figure B7.2.3 Areas of conservation significance, wetlands, and freshwater springs in the NT-Gulf.

The provision of appropriate employment opportunities also has a range of significant social and cultural benefits, including very substantial (multimillion dollar) reductions in government social security outlays.[6–8]

As outlined in this chapter and elsewhere in this book, such opportunities can only be realized through the development of robust business governance structures which build on and are accountable to local community needs.

REFERENCES

1. Aldrick JM, Wilson PL (1990) *Land Systems of the Southern Gulf region, Northern Territory.* Conservation Commission of the Northern territory, Darwin, NT.
2. McLean I, Holmes P, Counsell D (2014) *The Northern Beef Report – 2013 Northern Beef Situation Analysis.* Meat & Livestock Australia, North Sydney, Australia.
3. Pastoral Land Board NT, 2013–2014. Pastoral Land Board Northern Territory – Annual Report 2013–14. Pastoral Land Board, Northern Territory.
4. The Economics of Land Degradation (ELD) Initiative (2015) *The Value of Land: Prosperous Lands and Positive Rewards Through Sustainable Land Management.* ELD Secretariat, Bonn, Germany. Available from www.eld-initiative.org.

(Continued)

BOX 7.2 (Continued) LAND SECTOR OPPORTUNITIES IN THE NT GULF REGION

5. Walsh D, Cowley R (2016) Optimising beef business performance in northern Australia: What can 30 years of commercial innovation teach us? *The Rangeland Journal* **38**, 291–305.
6. Burgess CP, Johnston FH, Bowman DMJS, Whitehead PJ (2005) Healthy *Country*: Healthy people? Exploring the health benefits of Indigenous natural resource management. *Australian and New Zealand Journal of Public Health* **29**, 117–122.
7. Crowley, GM (2015) Trends in natural resource management in Australia's Monsoonal North: The conservation economy. The Cairns Institute, James Cook University, Cairns, QLD, Australia.
8. Social Ventures Australia (2016) Consolidated report on indigenous protected areas following social return on investment analyses. *Report to Department of Prime Minister and Canberra*, Australian Government, Canberra, Australia.

The ranger groups came under NLC administration at this time because senior landowners did not want either of the Indigenous organisations in the Gulf (Mabunji Aboriginal Resource Centre based at Borroloola and Mungoorbada Aboriginal Corporation based at Robinson River) to be involved in their caring for *country* projects. These organisations were deemed unsuitable because they were seen to have moved away from their primary objective – helping people remain living on their ancestral lands – and were viewed as having become administrative arms of government. Land owners expressed a number of concerns regarding transparency and accountability within these organisations. These concerns came to fruition in 2015 when both organisations went into administration, with Mabunji embroiled in corruption allegations. Because of their concerns about local governance, Garawa and Waanyi landowners asked the NLC to host their Ranger groups until they were in a position to establish a suitable local institution in the region to host their enterprise. The NLC assisted in the establishment of the Ranger groups by providing planning support and office space, a shipping container to store equipment, and by applying for funds from the NHT and Australian Government Natural Disaster Mitigation Fund to implement their community-based land management plans (Figure 7.5).

Figure 7.5 Senior Garawa Ranger Donald Shadforth undertaking aerial controlled burning. (Courtesy of M. Lawrence-Taylor.)

In 2007, the relationship between the NLC and the Garawa and Waanyi/Garawa Ranger groups became more formalised when the Australian Government introduced the Working on *Country* (WoC) program and began to shut down the CDEP program. The WoC program saw the Australian Government invest in Indigenous land and sea management and begin to pay wages to ranger groups in a fee-for-environmental service type arrangement.[8,16] When this occurred the seven Garawa and Waanyi WoC Rangers became employees of the NLC. These employees were managed by NLC located more than 1,000 km away in Darwin. The commitment of the NLC to Ranger groups over the past decade has been patchy.

Between 2012 and 2015, Garawa and Waanyi people experienced enormous challenges in keeping their ranger groups operating to implement their land management plans. Authority to make decisions and spend funds was held by staff in Darwin and exercised often with little consultation with the ranger groups or land owners. Simple decisions, such as filling a vacant Ranger position, could take more than 12 months. Grant funding opportunities vital for fire management were missed, leaving the Ranger groups with no operational funds to undertake burning. The neighbouring Queensland-based Carpentaria Land Council Aboriginal Corporation stepped in for a period to provide assistance. The inability to make their own decisions locally and in a timely manner has been a constant frustration for Garawa and Waanyi people actively involved in cultural and natural resource management projects.

Today, despite these administrative difficulties, the Garawa and Waanyi people operate two ranger teams with seven full-time staff, and up to 40 casual workers. Over the past decade they have taken control of wildfire, replacing the boom and bust cycle of wildfires with an early season mosaic-burning regime which has seen a considerable reduction in greenhouse gas emissions. After a decade of fire management however Indigenous people are yet to receive any income from carbon market opportunities. The ranger teams have also connected many young people with their *country*, bringing them to camps to participate in burning activities, planning meetings, and fauna surveys.[16] In June 2016, following a three-year planning process and a two-year wait, Waanyi and Garawa landowners declared over 10,000 km^2 of their land (the Waanyi/Garawa Aboriginal Land Trust) as the Ganalanga Mindibirrina IPA. Through the IPA, Waanyi and Garawa people aim to once again live on their ancestral land, to use two-way knowledge systems (Indigenous and scientific) to manage and protect their *country* and its biodiversity, and to expand their work activities. They see the Ganalanga Mindibirrina IPA as a way of engaging more young people with *country* and creating further employment opportunities, while also developing enterprises such as cultural tourism and participating in emerging carbon markets.[16]

Although wanting to, Gudanji and Marra people have not been able to establish any cultural and natural resource management projects to protect their *country*, assist with the maintenance of cultural and ecological knowledge, and create meaningful employment. There are a number of reasons for this. The majority of the Marra people live off their ancestral lands in Ngukurr and Numbulwar in southern Arnhem Land, and financial resources are in short supply to bring the wider landowner group together so that they can articulate their aspirations. There is also no local organisation whose role is to facilitate such action. Gudanji people, on whose ancestral *country* sits Glencore's McArthur River Mine, been divided by the politics of the mine. Glencore has been trying to establish a ranger group to fulfil one of their community engagement obligations of creating one new Indigenous business per year. Many Gudanji see this as an attack on their law and decision-making processes. It has long been a strategy of the mine to usurp Indigenous decision-making processes by isolating individuals from cultural groups[8,42] (Figure 7.6).

This painting is about how mining companies work in the Gulf region. They work out who they think the main Traditional Owners are for an area and then they swoop down like birds of prey, and grab them, and take them away to their nest. There they give them money to keep quiet and give the OK for the sacred sites to be damaged. The boomerang represents law in the country that's being disrespected. The bird with his tongue out is like the boss of the mine, he's laughing, saying, 'I got them, I got them and I'm bringing them now'. The other one says, 'Great I got the money ready'. The people standing on the hills and the plains represent our families who are worried cause their fathers and brothers have been taken away. The miners are trying to smash our law and how we want to organise and represent ourselves. They take our people away from our culture to make it easier for them to take our resources inside Country.

Figure 7.6 Birds of prey, Jacky Green, 2016.

7.6.1 All of Us Together

First time outsiders tried to push us off our country; they did it with guns and poison. This time they doing it again, and they still using poison.

Our fish, creeks, and waters are slowly being poisoned by mining companies. Some creeks in the Gulf country, mile after mile, got water like battery acid. There's no life, nothing but poison.

In other places fish and livestock been testing positive with lead. We losing our food, our way of life. Our river been diverted when the mining company dug a huge open-cut pit on our sacred site. They made huge waste rock pile too. And that's full of toxic waste that's gonna slowly leak acid into our waterways.

On top of this we got the fracking mob turn up and start drilling to suck all the gas out of our country. They got us real worried. They going to use so much of our water and damage our country, just like the rest of them miners.

Government and mining companies, they all in this together. They just like the one-eyed fella. He got one eye on the money while he covers the other one. He knows he's damaging our country, but he covers his eye so he don't see the destruction. He don't have to worry about it, as long as government and mining company getting their money. He just don't care.

They all poison our life, just like the first time.

This painting's about what's happening to us and our country right now. It's not just us blackfellas that are real worried, but whitefellas too. We all gotta be lined-up together, black and white, to protect our country. It's what gives us life (Figure 7.7).

Although Indigenous groups have been remarkably innovative and successful in developing Ranger groups in a very remote and challenging region of Australia, they continue to face significant challenges.[41] These include making their voice heard in regional planning processes; developing

Figure 7.7 One eye on the money, Jacky Green, 2016.

community-based management plans for Gudanji and Marra *Country*; and growing and diversifying their cultural and natural resource management activities. Many communities of traditional owners have identified the lack of Indigenous governance that allows all the groups to work together across the region to assist them with land and sea management as a weakness, constraining their ability to take full advantage of emerging opportunities.

In 2012, to address these challenges, some Indigenous groups, as part of the Ganalanga Mindibirrina IPA planning process, started developing a set of rules for a future regional governance institution to operate as a social enterprise. Its aim would be to strengthen their decision-making processes, while consolidating and growing their cultural and natural resource management initiatives so that they can achieve their long-term life project. Indigenous groups spoke of wanting to ensure that their future organisation could not be captured by an individual (staff, director or member of the company), that the organisation would be both accountable and transparent to its members, and that any future income from activities such as carbon farming and biosequestration be invested back into the company to grow and diversify its activities.

Through a series of planning meetings Indigenous people drafted a constitution for a company limited by guarantee to operate as a community-based enterprise under the Corporations Act (Cth) 2001. It is envisaged that while the company is established under Australian law, its members and directors will draw on Aboriginal custom – norms of behaviour, local conventions, and codes of conduct – in making their decisions.[44] It is through the community-based enterprise that Indigenous peoples in the Gulf aim to take more control of their caring for *country* activities (Box 7.3).

BOX 7.3 SOCIAL ENTERPRISE AS A MODEL FOR DEVELOPING ABORIGINAL LANDS IN THE SOUTHWEST GULF *COUNTRY*

SEÁN KERINS

CENTRE FOR ABORIGINAL ECONOMIC POLICY RESEARCH
AUSTRALIAN NATIONAL UNIVERSITY, CANBERRA

Traditional Owner communities across the southwest Gulf region hold significant natural and social capital, as evidenced by their land and sea *countries* as well as their kinship and ceremonial relationships.

Some groups like the Garawa, Waanyi, and Yanyuwa have managed to utilise both their natural and social capital to underpin significant collaboration between communities of Traditional Owners, Aboriginal Rangers, governments, and the environmental and science communities.

Over the past decade communities of traditional owners have implemented significant community-based development projects across the region. Underpinning these initiatives are small governance institutions, such as IPA management committees, whose role it is to implement Traditional Owners' plans of management.

Although these small-scale governance institutions are vital for community-based development, their responsibilities are limited, with final decision-making resting with external agencies that administer funding agreements. For example, the Northern Land Council, which is situated 1,000 km away from the southwest Gulf in Darwin, administers the Garawa and Waanyi Garawa Ranger programs, while the Mabunji Aboriginal Corporation, based in Borroloola, administers the Yanyuwa Li-Anthawirriyarra Sea Rangers.

Some communities of Traditional Owners have long voiced frustration with these agencies, complaining about the lack of transparency in both financial and administrative decision-making. The slowness of decision-making and the high turnover of staff are also seen as barriers to the continued growth of the region's community-based enterprises. Other communities of traditional owners have expressed a desire to utilise their natural and social capital in a similar way to other groups but have not received assistance to help them organise to develop management plans.

In 2012, long-frustrated by the lack of administrative support from the Northern Land Council – which is seen by some traditional owner communities as being preoccupied with facilitating outside interests at the expense of working with landowners on their own development aspirations – Indigenous groups set about developing their own regional governance institution.

What is a Social Enterprise?

Across North Australia, the Caring for *Country** movement has seen a number of Indigenous landowning groups enhance their social capital by forming new institutional arrangements to better cooperate in protecting and managing land and waters that they hold in common.

Some of these new institutional arrangements operate as social enterprises. They have enabled Indigenous people to build governance institutions amongst relatively small

* Kerins S (2012). 'Caring for *country* to working on *country*'. People on *country*, vital landscapes, Indigenous futures (Eds Altman J, Kerins S), Annandale: The Federation Press, 26–44.

(*Continued*)

BOX 7.3 (Continued) SOCIAL ENTERPRISE AS A MODEL FOR DEVELOPING ABORIGINAL LANDS IN THE SOUTHWEST GULF *COUNTRY*

land-owning groups on an economy of scale that enables the successful management of *country* with high biodiversity and cultural values in a more economically viable way.

Social enterprises are 'not based on utilitarian-economic models but rather an economic model in which resources provide for broader goals, economic, social, cultural, and political'.[*] They include the creation of jobs and the strengthening of social capital by supporting people who have been inactive back into the wider activities of the community.[†]

While social enterprises can have a profit motive, their primary aim is to provide social and environmental dividends to community members, and in some cases to the wider public. They rarely distribute financial profit to individuals, with any surplus being reinvested for the long-term benefit of the community.[‡]

An example of this type of cooperation can be seen in West Arnhem Land Plateau, where Indigenous landowning groups formed Warddeken Land Management Limited, a not-for-profit company, to protect and sustainably manage their land. Warddeken Land Management Limited is a social enterprise designed to deliver environmental services and create jobs for people living on *country* (see Box 5.5).

Warddeken Land Management Limited's constitution sets out a series of rules that define resource boundaries, membership, and directors' responsibilities, as well as what the company will do (the process for use of the resources). While the company is established under Australian law, its members and directors draw on Aboriginal custom (norms of behaviour, local conventions and codes of conduct) in making their decisions.[§] While operating in a very remote region of Australia, the social enterprise sustainably manages the co-owners' land and simultaneously provides social, cultural, economic, and environmental benefits to its members by tapping into national and global economic opportunities through, for example, a carbon farming initiative.[¶] These initiatives are examples of co-owners utilising their property to encourage labour and investment in their land, biodiversity and communities.

It is through social enterprise institutions that Indigenous peoples in West Arnhem Land Plateau are taking more control of their caring for *country* activities with the aim of 'breaking from the practice of relying entirely on government funding rounds for operational funds and move toward financial self-determination'.[**]

Building a Gulf Caring for *Country* Company

Inspired by Warddeken Land Management Limited, some traditional owner communities in the Gulf sought legal advice from an international law firm on the most appropriate way to begin building their own regional governance institution. Their aim was to continue their life project, to live on and manage their ancestral lands, to protect their *country* and livelihoods, and to determine their own future as Indigenous peoples.

Following a number of workshops, a company limited by guarantee was seen as the most effective institution to meet their needs. It was envisaged that the company would continue to

[*] Berkes F, Davidson-Hunt IJ (2007). 'Communities and social enterprises in the age of globalization'. Journal of Enterprising Communities: People and Places in the Global Economy, 209–21.

[†] Borzaga C, Defourny J (2001). The emergence of social enterprise. Routledge.

[‡] Pearce J (2003). Social enterprise in anytown. London: Calouste Gulbenkian Foundation.

[§] Cooke P (2012). 'A long walk home to Warddewardde'. People on *country*: vital landscapes, Indigenous futures (Eds Altman J, Kerins S), Annandale: The Federation Press, 155.

[¶] Russell-Smith J, Whitehead PJ, Cooke P (Eds) (2009). Culture, ecology, and economy of fire management in North Australian savannas, Rekindling the Wurrk Tradition, Collingwood: CSIRO Publishing.

[**] Ibid. Cooke P, 155.

(Continued)

BOX 7.3 (Continued) SOCIAL ENTERPRISE AS A MODEL FOR DEVELOPING ABORIGINAL LANDS IN THE SOUTHWEST GULF *COUNTRY*

build relationships with scientists, government, conservation and philanthropic organisations to help manage the region's ecosystems and biodiversity and at the same time create employment and enterprise opportunities for Indigenous landowners.

Some of the issues landowners wrestled with during their deliberations included how the proposed company would work with, and not against, customary governance institutions framed by the Aboriginal Land Rights (Northern Territory) Act 1976 (Cth) (ALRA); how the new institution could avoid the weaknesses seen in other local governance organisations, such as a lack of transparency in decision making and financial reporting and a lack of accountability of board members to their constituents; and how to avoid capture and undue influence by individuals and groups.

Landowners also stipulated that the governance institution should not distribute royalties. They described the current model of royalty distribution in the region as being divisive and providing no long-term benefits. They wanted to see any profit the company might make being invested back into the business of their social enterprise so that it could diversify its activities or be used to leverage external funds.

The proposed land area that would be managed by Traditional Owners with the assistance of the company is envisaged to comprise seven Aboriginal Land Trusts. These include the Garawa; Jandanku; Mambaliya, Rrumburriya, Wuyaliya; Mara; Narwinbi; Waanyi/Garawa; and the Wurralibi Aboriginal Land Trusts.

Land owners also identified a number of functions that they wanted their company to be able to perform to assist with the administration of their Ranger groups and enterprise development, including

- Acting as a host for all Aboriginal ranger groups in the south-west Gulf, but still allow the space for each group's cultural identity.
- Acting as a proponent for future carbon trading projects.
- Holding and administering Working on *Country*, Indigenous Protected Area, and other natural resource and cultural heritage management funds.
- Managing land and sea management data.
- Facilitating small enterprise development.
- Undertaking environmental fee-for-service work in the region.
- Performing Human Resource management.
- Being able to attract future investment from philanthropic and environmental organisations.

Next Steps

Although much work has been done to develop a constitution to frame the governance of the company much more needs to be done. Consultation meetings with all traditional owner communities whose ancestral lands may benefit from being managed by the company remain outstanding.

Consultations under the ALRA are a role for the Northern Land Council that is charged with the administration of the Act. While the Northern Land Council was initially supportive of the project to develop a regional governance organisation, it has not recently responded to requests for support and instead has consolidated its administrative position by taking control of Waanyi and Garawa Ganalanga-Mindibirrina Indigenous Protected Area funding and implementation of the Ganalanga-Mindibirrina Plan of Management.

One issue that communities of traditional owners would like to see change is for the Northern Land Council to give their development aspirations priority before facilitating the interests of outside groups who want access to Aboriginal owned and native title lands for their own profit.

Diversifying the funding base for ranger programs is a pressing issue for Indigenous groups. In August 2016, documents from the Department of Prime Minister and Cabinet suggest that WoC program may be under threat with the Minister of Indigenous Affairs wanting to overhaul [the WoC program] from a community-run to a top-down structure.[43] Indigenous groups in the Gulf region hope to diversify they funding base and formalise their relationships with, and draw on the skills, expertise and financial contacts of, private, conservation, and philanthropic organisations operating both nationally and internationally.

Relationships with conservation and philanthropic organisations have been an important development in the evolution of Indigenous community-based enterprises. This is because some recognise that heterogeneous and complex Indigenous communities in remote regions are not isolated but are embedded within larger systems and can play a role in problem solving, for example, combating climate change, minimising species extinction, and ensuring ecosystem maintenance, all things that underpin their life project. They also acknowledge that Indigenous people in remote regions maintain a competitive advantage in providing these environmental services. Collaborative relationships provide an opportunity for Indigenous groups to further develop new skills and new knowledge and to combine these with their existing knowledge systems to deal with a growing array of environmental threats. Importantly, these relationships create new funding opportunities to ameliorate the challenge of an overreliance on government funding and state domination of community-based projects.[44]

Many Indigenous people throughout the Gulf region hope that the work they have done in building Ranger groups and their ongoing effort to establish their social enterprise by way of a company limited by guarantee to care of their ancestral lands, create meaningful employment, and pass cultural and ecological knowledge between generations will continue to grow so that they can achieve their life project, to live on and manage their ancestral lands, to protect their *country* and livelihoods, and to determine their own future as Indigenous peoples.

REFERENCES

1. Roberts T (2005) *Frontier Justice: A History of the Gulf Country to 1900.* University of Queensland Press, St. Lucia, Queensland.
2. McGrath A (1987) *Born in the Cattle*, Allen and Unwin, Crows Nest, Australia.
3. Roberts T (2009) The brutal truth: What happened in the Gulf Country. *The Monthly*, November 2009: 42–51. Available at https://search.informit.com.au/documentSummary;dn=300817355796536;res=IELAPA.
4. Varacini L (2011) Introducing settler colonial studies. *Settler Colonial Studies* **1**(1), 4.
5. Trigger DS, Asche W (2010) Christianity, cultural change and the negotiation of rights in land and sea. *Australian Journal of Anthropology* **21**, 92.
6. Healy S (2006) Cultural resilience, identity and the restructuring of political power in Bolivia. *Paper submitted for the 11th Biennial Conference of the International Association for the Study of Common Property*, Bali, Indonesia. June 19–23.
7. Blaser M (2004) Life projects: Indigenous peoples' agency and development. In *In the Way of Development, Indigenous Peoples, Life Projects and Globalization* (Eds. M Blaser, HA Feit, G McRae), p. 26. Zed Books, London, UK.
8. Green J, Morrison J, Kerins S (2012) No more yardin' us up like cattle. In *People on Country Vital Landscapes Indigenous Futures* (Eds. J Altman, S. Kerins), pp. 190–201. The Federation Press, Sydney, Australia.
9. Scambary B (2013) My country, mine country indigenous people, mining and development contestation in remote Australia. *Research Monograph* **33**, 18–19.
10. Escobar A (2008) *Territories of Difference: Place, Movements, Life, Redes.* Duke University Press, Durham, NC, p. 12.
11. NT Government, ND McArthur River coastal floodplain, Sites of Conservation Significance, Department of Natural Resources, Environment, The Arts and Sport. Available at https://nt.gov.au/environment/environment-data-maps/important-biodiversity-conservation-sites/conservation-significance-list.

12. Bradley J, Yanyuwa families (2010) *Singing Saltwater Country*. Allen and Unwin, Sydney, Australia.
13. Pickering M (1994) The physical landscape as a social landscape: A Garawa example. *Archaeology in Oceania* **29**, 149–161.
14. Avery J (1985) The law people: History, society and initiation in the Borroloola area of the Northern Territory. Doctor of Philosophy thesis. University of Sydney, Sydney, Australia.
15. Gambold N, Kerins S (2015) *Ganalanga Mindibirrina Indigenous Protected Area, Plan of Management 2015–2020*. Darwin, NT.
16. Australian Government (2017) Gulf Fall and Uplands Bioregion. Department of Environment. Available at https://www.environment.gov.au/system/files/resources/a8015c25-4aa2-4833-ad9c-e98d09e2ab52/files/bioregion-gulf-fall-and-uplands.pdf.
17. Elkin AP (1961) The Yabuduruwa. *Oceania* **31**(3), 166–209.
18. Aboriginal Areas Protection Authority (1991) Mussolini Harvey. Second Annual Report for the period ended 30 June 1991. Aboriginal Areas Protection Authority, Darwin, NT.
19. Aikman A (2015) Aboriginal cash "siphoned" off by Northern Territory. *The Australian*, August 8.
20. Australian Bureau of Statistics (2015) *Prisoners in Australia*. Available at http://www.abs.gov.au/ausstats/abs@.nsf/Lookup/by%20Subject/4517.0~2015~Main%20Features~Prisoner%20characteristics,%20Australia~28.
21. Payer H, Taylor A, Barnes T (2015) Who's missing? Social and Demographic Impacts from the Incarceration of Indigenous People in the Northern Territory. Demography and Growth Planning, Northern Institute, Research Brief 5. Available at https://www.cdu.edu.au/sites/default/files/research-brief-2015-05.pdf.
22. Australian Bureau of Statistics (2011) Census of Population and Housing, Aboriginal and Torres Strait Islander peoples (Indigenous) Profile: Borroloola (SSC70030) and Mabunji – Mungoorbada Outstations (ILOC70500401). ABS, Canberra.
23. Australian Bureau of Statistics (2016) *Doomadgee Local Government Area*. Available at http://stat.abs.gov.au/itt/r.jsp?RegionSummary&region=32770&dataset=ABS_REGIONAL_LGA&geoconcept=REGION&datasetASGS=ABS_REGIONAL_ASGS&datasetLGA=ABS_REGIONAL_LGA&regionLGA=REGION&regionASGS=REGION.
24. Holmes J (2010) Divergent regional trajectories in Australia's tropical savannas: Indicators of a multifunctional rural transition. *Geographical Research* **48**(4), 342–358.
25. Northern Territory Department of Lands and Housing (1991) *Gulf Region Land Use and Development Study*. Northern Territory Government, Darwin, Australia.
26. Holmes J (1996) Diversity and change in Australia's rangeland regions: Translating resource values into regional benefits, *Rangeland Journal* **19**(1), 3–25.
27. Holmes J (1992) *Strategic Regional Planning on the Northern Frontiers*. North Australia Research Unit, Australian National University, Discussion Paper No. 4.
28. Kerins S (2014) Sick country: Poisoning garawa with mining and politics, *New Matilda*, 23 July. Available at https://newmatilda.com/2014/07/22/sick-country-poisoning-garawa-mining-and-politics.
29. Jones R (2013) Indigenous leaders want government help to stop copper sulphide leaks at Redbank mine. Australian Broadcasting Corporation. Available at http://www.abc.net.au/news/2013-09-18/redbank-copper-legacy-mine-problems-feature-story-ruby-jones/4965356.
30. ERIAS Group (2014) Independent Monitor Community Report, McArthur River Mine. *ERIAS Group*, South Australia. Available at http://mrmindependentmonitor.com.au/downloads/.
31. Kerins S (2015) Indigenous communities are losing out in the development of northern Australia. The Conversation, September 2. Available at https://theconversation.com/indigenous-communities-are-losing-out-in-the-development-of-northern-australia-46736.
32. Kerins S, Green J (2016) Indigenous country in the southwest Gulf of Carpentaria: Territories of difference or indifference? In *Engaging Indigenous Economy, Debating Diverse Approaches* (Ed. W Sanders). CAEPR Research Monograph No. 35, ANU Press, Canberra, Australia.
33. Kerins S (2009) The first ever Northern Territory homelands/outstations policy. *Indigenous Law Bulletin* **7**(14), 7–10.
34. Vanstone A (2005) *Beyond Conspicuous Compassion: Indigenous Australians Deserve More Than Good Intentions*. Address to the Australia and New Zealand School of Government, 7 December, Australian National University, Canberra, Australia.
35. Altman JC, Hinkson M (Eds.) (2007) *Coercive Reconciliation: Stabilise, Normalise, Exit Aboriginal Australia*. Arena Publications Association, Melbourne, Australia.

36. Kerins S (2012) Building from the bottom-up: Indigenous development initiatives in the South-west Gulf of Carpentaria, Australia. *Proceedings of Nga Pae o te Maramatanga Conference*, Auckland University, NZ.
37. Kerins S (2013) A key role for indigenous peoples in Australia's sustainable future. In *Black and White: Australians All at the Crossroads* (Eds. R Craven, A Dillon, N Parbury), pp. 287–302. Connor Court Publishing, Melbourne, Australia.
38. Polanyi K (2001) *The Great Transformation, The Political and Economic Origins of Our Time*. Beacon Press, Boston, MA.
39. Bradley J, Yanyuwa families (2007) *Barni-Wardmantha Awara, Yanyuwa Sea Country Plan*. Mabunji Aboriginal Resource Association, Borroloola, NT.
40. Yanyuwa families, Li-Anthawirriyarra Sea Rangers, Johnson S (2011) *Barni-Wardimantha Awara Yanyuwa Indigenous Protected Area Plan of Management and MERI Plan*. Mabunji Aboriginal Resource Association, Borroloola, NT.
41. Kerins S (2012) Caring for country to working on country. In *People on Country Vital Landscapes Indigenous Futures* (Eds. J Altman, S Kerins). The Federation Press, Sydney, Australia.
42. Northern Land Council (NLC) (1993) *Borroloola Community Development Planning Study Interim Report and Recommendations*. Northern Land Council, Darwin, NT.
43. Taylor J (2016) Leaked docs reveal secret Coalition plans for Indigenous Rangers. *Crikey*, July 15.
44. Kerins S (2013) Governing a black commons. *Indigenous Law Bulletin* **8**(8), 30–34.

CHAPTER 8

Governing North Australian Landscapes for a Better Future

Allan Dale

CONTENTS

8.1 INTRODUCTION AND HISTORICAL CONTEXT

Presented at the Australian Davos Connection Conference in Townsville in 2014, the multi-authored paper *Myth to Reality* explored more than 150 years of grand narratives regarding the future prospects for North Australia.[1] *Myth to Reality* highlighted that over the past 50 years, three big narratives have tended to reappear with great predictability.[2] One of these narratives is based on the perception that North Australia is a place of endless economic bounty and limitless resource development opportunity. The second derives from those who would like to see extensive conservation outcomes (Box 8.1). Both narratives are based on important national needs. There are significant resource development opportunities in the North, while at the same time the region is a largely intact bio-cultural landscape of immense international value (see Chapter 5). Both narratives, however, tend to discount major physical, climatic, economic, and social barriers and opportunities.

BOX 8.1 MYTH TO REALITY

Northern Australia [is] one Australia's most contested landscapes. Indeed, there is often a very sharp contest between different visions and realities about the North that means defining Northern Australia can be a contentious business.[1]

The third national narrative is more complex and relates to the way many non-Indigenous Australians have viewed and responded to explicit Indigenous policy narratives relating to their growing rights and interests in the North.[3] At one extreme, some policy actors have failed to understand the significance of hard-won land rights, leading to continued and facilitated appropriation of Indigenous interests in *country*, and engagement approaches that continue to marginalise traditional owners. Alternatively, other actors involved in Indigenous policy development may have not yet fully grasped the fact that the North's Traditional Owners have always governed their own affairs (see Chapter 6); with approaches that assume Indigenous acceptance of well-intentioned but ultimately assimilative national, state, and territory government policies.[4] Together, these views reflect a broader myth that Traditional Owners do not have significant ongoing control over much of the North Australian domain, despite having explicit rights and interests in 94% of the landscape.[5]

Together (Figure 8.1), these three national northern narratives, driven both from outside and from within the North (see chapter 3), have created the foundations for grand policies, plans, and even grander failures in the more recent history of northern development.[1]

Previous chapters in this book and historical detail outlined in *Myth to Reality* suggest that if northern development is to deliver genuine economic opportunity, it will require access to multiple knowledge sets, collaborative capacity building across several sectors and cross-governmental mobilisation of effort within North Australian regions. This must be coupled with serious analysis of the global comparative advantage of the most prospective economic opportunities to be developed. The history of grand northern visions based on policy myths rather than well-informed reality has generally resulted in conflict,[6] economic failure,[7] and a continuation of a boom-bust economy (see Chapter 4). The clear result of all three narratives has been persistent underdevelopment. Indeed, Lyndon Megarrity has shown that political promises for northern development not based on

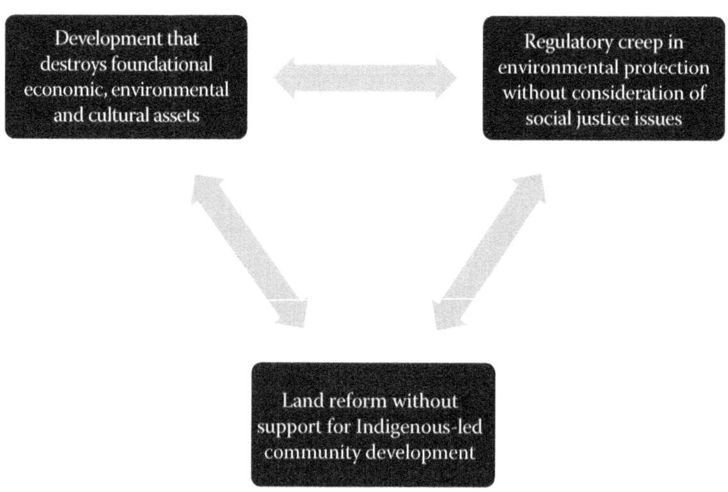

Figure 8.1 Three competing policy narratives influencing the North Australian landscape.

economic and social reality generally falter on the altar of economic austerity once the shorter-term political commitments made during election campaigns are assessed in the cold hard light of day.[8]

Very importantly, however, as suggested by the 'White paper on the development of Northern Australia',[9] there are great opportunities, and at the same time many complex issues to be identified, analysed, and resolved. Significant trade-offs will need to be negotiated and real partnerships established between development, Indigenous, and conservation interests. If we do this, the genuine opportunities in targeted agriculture,[10] tourism, mining, fishing and forestry, carbon, conservation and ecosystem services, tropical knowledge, and other human services will grow. There is, however, a real need to ensure this effort is underpinned by more robust evidence,[11] engagement, and improved governance of the North, all with a focus on building human equity[12] and regional liveability.

8.1.1 More Recent Approaches to Northern Development

Myth to Reality suggested that in the last 10 years there have been at least three new Australian Government-led efforts to revitalise northern development. First came the Australian Coalition Government's formation of a Northern Development Taskforce in 2007 (the Heffernan Committee).[13] This was followed by the Australian Labor Government's formation of the stakeholder-based Northern Australia Land and Water Taskforce (supported by CSIRO's Northern Australia Land and Water Science Review[14]). The Land and Water Taskforce's efforts were complemented by the establishment of the Office of Northern Australia, the North Australian Ministerial Forum (NAMF), and the commissioning of several key pieces of work informing the deliberations of the Forum through the Northern Australia Expert Advisory Panel to provide in-depth analyses of issues such as prospects for the northern pastoral industry[15] and tenure reform.[5] The evidence emerging from these landmark processes and studies, all of which have drawn on an ever-growing knowledge base and wide engagement, have identified significant growth prospects for major industries, as well as the attendant impediments and enablers. In effect, while the prospects for both development and extensive conservation are good, the operational focus on building the evidence and engaging well will remain critical.

The new Coalition Government's latest white paper processes[16] and the aligned Northern Australian Strategic Partnership[9] and new Ministerial Forum[17] present an additional new opportunity for national governance reform in the North. While these processes have to some extent been engaged (via a parliamentary committee) and evidence-based (via the white paper), they have not yet established stronger frameworks or actual processes for durable knowledge building, engagement, and adaptive management in the governance of the North. In particular, very little progress has been made with respect to any meaningful and durable settlement or dialogue between governments with Traditional Owners (see Chapter 2). This is most starkly outlined in Peter Yu's recent speech at the Developing Northern Australia Conference in Darwin in June 2016[18] (also see Box 8.2 and Chapter 1).

BOX 8.2 PETER YU'S 2016 NORTHERN DEVELOPMENT SPEECH

The fundamental problem in the relationship between Indigenous society and governments is that government policies and strategies that relate to northern development are underpinned by an entrenched government belief that Indigenous commitment to traditional culture and communal ownership of land is incompatible with economic development. A pervasive mantra … is that native title issues must be 'resolved'. What they mean by this word 'resolve' is extinguishing or winding back Native title rights, so that in government's view, economic development can proceed. 'Repugnant' captures the pervasive Aboriginal view of extinguishment.[18]

Hence, while early foundations for better pan-northern governance may be in place, real innovation is needed in how to design and institutionalise a better governance system.

8.2 GOVERNANCE CHALLENGES FACING IMPROVED LAND SECTOR OUTCOMES

The Australian Government's recent white paper on the Development of Northern Australia[9] explores some of the economic opportunities available to the North. The white paper, however, does not focus attention on the North's Indigenous development and environmental sustainability challenges. With all three of these key themes in mind, my recent book *Beyond the north-south culture wars*[2] explored the opportunities and risks ahead for the North given the well understood mega-trends facing our global future.[19] *North-south culture wars* speculated about how the future of the North might look if some of the critical mythologies and cultural divides that have been playing out within the North Australian landscape are not resolved or reconciled. It further posited that depending on how successful the nation is in charting this next phase of northern development, two very different scenarios could emerge. If future decision making about the North goes spectacularly wrong, then a failed state scenario is possible. However, with structured policy and engaged decision making based on sound evidence, then a much brighter future is possible. While this unfolding opportunity represents many possible futures, it could look like the emergence of a stable alliance of dynamic regional economies across North Australia. These scenarios have been fleshed out further elsewhere,[20] but are important to understand in order to help set the scene for discussing governance reform.

Achieving the second of these two scenarios will depend almost universally on the health of the over-arching system of governance across North Australia; from the pan-northern to the local and even property and paddock scales. My 2013 discussion paper, 'Governance challenges affecting northern Australia', detailed the current strengths and weaknesses of our governance system across all of these scales and their impact on the future of North Australia.[21] 'Governance challenges' considered that, because North Australia has such national significance, the nation should be actively debating the overall health of its governance system. In evidence, gleaned from the work of *Remote focus* in 2012, it seems clear that North Australians are far from happy about the current governance system (Box 8.3).[22]

Judging by the emerging social, economic and environmental outcomes projected for the North, its governance system appears to remain on a knife's edge. Some emerging features of the governance system and the outcomes being secured should be celebrated and expanded. Equally, however, some radical improvements are needed to avoid the failed state scenario. Indeed, if North Australia is to secure a prosperous future based on dynamic regions, Australia will need to spend the next decade getting the governance systems and consequent policy and delivery settings right.

BOX 8.3 REMOTE FOCUS ON GOVERNANCE

Governments are recognising that for large parts of remote Australia, the notion that the market will deliver positive outcomes – in services, settlement patterns, environmental management, and wider social inclusion – is a fiction. The national interest in remote Australia therefore needs clarity of purpose and resolute and continuing engagement by governments at all levels, accompanied by significant systemic reform.[22]

Before possible reforms can be explored, however, building on my 'Governance challenges' discussion paper, it is important to present an overview of how the North's contemporary governance system actually operates.

1. *A nationally fragmented governance framework*: North Australians experience similar cultural, historic, economic, climatic, environmental, and social conditions. There is, however, little governance connectivity *across* the North, and by and large, different governments manage common issues in isolation. Canberra's internal relationships and relationships with other jurisdictions tend to be compartmentalized and fragmented. In the last decade though, more progressive points of integration have been emerging between the Western Australia (WA), Queensland (QLD), Northern Territory (NT), and Australian Governments. As one of several historical attempts to do this, the NAMF, established by the Commonwealth around 2007, commenced a higher level of cross-jurisdictional dialogue. Within the context of implementation of the 'White paper on developing Northern Australia', the Commonwealth replaced this with a Strategic Partnership, bringing the Prime Minister, the NT Chief Minister and the WA and QLD Premiers together behind Council of Australian Governments (COAG) meetings.[13] It also established and new Office of Northern Australia explicitly to build strategic governance linkages. Linked to this, Regional Development Australia (RDA) boards[23] across the North have also been developing stronger alliances that could help shape a wider, more connected North Australian agenda aimed at raising agenda via the Strategic Partnership framework.

   At the same time and in parallel with governmental efforts, Indigenous people have had historically strong networks and connections across the North. In a contemporary sense, however, stronger cross-northern governance-related alliances are only just emerging. Traditional Owners are the torchbearers on this front, with informal but developing institutions like the North Australian Indigenous Land and Sea Management Alliance (NAILSMA).[24] Similarly, across the North, alliances have also been emerging in several not-for-profit and industry sectors such as the community-based natural resource management (NRM) and private sectors (the Northern Alliance).[25] Industries as diverse as the pastoral, fishing, and mining sectors have several informal alliances that peak and trough. Some major national institutions, like the Commonwealth Scientific and Industrial Research Organisation (CSIRO) and several universities with footprints in the north also have established cooperative arrangements that enable them to deal with North Australian issues more cohesively.[26] Hence, while the governance frameworks across the North are currently fragmented, it is possible to consider positive signs that the strengthening of cross-North Australian alliances can link into growing national and cross-jurisdictional policy debates. Governments will need to organically support these emerging alliances in a progressive and structured way if we are to see a growing conversation about cohesive landscape-scale outcomes.
2. *Centrally controlled government policy, programs, and investment*: Despite an emerging national policy cohesion in the North, 'Governance challenges' stresses that, with regard to landscape-scale policies and programs, the vast majority of government-based policy making, program design and budget development remains embedded in the south. This plays out differently in WA and QLD compared to the NT, so both circumstances deserve attention. In Queensland and WA, the locus of political decision-making sits in Brisbane and Perth. The level of political representation in the North of these states is restricted due to limited population and this is exacerbated in northern WA when compared northern Queensland. In both states North Australian interests are not strong in departmental executives, intra-governmental committees, and in Cabinet and Treasury deliberations. Day in and day out, major decisions do not always fully account for the North's differences. This traditionally leads to a high level of frustration both among North Australian communities and conflict within North Australian landscapes.[27]

   The situation in the NT is different, with it being North Australia's only nascent state, with its capital firmly embedded within the North. The NT, however, is not an economically self-sufficient state, and unlike QLD and WA, it is dependent on the Australian Government. The Australian Government, however, also has other major governance responsibilities in all three jurisdictions, including Defence and unusually large responsibilities for Indigenous affairs and the environment. The resultant and generally centrist approach to policy and program making, combined with the cross-government and cross-departmental fragmentation of effort can drive conflict between competing interests in the North. It also results in less innovative policy outcomes developed in genuine partnership with northern communities, particularly in the extensive pastoral and Indigenous land estate.

3. *Large corporates and northern Australia*: Due to its remoteness, terrain, and climate, North Australia is an expensive place to do business, and its major productive assets are in public hands, despite significant Traditional Owner rights and interests existing in these landscapes. As such, society's big corporate and non-profit institutions (whether they be in the development or conservation space) are often best placed to raise the finance required to successfully develop or enact conservation outcomes within these landscapes. This often means that the biggest decisions about the future of northern landscapes are often made in corporate board rooms beyond the region. The scale of many development and conservation projects also tends to mean they are of state or even national significance, requiring centralised decision making about approvals. This means corporate and government approval decisions, no matter how significant to regional communities, largely happen outside of the North Australian region.
4. *Cross-regional and sectoral fragmentation*: 'Governance challenges' also stressed that there can be quite fragmented approaches to governance among regions within the North, meaning innovative approaches about landscape scale management can be slow to disseminate into other regions. Isolation and competitiveness among North Australian regions, however, has tended to soften in the last decade, and institutions like NAILSMA and the RDAs are keen to foster cross-region sharing.

   Equally, there has also tended to be isolated development in the capacity of key economic sectors across the three jurisdictions as well. The pastoral industry, for example, has three very separate cattle-based agri-political institutions across the North. While some cross-sector industry alliances have emerged (e.g. the Northern Australian Beef Research Committee), these tend to be limited in scope. A lack of sectoral capacity in the North has led to national peak bodies taking a higher profile role than might have been the case otherwise in major discussions about the future of northern landscapes. In the conservation sector, for example, national groups such as the Australian Conservation Foundation, the Wilderness Society, the Nature Conservancy, and the World Wide Fund for Nature play significant cross-northern roles. These groups themselves, however, can be divided with respect to North Australian issues, as experienced in the 'Wild Rivers' debates in Far North Queensland. Such organisations tend to have out-posted positions in the North, but their extensive membership base and financial resources are derived from southern Australia.
5. *Locality and capacity*: Local governments play a much more significant role in landscape-scale planning in the south of Australia. Alternatively, the vast majority of local governments across North Australia cover vast territories and have tiny populations and a paltry rates base, limiting their capacity to cover basic asset maintenance and civic services. Alongside this, these Councils face significant complexity in both large and small development decisions relating to land use change. Native title considerations are ubiquitous. Widespread issues of international, national and state significance often trigger major environmental assessments and additional regulatory complexities. The difficult operating environment and annual wet season infrastructure damage bill further limits capacity. Indigenous Councils in QLD and the NT are also recent additions to the governance landscape. They have emerged in different forms from their origins as church or government-run missions or reserves and many have an even more limited income base than mainstream rural and remote councils. This means equally under-resourced Traditional Owner and pastoralist-based organisations often need to do the heavy lifting with respect to the negotiation of external development and conservation interests within and across the Northern Australian landscape.
6. *Indigenous governance*: 'Governance challenges' detailed the foundational Traditional Owner-based governance systems that drive and guide Aboriginal and Torres Strait Islander interests across the North. This first plays out through the cultural institutions (families, clans and tribal groups) that dominate decision-making. Additionally, these informal institutions often set up more formalised Traditional Owner organisations under particular legislative frameworks (e.g. land trusts, prescribed body corporates, management trusts, joint management boards, incorporated associations, and commercial companies). Sometimes the governance needs of Traditional Owners are scaled-up at the regional and even state/territory-wide level. Infused with this is the role of statutory land councils; regional or supra-regional organisations supporting the

determination of native title; and other land and sea rights under Commonwealth legislation. Either way, in the future, Traditional Owner organizations are likely to be significant economic institutions with an interest and responsibility in ensuring the effective management of North Australian landscapes (see Chapter 7).

8.3 PATHWAYS TO A BETTER FUTURE FOR THE NORTHERN AUSTRALIAN LANDSCAPE

Given these governance challenges, some of the primary pathways to achieving improved social, economic and environmental outcomes in the northern landscape can be further developed. Figure 8.2 articulates some key structural components of an emerging system that must function effectively for improved governance outcomes.[28] These structural components range from the national (e.g. COAG) to the property or Traditional Owner estate scales; effectively operating like nested substructures within a complex system. All levels, however, need to be viewed as part of an integrative whole; as a package of potential systemic governance reforms rather than as isolated reform actions. In doing so, I again build further on preliminary ideas raised in my original 2013 'Governance challenges' paper.

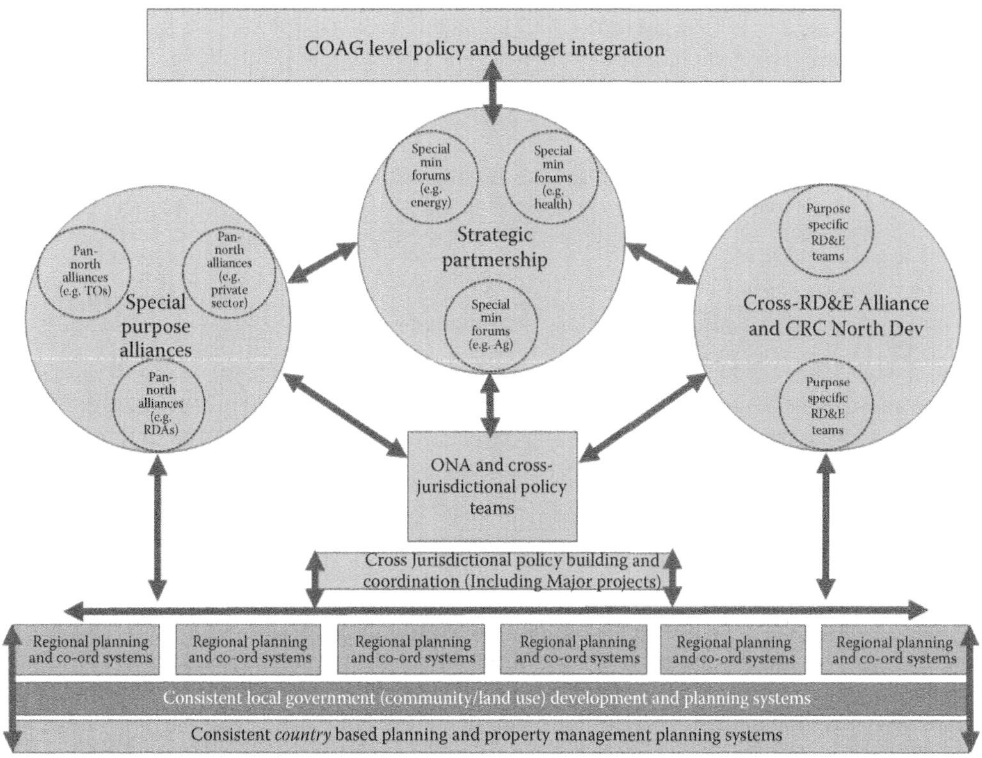

Figure 8.2 A broad overview of the key structural elements of an improved governance system for North Australia.

8.3.1 The Pan Northern Scale: Bilateral Policy and Investment in Northern Australian Priorities

The most significant policy initiative in the 'White paper on developing Northern Australia' could be viewed as the formalization of the Strategic Partnership arrangements.[9] It is the place where significant pan-northern issues could be identified and agreed, and where serious cross-northern engagement key sectors and evidence building might drive effective policy building. This highest-level governance mechanism, however, is currently underutilized for building fresh bilateral policy and strategy effort. It remains more administratively focused on implementing key initiatives committed to within the original 'White paper'. It could, however, become a more effective vehicle for overseeing continuous improvement and adaptive management. Aligned with this governance innovation, however, has been the establishment of the Office of Northern Australia (based in Darwin), together with its intra/inter-governmental coordinative responsibilities and mechanisms.

Importantly, with this broad (but underutilized) government-based institutional architecture in place, the possibility of North Australians identifying, leading, and progressing major policy initiatives into governments will be very strong, but will rely on the following:

- Grassroots identification and advocacy of the most significant strategic policy issues
- Strong debate and discussion about these issues within and among key sectors (e.g. the Indigenous sector, the pastoral sector, etc.)
- The coming together of more facilitated and collaborative cross-sectoral dialogue to identify common North Australian problems and potentially shared solutions
- The injection of structured input and evidence building support into policy debates from across the North's research and development sectors (including CSIRO and the key universities with a footprint within or major exposure to the North) and the emerging Collaborative Research Centre for Developing Northern Australia
- Some point or locus of integration across these policy-building efforts, including explicit partnership building between the North's major stakeholders, the WA, NT, and QLD Governments with the Commonwealth
- Brokerage of these emerging policy (and associated investment packages), either through portfolio specific Ministerial Forums (e.g. via Health Ministers on significant health related issues), and/or through the Office of Northern Australia and its associated inter- and intra-governmental administrative linkages
- The development of cohesive policy and budgetary responses via typical COAG-style framing, leading to joined up governance and shared budget influence

Through such approaches, key social and economic issues of importance to North Australians could be powerfully framed, resolved, and progressed through these sort of high-level governance arrangements. Classic examples of pan-northern issues that *could* be progressed in this way might include (1) building indigenous-led approaches to northern development and landscape management (see Chapter 2); (2) building a strong future for the North Australian beef industry, where economically and environmentally appropriate; (3) halting the dramatic decline of small mammals across tropical savannas; (4) harmonizing and improving regional land use and infrastructure planning and major project approval systems; (5) the strategic allocation of water resources; (6) the harmonization of tenure systems across jurisdictions; or (7) the cohesive development of an ecosystem services economy based within a supportive cross-jurisdictional policy framework (see Chapter 5).

Doing so, however, will take explicit cross-governmental support and investment in building governance capacities at all levels, but particularly through alliance building. While the basic frameworks now exist for that to occur, much more needs to be done to facilitate such serious engagement between North Australian interests and governments.

To inform a more nationally cohesive pan-northern policy agenda (as well as more specific jurisdictional, and regional and local innovation), a much stronger pan-northern RD&E alliance is particularly needed. Integrated science investment by the Commonwealth in North Australia has traditionally been patchy and generally poorly connected to policy building and delivery efforts. Such science-rich approaches have primarily been progressed through now defunct (and spatially-restricted) collaborative research centres (CRCs) and more recently, the Tropical Rivers and Catchment Knowledge (TRaCK); Marine and Tropical Research Facility (MTSRF); National Environment Research Program (NERP); and now National Environment Science Program (NESP) research and funding programs. The CRC program particularly showed promise in developing a strong, place-oriented and user-influenced framework for the integration of knowledge in decisions about the northern landscape.

If each of these public research investment frameworks had been able to build upon previous successes then, North Australia now may have had a much stronger knowledge base to inform policy development, land use planning and major project decisions. While the local delivery agents of these programs have previously attempted to secure real strategic integration and stakeholder ownership of the science generated, the basic design of the funding programs has tended to be centrist. Key design limitations have included the lack of genuine partnership in setting and implementing research priorities, and tensions between the role of Commonwealth departments and local science brokers in the North. Importantly, despite their critical roles in science and regional land use planning, state and territory governments have been generally excluded from program planning and delivery. This results in tenuous links between science, policy, and land use decision making.

The emerging CRC for Developing Northern Australia[29] now again presents a more cohesive opportunity to address these concerns and embed evidence building in policy development. If well embedded within the wider and emerging pan-northern governance system, the CRC could help provide the evidence required for more inclusive, collaborative and innovative approaches to decision making from pan-northern to local scales.

8.3.2 The Regional Scale: Strategic Frameworks for Regional Planning and Development

Limited bilateralism and bipartisanship in regional land use planning has undermined credibility in the practice in North Australia. Regional planning should be about providing everyone in the community with enhanced information as a basis for evidence-based decision making.[30] To enhance certainty for Indigenous, environmental and development-oriented stakeholders and investors, all jurisdictions in the North need to revisit the purpose of regional planning in highly contested landscapes and its ability to deliver long term security to competing interests. There is an opportunity to progress some sense of trilateral agreement about the role and scope of regional land use planning in contested regions, and this key area of policy effort could be advanced through the COAG-like Strategic Partnership. There is potential value in these arrangements creating a more integrative policy framework for northern development; building the equitable foundations for regional framework agreements and plans that are rights based (see Chapter 7) and that could deliver more stable, adaptive, and integrative forms of decision making within regions.

Within such frameworks, greater coordination and engagement could occur with respect to environmental regulation, particularly in the context of vegetation management, water quality improvement (e.g. in Great Barrier Reef catchments), the development of regional Native Title framework agreements, vegetation management and world heritage management, climate adaptation, and so on (Figure 8.3). The same needs to occur in planning for the allocation and health of North Australian catchments, surface, and groundwaters. Key, nationally agreed principles for water allocation and management have already been forged under the National Water Initiative. With some exceptions, however, these principles are not universally accepted, adapted, and adopted in water-based decision making in North Australia, potentially exposing northern catchments to the same type of

Figure 8.3 Not planning for climate change and the reconciliation of competing resource use interests in land could weaken development prospects for the North. (From Mohrmann, D., Yasi – Cardwell Marina, 2011, https://flic.kr/p/9fMV7d.)

fragmented decision making that has resulted in the degradation of systems like the Murray Darling Basin. With some catchment systems already fully allocated in the north-east of the nation, it is important that good water governance is fully institutionalized soon. The fundamental inclusion and resolution of Indigenous interests in water is also key.[31]

More importantly, the way in which broader scale natural resource allocation processes have been run at the state/territory level has undermined the trust that pastoral and other communities have in governments and people from the south in general.[32] Governments, however, can rebuild trust with these important northern communities, and doing so will require policy makers conferring some respect on the people in regions who actually manage North Australia's vast land and seascapes. *North-south culture wars* suggests that this means governments and industries sitting down together to explore what it is that both parties value and to build a common understanding of the problems at hand. There is a need to embed science in problem definition in a clear way so that a common understanding of the key issues can emerge. From here, both parties can co-design and jointly monitor implementation of the best approaches. Some great successes via the application of joint management, for example, have been achieved in securing sustainability in the northern prawn fishery,[33] but such cases tend to be the exception.

Regionally focused land and resource use planning approaches have the potential to set the scene for very effective and clear, bilaterally agreed, and stable infrastructure planning in the North, consistent with the intent of the Northern Australian Infrastructure Audit.[34] Such regional frameworks would also provide an improved approval-system for major project facilitation in the North. These systems should be about facilitating development in the right places while also securing agreed environmental and social standards and benefits at the landscape and local scales. Project approval processes should not be about bolstering economic aspects of development at all costs, nor becoming an impassable barrier. Major project approval systems have increasingly become too politicized and lack a clear underlying philosophy that drives their operational culture.[35] Regular staff changes in approval agencies and inconsistent decision-making can send mixed signals to investors without real environmental gains. Development approval systems could be much more efficiently guided by regional planning frameworks, integrated across all spheres of government, and deeply engaged within the communities they service.

While regional land use and infrastructure planning needs to be driven by the Commonwealth and particularly state/territory governments, the regular intervention of remote governments in the affairs of regions could be avoided through genuine attempts to build an effective scale and capacity for the operation of regional integrative institutions (e.g. WA's Planning Commissions) and local governments within the North Australian landscape (including strong voluntary regionalism among local governments). Some commentators have called for stronger council-based devolved regionalism and more regionally focussed policy/support from the state-NT Government administrations.[36] There is no alternative, however, to empowering northern regions to set the direction for and to manage and to monitor progress towards their own economic destiny.

North-south culture wars refers to strong regional determinism and decision making within a wider national and global context as endemic regionalism. Strong endemic regionalism is essential to securing a stable future for the North, but it needs national and state policy frameworks that foster it. It also needs to be explicitly resourced and supported by trilateral (local, state, and Australian) government commitment. A shift to endemic regionalism requires a shift from government and program-driven planning around the typical three or four-year political cycles and requires great steps to be taken in improving the planning capacity within a region's community and industry sectors, particularly at least across the regional development, natural resource management and human services sectors. Regions need to increasingly become more self-reliant in their own planning and decision-making and involve all players with a stake in the future (i.e. combining government, community, and private sector interests). This generally means:

- The need for regionally based integrative institutions that bring the parties together to find negotiated solutions to regional problems, building cross-sector trust, and supporting sector-based planning capacity.
- Well-structured negotiation between governments and communities over major regional problems can reduce significant and long-standing conflicts almost immediately and more strategically influence annual government budget cycles.
- State of the region reporting systems, which can harvest local technical expertise with regards to where a region is headed and the success of agreed actions.

Major change is required in the way the nation supports planning and decision making to achieve regional prosperity and sustainability. This change, however, is not one that will be too costly in the North; in fact, it could actually save limited human resources and be cheaper than our current fragmented and conflict-based regional planning system. Endemic regionalism simply requires the collective will of federal, state, and local government, industry, and communities to agree on what constitutes endemic regionalism and to act upon it. Both the Commonwealth and states have advanced in this field, but stronger bilateral approaches are needed as well as a more consistent approach across the North.

With both strategic land use planning and strong endemic regionalism in mind, along with Chapter 5, *North-south culture wars* reminded readers that, as Indigenous and pastoral lands cover the vast majority of North Australia's landmass, vast natural landscapes lend themselves well to being packaged and presented internationally within emerging regulated and voluntary ecosystem services markets. Australia is doing little to establish a sound national or North Australian policy and delivery framework for getting front-row access to these markets. Some change is in play though within the nation's emerging ecosystem service trading systems (e.g. carbon).[37] Consequently, the nation can't continue to just regulate the North's land managers to deliver outcomes on behalf of consumers in the south. Importantly, the cost of securing ecosystem services in the North is not prohibitive.

8.3.3 Local and Community Scales: Subsidiarity via Strong Local and Community Governance

Ultimately, real subsidiarity in the northern landscape will rely on continuing to build the democratic and operational capacity of local and community governments within the North Australian landscape (including strong voluntary regionalism). The different powers of local governments between WA, the NT, and QLD also need to be narrowed, along with the capacity differential between councils within the jurisdictions (for example, between Indigenous and non-Indigenous councils and rural versus urbanised councils). *North-south culture wars* previously considered how this could be assisted through increasing support for devolved and cooperative regionalism among local councils at appropriate scales. Together, stronger regional institutions and local government would create more enduring regional leadership. The international, national, state, and territory significance of cultural and conservation values in northern landscapes, however, would mean the Commonwealth, state, and NT Governments would still need to be trilateral players in any new approach.

In both Indigenous and non-Indigenous local and community governments, Australia's centralised policy and fragmented program delivery frameworks continue to eschew place-based or community-led approaches to local development. The only hope of bridging the gap facing northern Indigenous communities, for example, is through the development of genuine, long-term, and purposeful partnerships with a much stronger focus on building Indigenous social and economic capital (Box 8.4). A sound and durable community development philosophy and method is only just starting to emerge but is not common practice in remote communities. Neither strategies seeking to develop utopian Indigenous communities nor strategies seeking to close remote outstations will work.

BOX 8.4 FROM THE GRASSROOTS UP: THE BROOME SOCIAL INVESTMENT MODEL

PAUL LANE

KIMBERLEY INSTITUTE, BROOME

The 'Broome Model' aims to leverage and holistically coordinate social investment in the culturally cohesive region of Broome, the Dampier Peninsular, and Bidyadanga, Western Australia. The model has been devised by the Kimberley Institute, a peak advocacy organisation for Aboriginal peoples of the region. The approach brings corporate and philanthropic investors into grassroots organisations through new social investment mechanisms including bonds, grants, and venture capital that are tied to strong indicators and performance targets. In this model, government becomes an underwriting supporter not a top-down director.

The model changes the dynamic of solving community problems. While several aspects of the Broome Model have been attempted individually and within some spheres of activity, it takes a whole of community approach based on rigorous understanding of the nature of local problems and challenges. The collective partnership also includes the Native Title holders of the area, being the Yawuru, through their Prescribed Body Corporate and their commercial

(*Continued*)

BOX 8.4 (Continued) FROM THE GRASSROOTS UP: THE BROOME SOCIAL INVESTMENT MODEL

entity, Nyamba Buru Yawuru Limited. It provides a useful model for social investment for North Australia and across the continent in general.

The Broome Model addresses Indigenous socioeconomic disadvantage, including:

- High incarceration rates
- Juvenile antisocial behaviour including poor school attendance
- High incidence of substance abuse
- High incidence of poor physical and psychological health
- High rates of long-term unemployment
- High incidence of domestic abuse or dysfunction

These problems are addressed by an alliance of Broome Aboriginal NGOs that are culturally appropriate providers of:

- Employment and training services
- Primary and mental-health services
- Language and cultural well-being services
- Substance-abuse services
- Housing solutions services
- Men's and women's outreach services

One of the motivations for this coordinated approach has been volatile annual government funding cycles which challenge local capacity to address practical problems over short-, medium- and long-term timeframes. The model is more efficient than a cyclical top-down investment model since it is tied to the delivery of outcomes, saves government money, rewards good performance and provides long-term stability and professional development for organisations on the ground.

How Does It Work

To work, the grassroots community partners have to be able to define and show that key performance targets have been achieved. Once a priority has been identified, a social investment strategy is designed over the medium and long term to support a package of services provided by the partnership. The community impact partners then negotiate with governments and social investors on the amount of investment required, the dividend proposed by government as underwriters, and the specification of the services and outcomes to be delivered. As the lead coordinating group, the Kimberley Institute negotiates with government but brings in a range of different sources of funding including social bonds, social debt, philanthropic grants, contributions from participant organisations, and in-kind contributions.

Investment Partnership—The Way of the Future

The Broome Model creates a linkage between the public/corporates/community for the delivery of social outcomes in any particular region – a new way of delivering social outcomes through partnerships. The model provides an avenue for companies involved in developing the Kimberley to invest in Kimberley community service delivery in a structured manner on a long-term basis – which in turn builds strong partnerships between industry partners and the communities where they are focusing their activities. Payments for delivered outcomes may allow some corporates to achieve both social and commercial returns, as well as enabling industry partners to monitor and evaluate social impacts resulting from their activities in the region.

(Continued)

BOX 8.4 (Continued) FROM THE GRASSROOTS UP: THE BROOME SOCIAL INVESTMENT MODEL

Our focus for potential social investors is corporates, philanthropic institutions, and individuals and asking them to invest in new models dealing in areas of intransigent social problems that to date have proved beyond the remit of governments to deliver meaningful change. Ideally, government can be brought in to underwrite part or all of the investment and in so doing provide some return to investors.

Developing the Broome Model was never going to be a simple process, and as is inevitable with developing ideas and models largely from scratch, there are always going to be the unexpected opportunities and challenges. Capacity building in community organisations takes time and resources, and concerted efforts have been required to better inform and engage with government agencies.

Both collective community development and a focus on building the needs and the capacities of the individual are needed, and both of these approaches must be combined into one cohesive approach.

8.3.4 Property Scales: Stronger Governance of Properties and Traditional Owner Estates

North-south culture wars stressed that throughout the world, capitalism's success in poverty reduction is based on the introduction of transparent land title systems. Clearly defined and secure property rights in Indigenous lands would enable investment capital to be raised and private enterprise to flourish. Emerging markets can then also be well regulated to avoid or manage environmental and social downsides. There is a common view, however, that various forms of communal tenure are the antithesis of individual property rights and the operation of effective markets. Simply transitioning from communal to individual rights-based systems, however, is not essential.[38] It is the definition and management of the rights of parties with an interest in communal systems that is crucial. With clearly defined and stable communal rights, Traditional Owners can create the corporate structures needed to manage development and the consequent wealth generated. Alternatively, Traditional Owners may also seek to devolve land titles to third parties in areas of strategic economic importance in their communal estates (e.g. long-term leases). Clearly defining property rights and separating ownership from management in such ways *can* create the basis for marketable property rights and economic development (Box 8.5).

While lease-based arrangements on traditional lands have been part of Commonwealth, state, and NT intervention strategies, the primary focus has tended to have been aimed at securing essential facilities for the operation of the intervention itself rather than as a bold new strategy for solving the impoverished conditions facing people in remote Indigenous communities. Only recently (in the NT and QLD) have Traditional Owners been better supported by governments in their decisions relating to the creation of such leases (including leases for home ownership within communities). Some of these experiences, however, have also shown what might be possible, particularly through some of the specific strengths of the Commonwealth's Aboriginal Land Rights Act in the NT.[5]

There are several things that need to be done well if governance reforms associated with land (and increasingly water) tenure is to deliver real opportunity for communal lands and remote communities[39]:

- *Effective planning for communal lands and adaptive learning*: Independent *country*-based planning support for Traditional Owners to think about the wider aspirations they have for their estates and focused land use planning attention on the areas required for more intensive economic development

BOX 8.5 WHAT IF NATIVE TITLE WAS LIKE OTHER PROPERTY INTERESTS?

JEREMY DORE

ABORIGINAL CARBON FUND, ALICE SPRINGS

Since Mabo filled the national land rights void with Native Title, it hasn't gone too well. Instead of the seed of a new idea nurtured to grow and flourish, Native Title has been more like a transplanted tree that never quite takes alongside the others. Through all the reviews and introspection, Native Title has always been vulnerable and subordinate to other interests.

Take the case of a new industry, carbon farming. Carbon farming is about managing land better, as measured by carbon fluxes. In terms of land requirements, you need the right to carry out your project activities and, in some cases, you need the consent of others with an interest in the land, including the carbon rights holder.

States and territories control land. They have leaned in favour of pastoralists, being flexible to allow new kinds of management practices, such as burning, under their lease conditions. Plenty of farmers have taken this up. At the same time, they have leaned against Native Title holders: flexibility for farmers to expand their businesses has not required any discussion with Native Title holders. And when it comes to issuing carbon rights, Native Title holders either can't get them, or they certainly don't get a say when they are issued to someone else. The Native Title tree is pruned and kept neatly in its place.

But carbon farming is a federal scheme. And, here, legislators applied different thinking: rather than ask how does this work with Native Title, legislators asked how can we give Native Title holders the same opportunities as others? The Minister advanced this thinking in the second reading speech: 'The Carbon Farming Initiative includes a number of provisions to ensure Indigenous Australians can effectively participate and take up these opportunities'.

The bill contained special provisions deeming or assuming that exclusive possession Native Title holders have the right to carry out the project and the carbon right when applying for a project; much the same as a freehold owner, who could also assume to hold these rights if not granted to someone else.

In short, Native Title was dealt with in the carbon farming bill, rather than left to the uncertainties of the Native Title Act. And Native Title holders certainly got a better deal.

Native Title carbon farming projects have flourished. The Kimberley projects, relying heavily on their Native Title, have led the way to show that these provisions work. All up, Indigenous carbon projects have delivered nearly 2 million tonnes of carbon abatement at around $30 million in value, with a further $60 million contracted in future.

Ian Munro, former CEO of a large Indigenous organization in Arnhem Land, has said: 'Money generated from the sale of carbon credits has led to the largest non-government investment in Indigenous land and sea management in northern Australia … ever'.

Native Title has played a key part in unlocking an industry that fits with Indigenous aspirations for *country*.

The carbon industry is one example where Native Title has been treated fairly compared to other interests. Will it set an example for others to follow? Is this the start of a new approach to Native Title? It is interesting that the approach has not attracted much attention.

In time to come, could Native Title be equal to other interests and not vulnerable and subordinate to them? Maybe one day Native Title can be a native tree that sits comfortably along with all the others (Figure B8.5.1).

(*Continued*)

BOX 8.5 (Continued) WHAT IF NATIVE TITLE WAS LIKE OTHER PROPERTY INTERESTS?

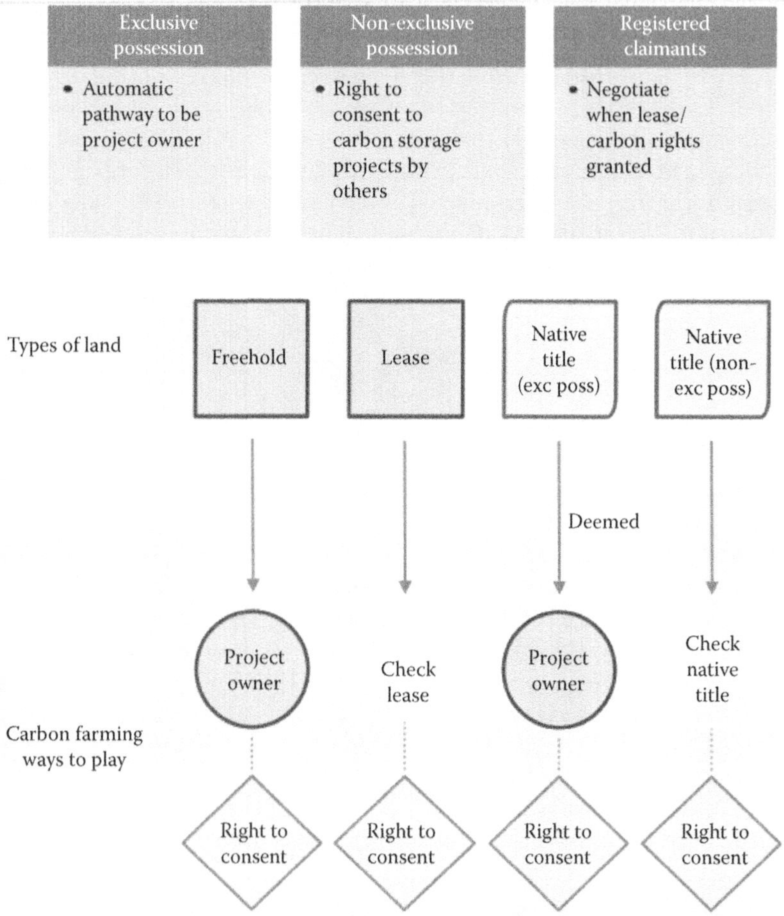

Figure B8.5.1 Indigenous native title rights and pathways.

- *Effective systems for recording and managing property rights*: The desired legal rights of Traditional Owners need to be defined well and information systems concerning land boundaries, community title, beneficiaries, lease boundaries, and lease conditions all need to be carefully resolved and recorded.
- *Supporting Indigenous capacity for managing communal wealth*: Several levels of deep and extended support are needed to assist Traditional Owners through any land reform and associated wealth management processes.
- *Institutional investment in land and sea management*: Delivering on the improved management of Indigenous cultural values means supporting strong Traditional Owner-based land and sea management institutions across the north (Chapter 4).

At the heart of this, at the Australian Government level, both political parties continue to see their key 'closing the gap' strategy as comprising welfare or supporting the integration of Aboriginal and Torres Strait Islander people into the mainstream economy. There has been little focus on supporting Traditional Owners to build a durable capacity to plan for their own future use and enjoyment of *country* to generate wealth.[4]

The other major land holding system in the North rests within the Crown-based land estate (including pastoral properties, national parks, the defence estate, and so on). While it may seem trivial in the bigger scheme of governance issues facing the North, reform in the relationship between governments and individual land managers across North Australia is a priority. This is partly because the vast bulk of the North Australia landscape is managed by a relatively small number of pastoral enterprises and Traditional Owner estates. Secondly, if southern Australia wants particular economic, environmental and cultural outcomes from the northern landscape, then the relationship between governments and land managers must change substantially. Relationships once focused on supporting rural enterprise have shifted to relationships almost exclusively focused on regulating land management.

A few key areas of reform could substantively improve this over time. Industry and government could reach agreement about simpler approaches to property management planning that starts with longer term enterprise profitability and simplifies the expression and management of the existing melting pot of regulatory obligations. Enhanced regionalised service frameworks could broker a wider range of government, not-for-profit, and business services to support enterprise viability at property-scale, including a stronger focus on landscape-scale collective action across properties. Finally, there could be significant consolidation of regulatory arrangements affecting property management. Such reform in turn could enable property management to be better guided by more regional strategies for industry development and stable land use planning frameworks.

8.4 CONCLUSION

This chapter suggests that while complex and multi-layered, the health of the governance system across North Australia essentially determines economic, social, cultural, and environmental outcomes in the landscape. The North's governance system is constituted by disparate activities at a variety of scales and many complex, often poorly connected governance domains (e.g. Indigenous policy versus employment policy). It also suggests that while great progress can be made in some governance areas, great stagnation or even regression can simultaneously be occurring in others. Dysfunction in key aspects of the North Australian governance system is particularly associated with Indigenous development, regional land use planning, natural resource management, and the human services sectors. However, at the same time there is reason to be buoyed by increasing capacity among key Indigenous institutions, the growth in policies supporting new ecosystem service markets (e.g. the nation's current Emissions Reduction Fund) and the emergence of new science alliances across the North. Of note also is improving (but unstable) cross-jurisdiction cooperation between the Commonwealth and the QLD, WA, and NT Governments, and the emergence of stronger pan-northern sectoral alliances and cross-sectoral communication.

Peter Yu's speech at the 2016 Developing Northern Australia Conference (reiterated in chapter 1) reminds, however, that it is the continuing gulf between mainstream Australian and Indigenous systems of governance that represent the widest disconnect in North Australia. While this is well represented in the generally poor performance of Indigenous policy and programs, it is perhaps most symbolically represented by the lack of cohesive vision in the white paper concerning the role of Indigenous people as the landscape's most significant and enduring custodians and the lack of clear pathways and strategies in place for supporting the development aspirations of Traditional Owners across the northern landscape. In this context, Yu's calls for a more negotiated revamp of the white paper's focus is both measured and appropriate. Maybe this means that the white paper should increasingly be viewed as representing a starting point for a more cohesive policy and delivery dialogue among governments and the people of North Australia; a starting point for more engaged and evidence-based governance.

This sort of thinking could lead to a significant tone of growing optimism within the context of the current North Australia development agenda. The release of the 'White paper on developing Northern Australia' does represent an attempt to set clearer strategic agendas in a more engaged, informed and connected way across all four jurisdictions; albeit one that is not yet very well engaged with the major sectoral and regional interests in the North. Governments alone will not be able to make this system of northern development work effectively. Its success will ride on the way local and regional institutions, cross-northern alliances and the research sector step up to ensure these new decision-making frameworks deliver real outcomes across the North. In short, those in North Australia need to make this new governance architecture work for and on behalf of Northern Australians. Governments, however, need to genuinely partner North Australians in that journey.

ACKNOWLEDGEMENTS AND LIMITATIONS

To meet the needs of this book, this chapter intentionally draws upon and updates text from two previous informally published papers: (1) 'From Myth to Reality' and (2) 'Governance challenges in Northern Australia'. It also updates and expands upon some key concepts originally published in 'Beyond the north-south culture wars'. In this respect I would like to thank other 'Myth to reality' contributors, including Chris Cocklin (James Cook University), Michael Douglas, Alistar Harris and Peter Davies (University of Western Australia), Ruth Wallace (Charles Darwin University), Sharon Bell and Andrew Campbell (both formally of Charles Darwin University), Andrew Johnson and Brian Keating (formerly CSIRO), and Peter Stone and Andrew Ash (CSIRO). Particular thanks must go to Jeremy Russell Smith and Kamal Sangha of Charles Darwin University in leading the charge on progressing the concept of this book, and to thought leaders in that process, including Melissa George, Paul Lane, Howard Pederson, Peter Yu, and Peter Whitehead. Thanks must also go to Ruth Wallace's support for this broader Northern Australian governance research agenda (via the Australian Government-funded Northern Futures Collaborative Research Network). I also thank Sandra Harding, Vice Chancellor of James Cook University, in supporting this research.

REFERENCES

1. Dale AP, Campbell A, Douglas M, et al. (2014). 'From myth to reality: New pathways for northern development'. Paper presented at the Northern Development Summit: creating the future Australia (Australian Davos Connection (ADC) Forum). 26–28 June 2014, Townsville, QLD, Australia.
2. Dale AP (2014). *Beyond the north-south culture wars: Reconciling northern Australia's past with its future*. London: Springer.
3. Stephen A, Oppermann E, Turnour J, et al. (2015). 'Identifying tensions in the development of northern Australia: Implications for governance'. *Journal of Economic and Social Policy* 17(1), 1–21. Retrieved 11 May 2018, from http://epubs.scu.edu.au/jesp/vol17/iss1/5.
4. Moran M (2016). *Serious whitefella stuff: When solutions became the problem in Indigenous affairs.* Melbourne: Melbourne University Press.
5. James Cook University, Commonwelath Scientific and Industrial Research Organisation (2013). *Land tenure in Northern Australia: Opportunities and challenges for investment.* CSIRO, Brisbane.
6. Davidson B (1972). *The northern myth: Limits to agricultural and pastoral development in tropical Australia* (3rd ed.). Melbourne: Melbourne University Press.
7. Greiner R (2000). 'The northern myth revisited: A resource economics research response to renewed interest in the agricultural development of the Kimberley region'. Paper presented at the 44th Annual Conference of the Australian Agricultural and Resource Economics Society, 22–25 January, Sydney.
8. Megarrity L (2011). 'Necessary and urgent? The politics of northern Australia, 1945–75'. *Journal of the Royal Australian Historical Society* 97(2), 136–60.

9. Australian Government (2015). *Our North, Our Future: White paper on developing Northern Australia.* Australian Government. Retrieved from http://northernaustralia.dpmc.gov.au/.
10. Deane P, Hogan W, Pervan M (2016). *Molehill to mountain: Agriculture in northern Australia.* ANZ Research In Depth, issue 1, November 2016.
11. Ash A (2014). *Factors driving the viability of major cropping investments in northern Australia: An historical analysis.* Commonwealth Scientific and Industrial Research Organisation, Brisbane.
12. Stoeckl N, Esparon M, Farr M, Delisle A, Stanley O (2013). The great asymmetric divide: An empirical investigation of the link between indigenous and non-indigenous economic systems in Northern Australia. *Papers in Regional Science* 93(4), 783–801.
13. The Coalition (2013). The Coalition's 2030 vision for developing northern Australia. Canberra: The Liberal National Party, Australia.
14. Commonwealth Scientific and Industrial Research Organisation (CSIRO) (2009). *Water in northern Australia: Summary of reports to the Australian Government from the CSIRO Northern Australia Sustainable Yields Project.* Australian Government National Water Commission. Retrieved from http://www.csiro.au/resources/NASY-Summary-report.
15. Gleeson T, Martin P, Mifsud C (2012). *Northern Australian beef industry: Assessment of risks and* opportunities. ABARES report to client prepared for the Northern Australia Ministerial Forum (ABARES project 43220), Canberra. Retrieved from http://www.industry.gov.au/ONA/Reports-and-publications/Documents/20120621-abares-final-report.pdf.
16. Joint Select Committee on Northern Australia (2014). *Interim report.* Australian Government, Canberra.
17. Canavan M (2017). Federal plan for jobs and investment in the NT. 31 March 2017. For further information see: http://minister.industry.gov.au/ministers/canavan/media-releases/federal-plan-jobs-and-investment-nt.
18. Yu P (2016). Keynote Address Northern Development Conference Darwin. Association of Sustainability in Business, 20 June 2016, Darwin.
19. Hajkowicz S, Cook H, Littleboy A (2012). *Our future world: Global mega-trends that will change the way we live.* CSIRO, Brisbane.
20. Dale AP, Campbell A, Douglas M, Robertson A, Wallace R, Davies P (forthcoming in 2017). New Pathways for the Governance of Northern Development. In *Rethinking Northern Australia Development* (Eds. R Wallace, S Harwood, R Gerritsen, et al.). ANU Press, Canberra.
21. Dale AP (2013). *Governance challenges for northern Australia.* Working Paper. The Cairns Institute, James Cook University, Cairns.
22. Walker B, Porter D, Marsh I (2012). *Fixing the hole in Australia's heartland: How government needs to work in remote Australia.* Retrieved from http://www.desertknowledge.com.au/Files/Fixing-the-hole-in-Australia-s-Heartland.aspx.
23. See http://rdafcw.com.au/nth-rda-alliance/.
24. See http://www.nailsma.org.au/.
25. See http://www.townsvilleenterprise.com.au/news/06-11-2015-Expansion-of-Northern-Austral-384.
26. Babacan H, Dale A, Andrews P, et al. (2012). *Science engagement and tropical Australia: Building a prosperous and sustainable future for the north.* Kingston, ACT: Department of Industry, Innovation, Science, Research, and Tertiary Education. http://www.science.gov.au/scienceGov/Documents/Tropical RegionEWGreport.pdf.
27. Skilton N, Adams M, Gibbs L (2014). 'Conflict in common: Heritage-making in Cape York'. *Australian Geographer* 45(2), 147–66. doi:10.1080/00049182.2014.899026.
28. Dale AP, Vella K, Potts R (2013). 'Governance Systems Analysis (GSA): A framework for reforming governance systems'. *Journal of Public Administration and Governance* 3(3). doi:10.5296/jpag.v3i3.4385.
29. See https://www.business.gov.au/assistance/cooperative-research-centres-programme/crc-for-developing northern-australia.
30. Sipe N, Vella K (2017). *Australian handbook of urban and regional planning.* New York: Routledge.
31. North Australian Indigenous Land and Sea Management Alliance (NAILSMA) (2013). *Indigenous people's right to the commercial use and management of water on their traditional territories: An indigenous water policy position.* Policy Paper 017/2013, NAILSMA Indigenous Water Policy Group. Retrieved from http://nailsma.org.au/hub/resources/publication/indigenous-peoples-right-commercial-use-and-management-water-policy.

32. Productivity Commission (2003). *Impacts of native vegetation and biodiversity regulations*. Productivity Commission Draft Report. Australian Government, Canberra.
33. Taylor M, Die D (Eds.) (1999). *Northern prawn fishery: 1997 and 1998 fisheries assessment report*. Canberra: Australian Fisheries and Management Authority.
34. Infrastructure Australia (2015). *Northern Australia audit: Infrastructure for developing north: Report January 2015*. Sydney: Infrastructure Australia. Retrieved from http://infrastructureaustralia.gov.au/policy-publications/publications/files/IA_Northern_Australia_Audit.pdf.
35. Dale AP, Pressey R, Adams VM, et al. (2014). 'Catchment-scale governance in northern Australia: A preliminary evaluation'. *Journal of Economic and Social Policy* 16(1), Article 2. Retrieved from http://epubs.scu.edu.au/jesp/vol16/iss1/2.
36. Hall L (2008). 'Territory politicians counter failed state claim'. *The Australian*, October 26, 4.
37. CSIRO (2012). The emerging carbon economy for northern Australia, CSIRO Sustainable Ecosystems. http://www.regional.gov.au/regional/ona/files/20121128-emerging-carbon-economy-discussion-paper.pdf.
38. Wensing E, Taylor J (2012). *Secure tenure for home ownership and economic development on land subject to native title*. Australian Institute of Aboriginal and Torres Strait Islander Studies. http://www.aiatsis.gov.au/research/documents/WensingTaylorDP_web_001.pdf.
39. NAILSMA (2012). *Towards resilient communities through reliable prosperity: First Indigenous experts forum report, NKS 013/2012*. http://www.nailsma.org.au/hub/resources/publication/towards-resilient-communities-through-reliable-prosperity-first-indigenous.

Index

Note: Page numbers followed by f and t refer to figures and tables respectively.

O

P

Q

For Product Safety Concerns and Information please contact our EU representative GPSR@taylorandfrancis.com Taylor & Francis Verlag GmbH, Kaufingerstraße 24, 80331 München, Germany

Batch number: 10732790

Printed by Printforce, the Netherlands